Biotechnology in India - Reworking A Strategy

George John • Amulya K. Panda
Editors

Biotechnology in India - Reworking A Strategy

Springer

Editors
George John
Department of Biotechnology
Government of India (Formerly)
New Delhi, Delhi, India

Amulya K. Panda
National Institute of Immunology
(Formerly)
New Delhi, Delhi, India

ISBN 978-981-97-0895-6 ISBN 978-981-97-0896-3 (eBook)
https://doi.org/10.1007/978-981-97-0896-3

This Springer imprint is published by the registered company Springer Nature Singapore Pte Ltd.
The registered company address is: 152 Beach Road, #21-01/04 Gateway East, Singapore 189721, Singapore

Preface

India set up a department for biotechnology at the national level very early on. The initial years were devoted to building the much needed infrastructure and manpower in biotechnology where very little existed. Bold initiatives were taken through the setting up of institutes, manpower development, R&D programmes, designing of a regulatory framework and a proactive policy for promotion of biotechnology. Today, the country has a significant presence at the global level, especially by emerging as a vaccine hub.

This book captures the sectoral developments in the country, most of which was achieved through the pioneering role played by the Department of Biotechnology, Government of India. Biotechnology as a tool in enhancing agriculture production, including animals as well as the marine sector, has been examined through interesting narratives. The section on health including AI has shown the impactful role of India worldwide, especially through its vaccines. Industry–academia collaboration, industrial biotechnology, lignocellulose utilisation and biopharma have brought out the applications of biotechnology. Human capital will play a vital role in this journey. The strong underpinnings of a regulatory mechanism cannot be overemphasised as the deployment of new tools in genetic modification will grow.

The cross-cutting theme of the chapters is the strong potential of biotechnology and the emerging India story. India has been formally in the biotechnology space for about 40 years since the setting up of the National Biotechnology Board (NBTB) in 1982 which became a department in 1986. This is an appropriate time to introspect, assess the successes and critically examine if enough has been done, and also, if things could have been better. The considerable churning in policies, especially with GM crops and related regulatory issues largely brought about by different perspectives should be factored in.

The unique feature of this book is that most of the authors have played a significant role in shaping biotechnology in India and their views are a blend of experience and forecasting. Their expositions have described the developments in India, assessed the present and emerging opportunities, and they have made recommendations for a reworking of the strategy.

Research institutes are critical platforms for furthering research and development agendas. National institutes and universities have played an important role. The Department of Biotechnology has pioneered by setting up 14 outstanding institutes. New-generation institutes like the Translational Health Science and

Technology Institute have a critical role to play. The decision to bring all institutes of the Department of Biotechnology under a single council will be administratively convenient and help converge efforts to drive excellence and innovation. The National Research Foundation and the BioE3-Biotechnology for Economy, Employment and Environment policy will ensure more funds for research in a speedy manner, foster innovation and technology development while capturing every opportunity for commercialisation.

New Delhi, Delhi, India

George John
Amulya K. Panda

Contents

Editors and Contributors

About the Editors

George John is a former Senior Advisor, Department of Biotechnology, Government of India, and former Vice Chancellor, Birsa Agricultural University, Ranchi. He has over 35 years of experience in various national and international organizations as a scientist, research manager, policy maker and institution builder. He holds a Master's and a Ph.D. from the University of Madras. He was a scientist with the Agricultural Research Service from 1978 to 1992 and later on with the Department of Biotechnology, Government of India, from 1992 to 2013. He served as the Vice Chancellor of Birsa Agricultural University, Ranchi, from September 2014 to September 2016. He made significant contributions to the fields of animal biotechnology, aquaculture, marine biotechnology, and public–private partnership. He served as a senior scientist at WorldFish (CGIAR), Cairo, from 2001 to 2006. He was involved in various national and international committees and also in the establishment of a new institute of animal biotechnology. Dr. John implemented the Small Business Innovation Research Initiative (SBIRI) for promoting public–private partnership and bio-entrepreneurship. He has published about 50 papers in peer-reviewed national and international journals, and book chapters.

Amulya K. Panda worked at the National Institute of Immunology, New Delhi, from 1990 to 2021 and retired as its Director in May 2021. He is a past president, Association of Microbiologists of India (2017); a Member, Scientific Board of Pharmacopeia Commission of India (2021–2024); and a Fellow of Royal Society of Chemistry (Cambridge) since 2020. He has a B.Tech degree from HBTI, Kanpur, a Master's degree from IIT Madras, and a Ph.D. from IIT Delhi. He was a visiting scientist at the Department of Chemical Engineering, University of California at Berkeley, USA, in 1994. His area of expertise lies in bioprocess engineering, encompassing fermentation, protein refolding, and the formulation of biologicals using nanoparticles. Dr. Panda's study reported the highest level of recombinant protein expression among the earliest Indian studies (1999). Novel protein refolding methods developed by him have gained worldwide acceptance for high-throughput recovery of bioactive proteins from inclusion bodies of *E. coli* (2005). He was awarded Young Asian Biotechnologist Prize by the Society for Biotechnology,

Japan, in 2004. He has published 165 research papers, 12 book chapters, and holds 35 patents. Currently, he serves as the Associate Director at Panacea Biotec Limited, New Delhi.

Contributors

Anil K. Angrish Department of Pharmaceutical Management, National Institute of Pharmaceutical Education and Research (NIPER), Mohali, Punjab, India

Rakesh Bhatnagar Laboratory of Molecular Biology and Genetic Engineering, Jawaharlal Nehru University, New Delhi, India

Kajal Chakraborty Central Marine Fisheries Research Institute, Kochi, India

Ghadiyaram Chakshusmathi Mynvax Private Limited, Bengaluru, India

Sambhavana Chauhan Department of Genetics, University of Delhi, South Campus, New Delhi, India

Astha Chhillar Department of Zoology, Maharshi Dayanand University, Rohtak, Haryana, India

Jagadeesh Gandla The Federation of Asian Biotech Associations, IIIT Campus, Hyderabad, India

Nirmalya Ganguli National Institute of Immunology, New Delhi, India

National Institute of Animal Biotechnology, Hyderabad, India

Prasanta Kumar Ghosh Department of Biotechnology, Ministry of Science and Technology, Government of India, New Delhi, India

Sompradip Publishers and Consultants, New Delhi, India

Himanshu Gogoi Laboratory of Molecular Biology and Genetic Engineering, Jawaharlal Nehru University, New Delhi, India

Translational Health Science and Technology Institute, Faridabad, Haryana, India

A. Gopalakrishnan Central Marine Fisheries Research Institute, Kochi, India

R. K. Jain ICAR-Indian Agricultural Research Institute, New Delhi, India

George John Department of Biotechnology, Ministry of Science and Technology, Government of India (Formerly), New Delhi, India

Divya Kandari Laboratory of Molecular Biology and Genetic Engineering, Jawaharlal Nehru University, New Delhi, India

Kawkab Kanjo Molecular Biophysics Unit (MBU), Indian Institute of Science, Bengaluru, India

Krishnamoorthy Kannan KK Biotech, Gurgaon, India

Biotechnology, GGSIPU, New Delhi, India

Nagaland Central University, Lumami, Nagaland, India

Swarnendu Kaviraj Gennova Biopharmaceutical Ltd., Pune, India

Ramesh Chander Kuhad Academy of Microbiological Sciences, Department of Microbiology, Maharshi Dayanand University, Rohtak, Haryana, India

D P G, Institute of Technology and Management, Gurugram, India

School of Biological Sciences Research, Sharda University, Greater Noida, Uttar Pradesh, India

Durgesh Kumar Gennova Biopharmaceutical Ltd., Pune, India

Krishan Kumar ICAR-Indian Institute of Maize Research, Delhi Unit, Pusa Campus, New Delhi, India

Subeer S. Majumdar National Institute of Immunology, New Delhi, India

National Institute of Animal Biotechnology, Hyderabad, India

Gujarat Biotechnology University, Gandhinagar, India

Satendra Kumar Mangrauthia ICAR-Indian Institute of Rice Research, Hyderabad, India

Raji Susan Mathew School of Data Science, Indian Institute of Science Education and Research, Thiruvananthapuram, India

Naveen Paluru Department of Computational and Data Sciences, Indian Institute of Science, Bengaluru, India

Amulya K. Panda Panacea Biotec Limited, New Delhi, India

Abhay H. Pande Department of Biotechnology, National Institute of Pharmaceutical Education and Research (NIPER), Mohali, Punjab, India

Manchikatla V. Rajam Department of Genetics, University of Delhi, South Campus, New Delhi, India

Pallu Reddanna School of Life Sciences, University of Hyderabad, Hyderabad, India

Brane Enterprises, Hyderabad, India

Vanga S. Reddy Plant Transformation Group, International Centre for Genetic Engineering and Biotechnology, New Delhi, India

Girish Sahni Panjab University, CSIR-Institute of Himalayan Biotechnology (IHBT), Palampur, and CSIR-Institute of Microbial Technology (IMTECH), Chandigarh, India

Sonu Saini Laboratory of Enzymology and Gut Microbiology, Department of Microbiology, Maharshi Dayanand University, Rohtak, Haryana, India

Department of Renewable and Bio-Energy Engineering, College of Agricultural Engineering and Technology, Chaudhary Charan Singh Haryana Agricultural University, Hisar, Haryana, India

Jameel Shaik Biomedical Engineering, HomeoWealth, Passion Life Skills Academy, Kadapa, India

Krishna Kant Sharma Laboratory of Enzymology and Gut Microbiology, Department of Microbiology, Maharshi Dayanand University, Rohtak, Haryana, India

Souvik Sen Sharma National Institute of Immunology, New Delhi, India

National Institute of Animal Biotechnology, Hyderabad, India

Ajay Singh Gennova Biopharmaceutical Ltd., Pune, India

Sanjay Singh Gennova Biopharmaceutical Ltd., Pune, India

R. M. Sundaram ICAR-Indian Institute of Rice Research, Hyderabad, India

V. V. R. Suresh Central Marine Fisheries Research Institute, Kochi, India

Ruby Tiwari Department of Genetics, University of Delhi, South Campus, New Delhi, India

Raghavan Varadarajan Molecular Biophysics Unit (MBU), Indian Institute of Science, Bengaluru, India

Neerja Wadhwa National Institute of Immunology, New Delhi, India

Phaneendra K. Yalavarthy Department of Computational and Data Sciences, Indian Institute of Science, Bengaluru, India

Part I

Agriculture Biotechnology

1 Transforming Indian Agriculture Through Genome Editing Technologies: Challenges and Opportunities

Satendra Kumar Mangrauthia, R. M. Sundaram, and R. K. Jain

Abstract

Precision breeding through genome editing is gaining momentum worldwide as it has the potential of making desired and specific changes in crop genomes in the shortest possible time. The progress of technology toward enhancing its scope and specificity has been tremendous in the past few years as evidenced by emerging tools of base editing, prime editing, and more efficient nucleases. Genome editing can improve several traits of crop plants including yield and nutrition, tolerance to biotic and abiotic stresses, resource use efficiency, resilience to a rapidly changing climate, etc. The possibility of using genome editing for enhancing the oilseed and pulses production to reduce the import burden of India will be discussed. The challenges of nonavailability of genome sequences of millet and minor pulses and oilseed crops, lack of information of gene function and efficient tissue culture and transformation protocol of crops and genotypes will be elaborated. Comparison of GM technology with genome editing will be made from an Indian perspective to retrospect and learn from past mistakes. This chapter will provide a strategic framework for efficient utilization of genome editing technologies to transform Indian agriculture in alignment with government goals of doubling farmers income, producing more crop per drop, improving carbon credits, and reducing the emission of greenhouse gases. Regulatory framework and issues of nontarget edits, IPR, infrastructure, and human resource will also be discussed.

S. K. Mangrauthia · R. M. Sundaram
ICAR-Indian Institute of Rice Research, Hyderabad, India

R. K. Jain (✉)
ICAR-Indian Agricultural Research Institute, New Delhi, India
e-mail: rakeshjain56@yahoo.co.in

G. John, A. K. Panda (eds.), *Biotechnology in India - Reworking A Strategy*, https://doi.org/10.1007/978-981-97-0896-3_1

1.1 Genome Editing and Its Tools

Genome editing can be defined as addition, deletion, or replacement of nucleotides/bases in a genome in a highly precise manner (El-Mounadi et al. 2020). It includes a set of molecular biology tools that enable researchers to make targeted modifications in genome in a most precise and efficient manner to decipher the gene function or to improve the genome structure and associated phenotype.

The genome editing tools work on the principle of DNA recognition and breakage at targeted site. Specific sequences of DNA can be recognized either by DNA binding proteins or single-stranded small RNA molecules (referred to as guide RNA or gRNA) (Doench et al. 2014). Based on DNA recognition by proteins or RNA, the genome editing tools can be classified into two groups: (1) Protein based such as meganucleases, zinc finger nucleases (ZFNs), and transcription-activator-like effector nucleases (TALENs) and (2) RNA based clustered regularly interspaced short palindromic repeats (CRISPR)/Cas systems (Khalil 2020). Meganucleases are endonucleases that usually recognize 12–45 bp DNA target site and create double-stranded break. Despite being very specific, meganuclease-based genome editing could not get much popularity because of the limitations of the absence of cleavage site in the target locus. Naturally occurring DNA recognition sites of meganucleases are very few in a genome that restricted its use. Although efforts were made to engineer the DNA recognition site of these enzymes and a few reports appeared, suggesting that its DNA recognition site can be altered by protein engineering, it could not be attempted in most of the laboratories because of the absence of protein engineering expertise. Further, DNA recognition and endonuclease sites overlap in these enzymes, which makes protein engineering more difficult. In the case of zinc finger nucleases, DNA binding domains of finger proteins and nuclease domain of *Fok*I (type IIS restriction endonuclease isolated from *Flavobacterium okeanokoites*) are fused together for site-specific DNA breakage. Similarly, TALENs represent the fusion of DNA binding domain of transcription activator-like effector (TALE) proteins and nuclease domain of *Fok*I. The bacterial plant pathogen *Xanthomonas* spp. secrete TALE proteins, which bind the specific DNA sequences to induce the susceptibility genes of host plant. Both ZFNs and TALENs were considered as very specific genome editing tools. However, these could not be utilized in common laboratories because of complexity in designing and arranging the zinc fingers or TALEs according to the target sequence of DNA (Zhang et al. 2018).

The limitations of DNA sequence recognition by proteins were overcome by the invention of CRISPR/Cas system where DNA sequences are recognized by small RNA called as guide RNA (Fig. 1.1). Since it is easy to design oligos for the complementary DNA target site, the CRISPR/Cas system became a very popular tool of genome editing (Hwarari et al. 2024). It can be implemented in a general laboratory without any need for specific equipment or infrastructure. The CRISPR/Cas is an adaptive immunity mechanism present in bacteria and archaea to suppress the invasion of foreign genetic elements (Barrangou et al. 2007; Doudna and Charpentier 2014). The system operates like RNA interference (RNAi) in eukaryotes where

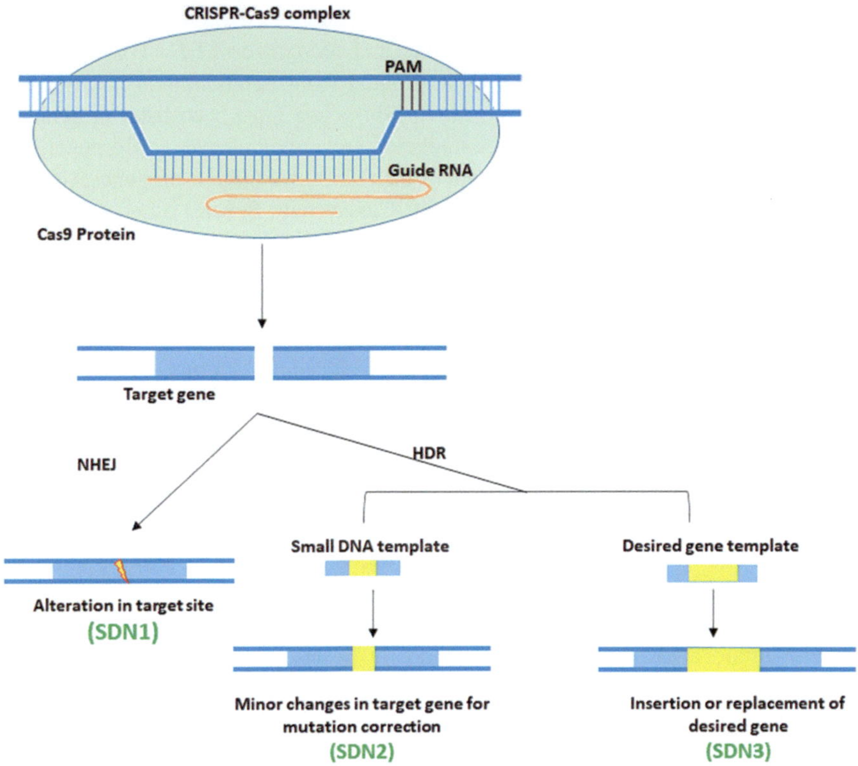

Fig. 1.1 CRISPR/Cas mechanism of genome editing. Guide RNA helps recognition of target DNA for editing while Cas9 nuclease produces PAM (protospacer adjacent motif) guided double stranded break at target site. The broken ends of DNA are repaired by either non-homologous end joining (NHEJ) or homology directed repair (HDR) mechanisms leading to either SDN1 (Site-Directed Nuclease1) or SDN2 or SDN3 genome editing

siRNAs guide the RNAi machinery (specifically argonaute endonuclease) to chop up the foreign genetic elements (Kola et al. 2016). The CRISPR arrays present in bacterial genome have unique features of multiple repeats separated by unique spacers and the cluster of many Cas genes. After invasion of foreign DNA, the Cas complex binds to foreign DNA by recognizing 2–4 bp short motif known as protospacer-adjacent motif (PAM). The Cas protein makes two double-stranded breaks in foreign DNA to release a DNA segment known as protospacer. The released protospacer is inserted between repeats of CRISPR arrays, thereafter referred to as spacer transcribed, to produce precursor CRISPR RNA (pre-crRNA), which is processed to produce crRNA. The processed crRNA along with Cas protein complex recognizes the PAM and complementary sequences of invading DNA, which is chopped by endonuclease activity of Cas protein (Jinek et al. 2012).

Among the Cas nucleases employed in CRISPR/Cas based genome editing, Cas9 and Cas12a have been most extensively used. *Streptococcus pyogenes* (spCas9) is the most widely used Cas protein that recognizes NGG as PAM motif. HNH and

RuvC nuclease domains of Cas9 protein perform dsDNA cleavage activity and produce blunt ends after digestion. Cas12a or Cpf1 recognizes TTTV PAM motif in target DNA (Zetsche et al. 2015). Cpf1 isolated from *Acidaminococcus*, FnCpf1 from *F. novicida,* and LbCpf1 from *Lachnospiraceae* have been used in genome editing. Unlike Cas9, Cpf1 produces staggered ends upon cleavage. Moreover, the smaller size of Cpf1 and no requirement of tracrRNAs for crRNA biogenesis makes Cpf1 an attractive nuclease in the CRISPR/Cas system. Several other variants of Cas9 protein such as dCas9 (deactivated Cas9 or catalytically inactive Cas9) and Cas9 nickase (nCas9) have been produced with an aim to explore various new applications of genome editing such as gene activation or suppression, base editing, and epigenomics (Bandyopadhyay et al. 2020).

Site-specific double-stranded DNA breaks produced by different genome editing tools (meganucleases, TALENs, ZFNs, CRISPR/Cas) are repaired by endogenous DNA repair systems present in eukaryotic cells. The broken ends of DNA are repaired by either template independent nonhomologous end joining (NHEJ) or template-dependent homologous repair (HR) pathways (Symington and Gautier 2011). As the name suggests, NHEJ does not require any template DNA for filling the broken ends of DNA; therefore, repair of DNA is error prone, leading to mutations in the target site. It can add, delete, or substitute the nucleotides at the target sequence, leading to frame shift mutation or introduction of stop codon or synthesis of new amino acids. NHEJ is utilized mostly in cases where gene knockout or knock-down is aimed. Conversely, the HR pathway requires a template to repair the broken ends; therefore, it is exploited in case of replacement of gene/specific domain/few nucleotides. The frequency of editing through NHEJ pathways is much higher than the HR pathway mediated genome editing.

Of late genome editing tools have witnessed several refinements to make the technology more specific and efficient. Multiplex genome editing has been employed to edit several genes/alleles together (Cong et al. 2013; Wang et al. 2017; Zetsche et al. 2017). This is particularly important in case of crops with polyploid genome. Also, multiplex editing can be useful in switching off the complete metabolic pathways or editing of quantitative traits. Similarly, base editing using CRISPR/Cas has become possible with the help of cytidine base editors (C:G to T:A conversion) or adenine base editors (A:T to G:C conversion) (Gaudelli et al. 2017; Komor et al. 2016). Base editing can be introduced without creating the double-stranded break in DNA. Therefore, deaminases (cytosine deaminase or adenine deaminase) are fused with dCas9 or nCas9, where Cas9 and gRNAs assist in landing the complex at target DNA sequence, while deaminating function of base editors helps in point mutations. Base editing is a powerful tool to alter the traits or phenotypes governed by single nucleotide polymorphism (SNP) (Henikoff and Comai 2003; Hess et al. 2017). Another such development in genome editing tools is the emergence of prime editing, which operates through nCas9, engineered reverse transcriptase, and prime-editing guide RNA (pegRNA). Primer editing does not require double-stranded DNA break (Vu et al. 2024). In comparison to other genome editing tools, primer editing is more flexible in terms of target DNA and the kinds of edits (Anzalone et al. 2019).

1.2 Genome Editing Applications

Among various applications of genome editing, its rightful use in agriculture can bring revolutionary changes (CAST 2024). It is a new addition to the toolbox of plant breeders, enabling them to make breeding more precise, specific, and faster. Developing a crop variety through cross breeding approach takes 8–10 years, while genome editing can deliver a new cultivar in 2–5 years (Gao 2021). Moreover, genome editing does not require several rounds of back-crossing to recover the genome of parent genotype. Other issues associated with cross breeding approach such as linkage drag can be easily addressed through genome editing. Several neglected crops can be brought under the mainstream of breeding. The crops can be bred for biotic (bacteria, fungi, viruses, insects, nematodes, etc.) and abiotic (drought, cold, heat, submergence, salinity, etc.) stress tolerance with an aim to develop new varieties for food security and achieving sustainable agriculture (Table 1.1). The different steps involved in the development of genome-edited plants are described in Fig. 1.2.

1.2.1 Genome Editing for Enhanced Yield and Nutrition

Enhancing crop productivity vis-à-vis nutrition enrichment is imperative to feed >7.99 billion people globally and >1.41 billion people India alone. Since independence, India witnessed tremendous progress in cereals and overall food production; however, feeding a continuous growing population is going to be a challenge in the years to come. Despite intense efforts of researchers and plant breeders, the productivity of major cereals has reached a stagnation, and significant gain or yield advantage is not seen during past several years. In addition to quantum-leap of agriculture produce, its quality and nutraceutical value needs be enhanced for curbing hidden hunger and making a healthy India. New breeding technologies such as genome editing can be smartly utilized for attaining the future goals of food and nutrition security. Yield genes in rice were edited to enhance its productivity. Wang et al. (2022) created a new allele of GRAIN SIZE ON CHROMOSOME 2 (*GS2*) gene, which encodes growth-regulating factor 4 (OsGRF4) in *Chinese indica* inbred X12. Using CRIPSR/Cas9, miRNA396 binding site in *GS2* was disrupted, resulting in increased expression of gene and >10% grain yield/plant. Genome editing of *KRN2* (kernel row number2) in maize and its rice orthologue *OsKRN2* resulted in ~10% and ~8% increased grain yield, respectively. There were apparently no other changes in agronomic traits of the plant. *KRN2* encodes WD40 protein that negatively regulates grain number in maize and rice (Chen et al. 2022). Gene editing of a newly identified gene *DUO-B1* resulted in increased grain yield in wheat without affecting other major agronomic traits (Wang et al. 2022). The gene encodes an APETALA2/ethylene response factor (AP2/ERF) transcription factor. Similar efforts of genome editing to enhance the yield of tomato and other crops have been made (Xia et al. 2021).

In addition to yield enhancement, biofortification through genome editing has also been achieved by tweaking few genes/transcriptions factors associated with

Table 1.1 List of crops for various traits developed through genome editing approaches

S. no.	Crop	Gene	Trait	Approach	References
Model crops					
1	*Arabidopsis thaliana*	CP gene	Resistance to Cauliflower Mosaic Virus	CRISPR/Cas9	Liu et al. (2018)
		eIF(iso)4E	Resistance to Turnip mosaic virus	CRISPR/Cas9	Pyott et al. (2016)
2	*Nicotiana benthamiana*	Short intergenic region, transacting replication initiation protein	Resistance to Bean yellow dwarf virus	CRISPR/Cas9	Baltes et al. (2015)
		43 sgRNAs targeting the BSCTV genome	Resistance to Beet severe curly top virus	CRISPR/Cas9	Ji et al. (2015)
Cereals					
3	*Oryza sativa*	OsERF922	Blast disease resistance	CRISPR/Cas9	Wang et al. (2016)
		OsSWEET14	Bacterial blight resistance	TALENs	Ting Li et al. (2012)
		OsSWEET13	Bacterial blight resistance	CRISPR/Cas9	Miao et al. (2013)
		OsSWEET13	Bacterial blight resistance	TALENs	Blanvillain-Baufumé et al. (2017)
		eIF4G	Rice tungro spherical virus resistance	CRISPR/Cas9	Macovei et al. (2018)
		Os09g29100	Bacterial leaf streak virus resistance	TALENs	Cai et al. (2017)
		ALS	Herbicide resistance	CRISPR/Cas9	Li et al. (2018a, b, c)
		BEL	Herbicide resistance	CRISPR/Cas9	Xu et al. (2014)
		C287	Herbicide resistance	Base editing	Shimatani et al. (2017)
		OsQQR	Trait stacking	ZFNs	Cantos et al. (2014)
		OsBADH2	Aroma	TALENs	Shan et al. (2015)
		OsMATL	Induction of haploid plants	CRISPR/Cas9	Li et al. (2017a, b)
		SlAGL6	Parthenocarpy	CRISPR/Cas9	Zhou et al. (2015)
		CSA	Photoperiod controlled male sterility	CRISPR/Cas9	Li et al. (2016)
		Hd2, Hd4, and Hd5	Earliness	CRISPR/Cas9	Li et al. (2017a, b)
		SP5G	Earliness	CRISPR/Cas9	Wang et al. (2016)
		OsPDS, OsSBEIIb	Nutritional improvement	Base editing	Li et al. (2017a, b)
		NRT1.1B gene	Enhance nitrogen use efficiency	Base editing	Lu and Zhu (2017)
		OsHAK-1	Low cesium accumulation	CRISPR/Cas9	Nieves-Cordones et al. (2017)
		OsNRAMP5	Low cadmium content	CRISPR/Cas9	Tang et al. (2017)
		Os SAPK2	Drought tolerance	CRISPR/Cas9	Lou et al. (2017)
		OsEPFL9	Leaf stomatal density	CRISPR/Cas9 and CRISPR/Cpf1	Yin et al. (2017)
		OsCDC48	Senescence and death	Base editing	Zong et al. (2017)
		SBEI and SBEIIb	Increased content of resistant starch	CRISPR/Cas9	Sun et al. (2017)
		LOX3	Enhanced storage time	TALENs	Ma et al. (2015)
		GW2, GW5	Improvement of grain weight	CRISPR/Cas9	Xu et al. (2016)
		Gn1a, DEP1, GS3, and IPA1	Improvement of grain number, panicle architecture, grain size, and plant architecture	CRISPR/Cas9	Li et al. (2016)
		PYLs	Improved growth and productivity	CRISPR/Cas9	Miao et al. (2018)

(continued)

Table 1.1 (continued)

S. no.	Crop	Gene	Trait	Approach	References
4	*Zea mays*	ARGOS8	Drought stress tolerance	CRISPR/Cas9	Shi et al. (2017)
		ALS	Herbicide resistance	CRISPR/Cas9	Svitashev et al. (2015)
		ZmIPK1	Herbicide tolerance and phytate reduced maize	ZFNs	Shukla et al. (2009)
		ZmGL2	Reduced epicuticular wax in leaves	TALENs	Char et al. (2015)
		TMS5	Thermosensitive male-sterility	CRISPR/Cas9	Li et al. (2017a, b)
		ZmMTL	Induction of haploid plants	TALENs	Kelliher et al. (2017)
		ZmTLP	Trait stacking	ZFNs	Ainley et al. (2013)
5	*Triticum aestivum*	PPO	Anti-browning phenotype	CRISPR/Cas9	Zhang et al. (2017a, b)
		TaMLO	Powdery mildew resistance	TALENs	Wang et al. (2014)
		TaMLO-A1	Powdery mildew resistance	CRISPR/Cas9	Wang et al. (2014)
		sgAlpha-1 sgAlpha-2	Low gluten content	CRISPR/Cas9	Sanchez-Leon et al. (2018)
6	*Sorghum bicolor*	k1C gene family	Increased digestibility and protein quality	CRISPR/Cas9	Li et al. (2018a)
7	*Saccharum officinarum*	COMT	Improved saccharification efficiency	TALENs	Kannan et al. (2018)
8	*Hordeum vulgare*	MP/CP, Rep/RepA genes and LIR	Wheat dwarf virus resistance	CRISPR/Cas9	Kis et al. (2019)
Legumes					
9	*Lotus japonicus*	Leghemoglobin 1/2/3, kinase (LjSYMRK), and Lotus histidine kinase 1 (LHK1)-interacting protein (LjCZF1-2)	Functional analyses of SNF (symbiotic nitrogen fixation) related genes	CRISPR/Cas9	Wang et al. (2016, 2019), Cai et al. (2018)
10	*Cicer arietinum*	4-Coumarate ligase (4CL) and Reveille 7 (RVE7)	Drought tolerance	CRISPR/Cas9	Bardhan et al. (2021)
11	*Vigna unguiculata*	Symbiotic receptor like kinase and VuSPO11-1	Symbiotic nitrogen fixation	CRISPR/Cas9	Juranić et al. (2020)
12	*Glycine max*	Phytoene desaturase, glutamine synthase, chalcone-flavanone isomerase, glucosyltransferase, DDM1 orthologs, MET1 orthologs, dicer-like3a, double-stranded RNA-binding2a/2b, fatty acid desaturase 2, and Squamosa promoter-binding protein-like	To increase nutritional value	CRISPR/Cas9	Bao et al. (2019, 2020), Al Amin et al. (2019), Curtin et al. (2018), Du et al. (2016), Jacobs et al. (2015), Li et al. (2015a, b), Michno et al. (2015), Sun et al. (2015)
		FAD2-1A, FAD2-1B, FAD3A	High oleic, low linoleic contents	TALENs	Demorest et al. (2016)
		ALS	Herbicide resistance	CRISPR/Cas9	Li et al. (2015a, b)
13	*Arachis hypogaea*	FAD2A and FAD2B	Increased oleic acid content	CRISPR/Cas9	Yuan et al. (2019)
Vegetable crops					
14	*Camelina sativa*	CsLOB1 promoter	Alleviated citrus canker resistance	CRISPR/Cas9	Jiang et al. (2017)

(continued)

Table 1.1 (continued)

S. no.	Crop	Gene	Trait	Approach	References
15	*Solanum tuberosum*	VInv	Minimizing reducing sugars	TALENs	Clasen et al. (2016)
		PP02	Decreased browning	CRISPR/Cas9	González et al. (2020)
		16DOX	Decreased steroidal glycoalkaloid content	CRISPR/Cas9	Nakayasu et al. (2018)
16	*Ipomoea batatas*	GBSSI	Decreased amylase content	CRISPR/Cas9	Wang et al. (2019)
		SBEII	Deceased amylopectin content; increased amylase	CRISPR/Cas9	Wang et al. (2019)
17	*Solanum lycopersicum*	GABA-TP1, GABA-TP2, GABA-TP3, CAT9 and SSADH	Reduced concentration of γ-aminobutyric acid	CRISPR/Cas9	Li et al. (2018b)
		SlGAD2 and SlGAD3	Increased GABA biosynthesis	CRISPR/Cas9	Nonaka et al. (2017)
		TMS5	Thermosensitive male-sterile	CRISPR/Cas9	Nekrasov et al. (2017)
		OsMATL	Induction of haploid plants	CRISPR/Cas9	Ortigosa et al. (2019)
		SP5G	Earlier harvest time	CRISPR/Cas9	Soyk et al. (2017)
		SlAGL6	Parthenocarpy	CRISPR/Cas9	Klap et al. (2017)
		SP, SP5G, CLV3, WUS, GGP1	Domestication	CRISPR/Cas9	Yao et al. (2018)
		ANT1	Purple tomatoes with high anthocyanin	TALENs	Čermák et al. (2015)
		CP, Rep genes	Tomato yellow leaf curl virus resistance	CRISPR/Cas9	Tashkandi et al. (2018)
		SlMlo1	Resistance against *Oidium neolycopersici*	CRISPR/Cas9	Nekrasov et al. (2017)
		SlAPX4	Elevation of ascorbate content in tomato fruits	CRISPR/Cas9	Do et al. (2022)
		MYB12	Pink tomatoes	CRISPR/Cas9	Deng et al. (2018), Zhu et al. (2019)
		CRTISO or PSY1	Orange tomatoes and yellow tomatoes	CRISPR/Cas9	Dahan-Meir et al. (2018)
		SlANT1	Purple tomatoes	TALEN and CRISPR/Cas9	Čermák et al. (2015)
		SGR1, LCY-E, Blc, LCY-B1, and LCY-B2	Increase in lycopene content	CRISPR/Cas9	Li et al. (2018c)
		SlDDB1, SlDET1 SlCYC-B	Increased carotenoid lycopene, and β-carotene	Target-AID	Hunziker et al. (2020)
		GGPI	Increased vitamin C	CRISPR/Cas9	Li et al. (2018b)
		SlANT2, SlAN2-like	Decreased anthocyanin content	CRISPR/Cas9	Yan et al. (2020), Zhi et al. (2020)
		FLORAL4	Increased phenylalanine-derived volatile content	CRISPR/Cas9	Tikunov et al. (2020)
		RIN	Decreased volatile organic compounds	CRISPR/Cas9	Ito et al. (2017), Zhi et al. (2020)
		L1L4	SSC, fiber, fructose ascorbic acid, total phenol, carotene, oxalic acid	ZFN	Gago et al. (2017)
		cycB	High lycopene content	CRISPR/Cas9	Zsögön et al. (2018)
18	*Theobroma cacao*	TcNPR3	Increased resistance to *Phytophthora tropicalis*	CRISPR/Cas9	Fister et al. (2018)

(continued)

Table 1.1 (continued)

S. no.	Crop	Gene	Trait	Approach	References
19	*Cucumis sativus*	EPSPS	Herbicide resistance	CRISPR/Cas9	Chandrasekaran et al. (2016)
		Cs-eIF4E	Resistance to Cucumber vein yellowing virus, Zucchini yellow mosaic virus and Papaya ring spot mosaic virus-W	CRISPR/Cas9	Chandrasekaran et al. (2016)
20	*Daucus carota*	F3H	Anthocyanin content	CRISPR/Cas9	Klimek-Chodacka et al. (2018)
21	*Agaricus bisporus*	Polyphenol oxidase gene (PPO)	Increase in shelf life	CRISPR/Cas9	Waltz (2016)
22	*Solanum melongena*	PPO4, PPO5, PPO6	Decreased browning	CRISPR/Cas9	Maioli et al. (2020)
Fruit crops					
23	*Citrus sinensis*	Susceptibility gene CsLOB1 promoter in citrus	Improvement of citrus canker resistance	CRISPR/Cas9	Peng et al. (2017)
24	*Musa* spp.	Viral ORF1,ORF2 and ORF3	Banana streak virus resistance	CRISPR/Cas9	Tripathi et al. (2019)
		Lycopene epsilon-cyclase	Increased beta carotene content	CRISPR/Cas9	Kaur et al. (2020)
25	*Citrus paradisi*	ALS	Herbicide resistance	CRISPR/Cas9	Jia et al. (2017)
26	*Malus pumila*	DIPM-1, 2 and 4	Resistance to fire blast disease (*Erwinia amylovora*)	CRISPR/Cas9	Malnoy et al. (2016)
27	*Punica granatum*	PgUGT84A23 and PgUGT84A24	Unique accumulation of gallic acid 3-0- and 4-0-glucosides	CRISPR/Cas9	Chang et al. (2019)
Other crops					
28	*Linum usitatissimum*	EPSPS	Herbicide resistance	CRISPR/Cas9	Sauer et al. (2016)
29	*Manihot esculenta*	EPSPS	Herbicide resistance	CRISPR/Cas9	Hummel et al. (2018)
		AC2 and AC3 genes encoding viral proteins Tr AP and Ren	CRISPR system fails to confer effective resistance against African cassava mosaic virus	CRISPR/Cas9	Mehta et al. (2019)
		Eif4e isoforms Ncbp-1	Resistance to cassava brown streak virus	CRISPR/Cas9	Gomez et al. (2019)
		MeCYP79D1	Reduced level of linamarin and evolved cyanide in leaves	CRISPR/Cas9	Juma et al. (2022)
30	*Brassica oleracea*	FRIGIDA	Early flowering	TALENs	Sun et al. (2013)
31	*Lactuca sativa*	–	Increased oxidation stress tolerance and ascorbate content	CRISPR/Cas9	Zhang et al. (2018)
32	*Brassica napus*	SFAR4 and SEARS, FAD2	Increased oleic acid content; decreased linoleic and linolenic	CRISPR/Cas9	Okuzaki et al. (2018), Karunarathna et al. (2020)
33	*Brassica oleracea* var. *alboglabra*	BoaCRTISO	Yellow color of Chinese kale with improved market prospects	CRISPR/Cas9	Okuzaki et al. (2018)
34	*Gossypium herbacium*	Gh14-3-3d	Resistance to *Verticillium dahlia* infestation	CRISPR/Cas9	Zhang et al. (2018)
35	*Theobroma cacao*	Non-expressor of pathogenesis-related 3 (TcNPR3)	Resistance to *Phytophthora tropicalis* fungal pathogen	CRISPR/Cas9	Fister et al. (2018)

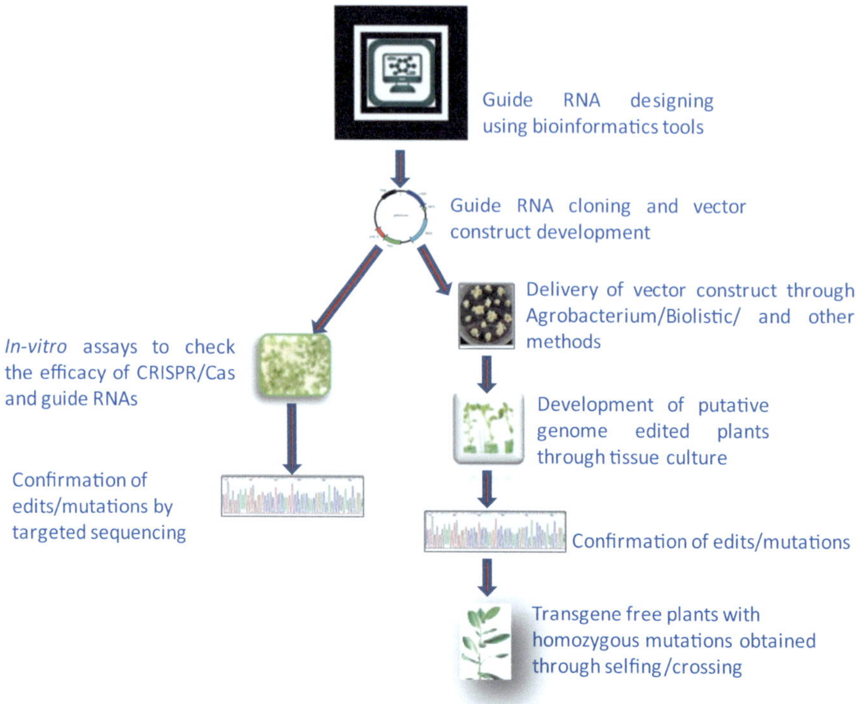

Fig. 1.2 Steps involved in the development of genome edited plants. The left side of flow-chart can be followed to check the efficacy of guideRNAs and CRISPR/Cas reagents, while the right side of flow-chart indicates the steps involved in development of transgene free genome edited plants

grain/fruit nutrient metabolism (Kumar et al. 2022). Gamma-aminobutyric acid (GABA) content in tomato fruit was enhanced to the extent of 7–15 fold by mutating glutamate decarboxylase (*SlGAD2* and *SlGAD3*) genes using CRISPR/Cas9 (Nonaka et al. 2017). GABA is a nonprotein amino acid, which helps in lowering blood pressure. The GABA rich tomatoes (Sicilian Rouge high GABA tomato) developed in collaboration with University of Tsukuba are being sold in market by a Japan-based company—Sanatech Seed (https://sanatech-seed.com/en/20201211-2-2/). Another exciting application of genome editing in tomato has been demonstrated recently wherein Li et al. (2022) developed biofortified tomatoes rich in Vitamin D3 by knocking out 7-dehydrocholesterol reductase (*Sl7-DR2*) gene through CRISPR/Cas9. The knockout of *Sl7-DR2* stops the conversion of 7-dehydrocholesterol (provitamin D3) to cholesterol, which eventually enters toward biosynthesis of α-tomatine in leaves and fruit. One medium size biofortified tomato developed by these researchers can provide Vitamin D equivalent to two medium-sized eggs, which can help in reducing the Vitamin D deficiency in vegetarian or vegan diets. Calyxt, Inc. (NASDAQ: CLXT) a consumer-centric, food- and agriculture-focused company based in USA launched high oleic acid containing soybean oil developed by TALEN genome editing. Natural soybean oil is rich in polyunsaturated fatty acids (linoleic acid and linolenic acid) and has low concentration of monounsaturated fatty acid oleic acid. By creating knockout alleles of fatty acid desaturase *FAD2-1A, FAD2-1B,* and *FAD3A* genes, the edited soybean showed more than 80% oleic acid in its oil (Demorest et al. 2016; Haun et al. 2014). In addition to known health benefits, monounsaturated fatty acid rich oil helps food last longer. Genome editing has also been used for other important traits such as low gluten wheat, low glycemic index rice, low cesium rice, etc. (Nagamine and Ezura 2022).

1.2.2 Genome Editing for Pest and Disease Resistance

Pests and diseases continue to be the major threat to agriculture productivity worldwide. In India, a significant portion of agriculture produce is damaged every year due to pests and diseases. In addition to the quantity, these biotic agents adversely affect the quality of produce and fetch lesser market price to the farmer/producer. While on one side these biotic stresses result in production loss, on the other side growers are forced to apply pesticides and chemicals that lead to environmental pollution and health hazards. Genome editing can be deployed to develop host plant resistance against major diseases (Karmakar et al. 2022). The resistant alleles can be created by tweaking a few nucleotides of genes already present in the crop genome. One such successful example was demonstrated recently by Oliva et al. (2019) wherein the researchers used CRISPR/Cas9 to edit the promoters of three *SWEET* genes (*SWEET11,13,* and *14*) in rice, leading to broad spectrum resistance against bacterial blight disease caused by bacterium *Xanthomonas oryzae* pv. *oryzae* (Xoo). *SWEET* genes in rice encode proteins involved in sugar transport. During the infection, Xoo secretes effector proteins, which bind to the promoter regions of *SWEET* genes to induce their expression. The induction of rice *SWEET* genes by bacterial effector proteins helps accumulation of more sugar, which is used as carbon source during bacterial multiplication. The CRISPR/Cas9 was used to mutate the effector binding elements present in *SWEET* gene promoters, which eventually help in developing resistance against bacterial blight disease. The *SWEET* genes of rice are also known as susceptibility genes (Antony et al. 2010). The powdery mildew disease of wheat caused by fungus *Blumeria graminis* f. sp. *tritici* (Bgt) was controlled by editing three *Mlo* (mildew resistance locus o) genes *TaMlo-A1, B1,* and *D1* by employing TALEN and CRISPR/Cas9 approaches (Wang et al. 2014). The genome editing tools have also been used against viral pathogens. The translation initiation factor eIF4E present in plant genome is used by virus for translation of its own proteins. CRISPR/Cas9 guided mutation in eIF4E of several plants resulted in virus resistance, for example, turnip mosaic virus (TuMV) resistance in *Arabidopsis*, and cucumber vein yellowing virus (CVYV), papaya ring spot mosaic virus-W (PRSV-W), and zucchini yellow mosaic virus (ZYMV) resistance in cucumber (Yin and Qiu 2019). Similarly, CRISPR/Cas9 based editing of translation initiation factor 4 gamma (*eIF4G*) gene of rice led to resistance against rice tungro virus disease (Kumam et al. 2022).

The utilization of genome editing for insect pest resistance development can help sustainable growth in agriculture by reducing the use of pesticides and enhancing crop productivity. A recent study by Lu et al. (2018) made a significant breakthrough in this direction. Production of serotonin, a neurotransmitter is enhanced in rice after the insect infestation. The CRISPR/Cas9 mediated mutation in rice cytochrome P450 gene (*CYP71A1*) suppresses serotonin production and leads to resistance against stem borer and plant hoppers. The gene encodes tryptamine 5-hydroxylase enzyme catalyzing the conversion of tryptamine to serotonin. In addition to plant genes, insect genes encoding olfactory receptors, pheromones, digestive enzymes, detoxifying enzymes, and other proteins involved in growth and development of insects have been targeted through CRISPR/Cas9 to demonstrate its effectiveness in controlling insect multiplication (Tyagi et al. 2020). However, the discovery of new plant target genes for CRISPR/Cas9 will be a safer and more pragmatic approach to develop insect resistant genome-edited crops. As far as biotic stresses are concerned, only a few well-characterized plant target genes are known that can be deployed in genome editing driven resistant crop development. It is necessary for funding agencies and researchers to intensify efforts toward understanding the host–pest/pathogen interactions and decipher the regulators associated in pathogen/pest recognition and establishment of disease or insect infestation. High-throughput genomic tools need to be combined with the advanced gene function validation tools, including CRISPR/Cas9 to unravel the potential targets for genome editing.

1.2.3 Genome Editing for Resource Use Efficiency and Abiotic Stresses

Agriculture resources, specifically nitrogen and phosphorus fertilizers and water are going to be the most essential inputs for crop growth and productivity in the years to come. Saving nitrogen fertilizers is more important from the climate change point of view as the excess use of N fertilizers is causing water pollution and increase in greenhouse gases (Menegat et al. 2022). Similarly, the sources of rock phosphate (precursor used for preparation of P fertilizers) are finite and the estimate says that if P fertilizers are being used at the current rate, the existing rock phosphate may not last for more than 100–150 years (Hellal et al. 2019). This situation is more alarming for country like India where most of the agricultural soils are either deficient or have moderate levels of available phosphorus (Dey et al. 2017). The Indian government pays huge subsidy on fertilizers, which is adding to the financial burden. Therefore, reducing the application of fertilizers by developing nutrient use efficient crops will not only make agricultural growth sustainable but also help in minimizing the adverse effects on climate. Recently, Li et al. (2018a, b, c) demonstrated the central role of the GROWTH-REGULATING FACTOR 4 (*GRF4*), a transcription factor, in coordinating the expression of multiple nitrogen-metabolism genes in rice, including DELLAs (aspartic acid–glutamic acid–leucine–leucine–alanine) encoding proteins. The semi-dwarfism trait of green revolution varieties was induced by DELLAs, which are negative regulators of gibberellins signaling. The accumulation of DELLAs in high yielding semidwarf varieties was also responsible for reduced nitrogen use efficiency of these varieties. A specific haplotype of *GRF4* showed desirable GRF4–DELLA balance, thereby enhancing nitrogen use efficiency and grain yield. Similarly, a specific haplotype of APETALA2-domain transcription factor *NGR5* (NITROGEN-MEDIATED TILLER GROWTH RESPONSE 5) was associated with high yield of rice under low nitrogen conditions (Wu et al. 2020). These specific haplotypes of *GRF4* and *NGR5* can be created in popular rice varieties using the base editing. CRISPR/Cas9 genome editing was used to ascertain the function of PHOSPHATE 1 (*PHO1*) homologs in phosphorus acquisition and transport in tomato (Zhao et al. 2019). A recent review elaborated the potential targets of genome editing for improving the nutrient use efficiency and nutrient stress adaptation in different crops (Sathee et al. 2022). To impart drought and salt stress tolerance in rice, the drought and salt tolerance (*OsDST*) gene encoding zinc finger transcription factor was edited to delete 366 bp of coding sequences of gene in the Indian rice cultivar MTU1010 (Santosh Kumar et al. 2020). The edited plants showed enhanced leaf water retention under dehydration stress and high level of salinity stress. Drought tolerance in tomato was obtained by developing CRISPR/Cas9 mediated mutation in nonexpressor of pathogenesis-related gene 1 (*NPR1*) (Li et al. 2019). *OsRR22* gene plays an important role in cytokinin signal transduction and metabolism, and its suppression or loss of function improves salinity tolerance in rice plants (Takagi et al. 2015). CRISPR/Cas9 mutation in *OsRR22* gene encoding a 696-amino acid B-type response regulator transcription factor showed salinity tolerance in rice (Zhang et al. 2019). Many QTLs and genes have been identified in various crops for different abiotic stress tolerance. Using genome editing or base editing for designing of favorable alleles in popular cultivars can make them tolerant to various stresses.

1.2.4 Genome Editing for Climate-Resilient Agriculture

Climate change is posing a serious threat to global food production. In view of this, researchers need to formulate a comprehensive strategy to make food production systems more resilient to weather extremes driven by climate change *vis-a-vis* reduction of greenhouse gases from cropping systems (Zabel et al. 2021). The adverse effects of climate will not only compromise the crop yield and productivity but also will have a large impact on society due to increased prices, hunger, unemployment, and poverty, specifically in Asia and Africa. It is a well-established fact that phenotype is the outcome of genotype × environment interaction. Since the environment cannot be controlled in the changing climate scenario, genotypes need to be designed to acclimatize the new environment for getting the best expression of genes and phenotypes. Temperature and precipitation are major environmental factors, which influence crop growth, development, and productivity (Krishnan et al.

2011). Specifically, germination, seedling, flowering, and grain filling are crucial growth stages of crop, which are highly vulnerable to weather extremes. The rise in global average surface temperature by 0.85 °C in the last century (IPCC 2013) and reports of regional-scale warming of over 1.5 °C in at least one season is an alarming situation for rice cultivation in various part of the world (Allen et al. 2018). During the growing season, an increase of 1 °C in the minimum temperature could decrease rice grain yields by 10% (Peng et al. 2004), while an increase of 1 °C in global mean temperature is estimated to reduce the global production of rice by 3.2 ± 3.7% (Zhang et al. 2017a, b), which varies at regional scales (Li et al. 2017a, b; Zhang et al. 2017a, b). Not surprisingly, the frequency of droughts, floods, cold, and heat stress in rice ecosystems has increased significantly (Hoegh-Guldberg et al. 2018), causing severe losses to crop production in recent times in different parts of the world. Therefore, analysis and drafting of appropriate strategies for mitigating the potential impact of climate change on crop production systems through designing smart genotypes by novel genome engineering tools is important for long-term food security and sustainable rice production.

One of the major impacts of climate change is the unpredictable rainfall during season and off-season. This causes flooding for short or prolonged durations. Though rice is a semiaquatic plant that can survive and flourish in standing water, the submergence of the complete plant for long affects growth and productivity immensely. The *Sub1A* gene controls submergence tolerance phenotype and encodes an ethylene response factor (*ERF*), which suppresses the GA-mediated shoot elongation through induction of GA signaling suppressors *SLR1* (slender rice 1) and SLR like-1 (*SLRL1*). The tolerant allele of submergence induced *Sub1A* gene has SNP at 556, which encodes Ser^{186} instead of Pro^{186} present in susceptible allele (Fukao and Bailey-Serres 2008; Xu et al. 2006). This amino acid substitution is crucial for phosphorylation of Sub1A by mitogen-activated protein kinase3 (MPK3), which is also induced upon submergence (Singh and Sinha 2016). The base editing *Sub1A* gene for inducing the desirable SNP in cultivars grown in flood prone areas can be a revolutionary step toward designing climate-resilient crops. Interestingly, gas films on hydrophobic leaves facilitate the exchange of CO_2 and O_2 for photosynthesis under submergence conditions. This phenomenon in rice is regulated by a gene leaf gas film 1 (*LGF1*), which controls the synthesis of C30 primary alcohol associated with the formation of gas films on submerged leaves (Kurokawa et al. 2018). Alleles with better expression of *LGF1* can further augment the phenotype obtained from SUB1A. *LGF1* can also reduce the water loss from leaf surface and help in developing rice cultivars for drought tolerance. In the rainfed rice cultivation, submergence and drought both can occur in a single crop growth season. Therefore, a combination of *LGF1* and *SUB1A* can be employed for such ecologies. It is interesting to note that *SUB1A* also imparts drought tolerance through augmenting the ABA response and activating the stress-inducible genes involved in retention of leaf water loss and lipid peroxidation (Fukao et al. 2011). In maize, knock-in genome editing was used to replace the promoter of *ARGOS8* gene with the native maize GOS2 promoter, which eventually enhanced the drought tolerance through increased expression of *ARGOS8*. This promoter replacement by genome editing helped enhance reproductive stage drought tolerance while maintaining yields in normal growth conditions (Shi et al. 2017).

Though, QTLs for heat stress tolerance have been identified by several research groups (Hill and Li 2022), the research toward dissection of QTLs to genes and their further structural and functional characterization has not been very encouraging. Homolog of *Arabidopsis* receptor-like kinase ERECTA (*ER*) and a thermo-tolerance 1 (*TT1*) encoding α2 subunit of the 26S proteasome were identified as important genes contributing to heat stress tolerance in rice (Li et al. 2015a, b; Shen et al. 2015). The *O. glaberrima* allele of *TT1* was shown to be more efficient in degradation of cytotoxic denatured proteins during heat stress. Another gene *OsDPB3-2* (LOC_Os03g63530) imparts heat stress tolerance in rice through positive regulation of dehydration-responsive element binding protein 2A (*DREB2A*). Wei et al. (2013) identified two SNPs in gene encoding zinc finger family protein (ZFP) (LOC_Os09g15430) to be associated with heat stress tolerance response. Genome editing is a powerful breeding tool that can be utilized for designing the genotypes resilient to climate change, as summarized in recent review papers (Biswal et al. 2019; Karavolias et al. 2021).

1.2.5 Genome Editing for Enhancing the Oilseed and Pulses Production

Enhancing oilseed and pulses productivity to attain self-sufficiency is one of the primary goals of Indian agriculture. During 2021–2022, India imported ~140.0 lakh tons of edible oil amounting to ₹ 1.57 trillion (https://www.business-standard.com/article/economy-policy/india-s-edible-oil-import-bill-rises-34-to-rs-1-57-trillion-says-sea-122111400927). Similarly, import of pulses in 2021-22 was 26.99 lakh tonnes amounting to ₹16,627 crores (https://www.thehindubusinessline.com/economy/agri-business/pulses-imports-up-over-9-per-cent-at-2699-lakh-tonnes-in-fy22/article65527375.ece). Despite numerous breeding programs and biotechnology interventions, productivity of oilseed and pulse crops has not increased to the desired level. Genome editing tools should be utilized to enhance the productivity of these crops. Homologue or orthologue of yield genes characterized in other crops can be edited in these crops. In addition to productivity, the oil quality and protein composition of pulses can be improved using CRISPR/Cas technology. Among the pulse crops, most of the genome editing effort has been made in soybean (Bhowmik et al. 2021). Under short day conditions, the double mutants of soybean *GmFT2a* and *GmFT5a* genes generated by CRISPR/Cas9 yielded more pods and seeds (pods 54.7 ± 12.6; seeds 106.3 ± 22.8) than the wild-type soybean (pods 11.9 ± 3.2; seeds 27.7 ± 6.4) (Cai et al. 2020). The double mutants are promising edited lines of soybean to enhance the productivity under short day conditions. In oilseed crops, genome editing has been mostly used for altering the fatty acid composition of oil, but the increase in content of oil through CRISPR/Cas has not been very successful due to complexity of the trait and limited understanding of molecular pathways associated with lipid biosynthesis (Park and Kim 2022; Subedi et al. 2020). Genome editing tools can be used for deciphering the potential regulators of oil content that can be later used as gene targets for trait improvement.

1.3 Genome Editing as a Powerful Tool to Attain Sustainable Development Goals (SDGs)

Among the 17 SDGs adopted by all the United Nation member states, genome editing can influence in achieving most of the SDGs either directly or indirectly. The second SDG (zero hunger) that includes food for all, end of malnutrition, doubling the agricultural productivity and incomes of small-scale food farmers, sustainable food production systems, and climate-resilient agricultural practices can be accelerated using genome editing tools. As described above, the crop genes associated with yield, nutrition, nutrient and use efficiency, and stress tolerance need to be tweaked to develop new generation cultivars meeting the aims of SDGs. The other SDGs such as clean water and affordable clean energy may also be partly met through the routes of genome editing. Viridos (formerly Synthetic Genomics, Inc.), a California-based biotechnology company is working on genome editing of microalgae to enhance the oil content for efficient biofuel production, which will significantly reduce the emission of greenhouse gases (https://www.viridos.com/technology/).

1.4 Genome Editing vs. GM Crops: Indian Perspective

Transgenic or genetically modified (GM) crops show huge potential to address longstanding issues of agriculture (Qaim 2020). The technology gives an opportunity to exploit genes of cross kingdom or species for the benefit of humanity. Even the *cis* genes fused with appropriate promoters produced exciting results (Srinivas 2023). Despite its proven success in other parts of the world, India could not reap the benefit of GM technology due to several issues, primarily the complex and exhaustive biosafety regulations and uncertainty of release of the product due to opposition by NGOs and also due to legal issues. The negative environment toward GM crops has discouraged Indian researchers. The huge funding spent on GM research and product development went in vain due to the unclear policies and the abovementioned issues. Except for the Bt cotton, which showed huge success (Kathage and Qaim 2012), India could not release any other GM crops for cultivation in farmers' field. With learning lessons from GM crops, India should make all possible efforts to harness the potential of genome editing technology for the benefit of farmers. The CRISPR/Cas and associated genome editing tools are clean and safe without any controversy. In fact, it is considered as the most precise breeding tool for improving the agronomically important traits (Wada et al. 2020). The technology (at least SDN1 and SDN2) does not leave any foreign nucleotides inside the plant genome which is considered as the biological mutagen for site-directed mutagenesis. In addition to precision, the technology can facilitate in bringing the new cultivars for field trials as early as in 2 years, which will save a lot of resources and money involved in new variety development (Gao 2021). Unlike GM crops where private companies had the upper hand, the genome-edited crops will be deployed more by public sector institutes because of the ease of technology and less resources required for product development. Moreover, genome editing can be utilized to improve the inbreds without the necessity to develop hybrids, which will save significant resources of farmers for purchasing hybrid seeds, and this will eventually help in doubling farmer's income. India should not lag behind in the global CRISPR/Cas market, which was USD 1095 million in 2021 and is expected to reach USD 9.6 billion by 2030 (https://finance.yahoo.com/news/crispr-cas-genes-market-size-120200436.html; https://www.globenewswire.com/en/news-release/2022/11/07/2549367/28124/en/CRISPR-and-CAS-Genes-Global-Market-to-Expand-by-26-82-Yearly-Through-2029.html). India's global position in CRISPR/Cas research is insignificant at present. Most of the patents and publications on genome editing research are from either China or USA (Cohen and Desai 2019). Until 2018, the number of papers on genome editing from USA were 898, China 824, and only 50 from India. Further, more than 98% of these 50 papers from India were review papers. Similarly, the CRISPR/Cas patent applications available in the public domain as on December 2018 were 872 from USA and 858 from China. India does not figure in patents as reported by Cohen and Desai (2019).

1.5 Regulatory Framework of Genome-Edited Crops

Regulation of genome-edited crops has been kept out of the ambit of strict biosafety regulations in most parts of the world including USA, China, Canada, Japan, China, Brazil, Israel, Australia, etc. (reviewed by Buchholzer and Frommer 2023) and more recently compiled by the FAO (2023). USA, Canada, and a few South American countries have exempted SDN1 and SDN2 edits from biosafety regulations, and these are considered as non-transgenic and equivalent to that of conventional breeds. Several other countries such as China, Japan, and Brazil follow a case-by-case approach with relaxed or minimum biosafety regulations. In India, the Ministry of Environment, Forest and Climate Change exempted the SDN1 and SDN2 genome-edited plants from rules 1989 through office memorandum (F.No. C-12013/3/2020-CS-III), dated 30th March 2022. The exemption has been given based on the Rule 20 of Rules, 1989 that empowers the ministry to exempt any genetically engineered organism. The Rule 20 says "The Ministry of Environment and Forests shall, wherever necessary, exempt an occupier handling a particular microorganism/genetically engineered organism from rule 7 to 11." Since the SDN1 and SDN2 genome-edited plants are free from foreign DNA, the Ministry of Environment, Forest and Climate Change has exempted these genome-edited plants from biosafety assessment, based on the recommendations of the Ministry of Agriculture and Farmers Welfare and Ministry of Science and Technology. All the SDN1 and SDN2 plants will be regulated by Institutional Biosafety Committee (IBSC) until the development of exogenous DNA free plants. Recently, the

Department of Biotechnology, Ministry of Science and Technology, Government of India released an office memorandum (File No. PID-15011/1/2022-PPB-DBT dated 04th October 2022), wherein Standard Operating Procedures for Regulatory Review of genome-edited plants under SDN1 and SDN2 have been given in detail.

1.6 Challenges

The last decade has been largely dominated by genome editing technologies and derived products. However, the application of genome editing for the benefit of farmers and consumers has still some concerns and challenges. The first and foremost requirement of genome editing for product development is the well-characterized genes and their regulation. Except for a few crops like rice and maize, the information of gene function is very minimal. This is a major bottleneck in the application of genome editing technology in millets, legumes, and oilseed crops. Even in rice, the promising genes for some of the priority traits like sheath blight, false smut, and yellow stem borer resistance lack this information, thus limiting the use of genome editing. Therefore, concomitant efforts toward functional genomics and basic research on characterization of genes is necessary to fully appreciate the benefits of genome editing technology. The sequencing and re-sequencing of genomes in several crops has seen tremendous progress in the past few years since the advent of high-throughput advanced sequencing technologies and platforms. However, the reference genome sequences of gold standard are not available in millets, minor pulses, and oilseed crops. In addition to genomics and gene characterization, well-standardized transformation and regeneration protocols of tissue culture is another major constraint in harnessing the benefits of genome editing technology. While some crops like rice and maize have been worked on extensively to develop reliable protocols, many legume and oilseed crops do not have standardized protocols for transformation and regeneration. Intensive research on developing efficient vehicles for transformation of CRISPR/Cas reagents in various crop plants will open avenues for gene-edited products development. The recently developed graft-mobile gene editing system for creating heritable transgene-free genome edits can be a revolutionary step in this direction (Yang et al. 2023). Here, the tRNA-like sequence (TLS) motifs facilitate the transport of Cas9 and gRNA from transgenic roots to grafted wild-type shoots. This method can give transgene-free genome-edited crops in just one generation. Further improvement on development of viral editing vectors and methods for direct delivery of active Cas protein plus gRNA complex will strengthen this research.

Off-targets or unintended changes in genome induced by genome editing tools is another area that attracts some attention while discussing the challenges of genome editing. However, it is not very significant in the use of genome editing for crop improvement or plant breeding. The unintended changes in genome induced by CRISPR/Cas tool are much lesser than the unintended variations induced by cross breeding or spontaneous mutations or the mutants derived from chemical treatment or ionizing radiation (Graham et al. 2020). Further, the off-targets can be minimized by designing specific guide RNAs through advanced bioinformatics tools and following appropriate computational algorithms. Moreover, the genome-edited derived crops are subjected to stringent phenotypic selections for 2–3 generations and other practices followed for development and release of commercial cultivars.

In addition to the above-mentioned challenges of product development through genome editing, the commercialization of genome-edited products seems to be more difficult. The complexity associated with CRISPR Intellectual Property can adversely influence the progress and adoption of the technology. The IP battle between University of California (UC) and Broad Institute of MIT and Harvard is not yet over, rather getting more complex (https://www.linkedin.com/pulse/crispr-intellectual-property-part-i-history-landscape-brian-dwyer). More than 12,000 CRISPR-related patent applications have been filed, a majority from USA and China. Discovery of several Cas protein variants and associated gene editing tools is further adding to the complexity of CRISPR/Cas IP landscape. Among these uncertainties, a satisfying scenario is that the license for use of CRISPR/Cas for agriculture is being given. The Dupont (now part of Corteva Agriscience) was provided license by UC to use CRISPR/Cas9 for agriculture. Recently, the International Rice Research Institute (Philippines) also acquired the license to use CRISPR/Cas9 through a tripartite agreement with the Broad Institute of MIT and Harvard, and Corteva Agriscience (https://www.irri.org/irri-licenses-crispr-technology-broad-institute-and-corteva-agriscience). While CRISPR/Cas tools democratize the benefits of new breeding technology to small and marginal farmers of developing and under-developed nations, with current legal and trade framework, the granting of social license may become major impediments in exploiting the full potential of technology for the benefit of humanity and addressing the global challenges of poverty, hunger, and malnutrition (Pixley et al. 2022).

1.7 Reworking of a Strategy

Constant research efforts toward discovery, engineering, and applications will shape the success of genome editing technology for the benefit of society. Discovery of novel nucleases having high specificity and efficiency is necessary to strengthen the genome editing toolbox. Currently, two Cas proteins (Cas9, Cas12a) have been predominantly used in development of genome-edited crops (Jinek et al. 2012; Zetsche et al. 2015). Cas9 recognizes G-rich PAM 5′-NGG-3′ (N = A, T, G, or C) that cuts 3–4 nucleotides upstream of PAM and produces blunt end. Cas12a (Cpf1) recognizes T-rich PAM 5′-TTTV-3′ (V = A, G, or C) that cuts 18–23 bp downstream of PAM and produces cohesive ends. Most recently, CRISPR-CasΦ discovered from huge bacteriophages further empowers the toolbox of genome editing (Pausch et al. 2020). This small size (~70 kDa) dsDNA cutter produces staggered ends through recognition of 5′-TBN-3′ PAM sequence (B = C, G or T).

Engineering of the nucleases to improve their efficiency and specificity is another priority research area in the field of genome editing. Cas9 has been engineered to produce dCas9, nCas9, and several other variants to repurpose its applications. These proteins have been engineered and codon optimized to enhance their efficiency in a crop-specific manner. Recently, the efficiency of CasΦ was enhanced by producing two new variants, vCasΦ and nCasΦ. These variants of CasΦ showed high editing efficiency without any off-targets in Arabidopsis (Li et al. 2023).

The research toward effective applications of CRISPR/Cas systems and the engineered variants need to progress in parallel. The effective applications include research for more efficient delivery mechanisms, gene–phenotype interactions, and crop genomics. Recently, researchers from Purdue University, USA, filed a patent for novel Agrobacterium strains that can deliver a transgene without integrating it into the plant nucleus, hence avoiding the step of transgenic plant development (https://www.purdue.edu/newsroom/releases/2023/Q2/purdue-biology-innovation-allows-the-introduction-of-valuable-traits-in-plants-without-creating-transgenic-plants.html). Such innovations on delivery mechanisms will ease and fasten the production of genome-edited crops. The application scientists need to foresee the use of genome editing for addressing the current and future challenges of agriculture.

1.8 Conclusion and Future Prospects

Agriculture has been the backbone of the Indian economy and its sustainable development. In addition to feeding more than one billion people, India is one among the largest exporters of agriculture produce and commodities feeding several other nations. In fact, agriculture products have been an important component of Indian diplomacy. With the shrinking agriculture inputs (land, water, labor) and finite nutrient sources (particularly phosphorus and potassium), the sustainable growth of agriculture will largely rely on effective utilization of advanced technologies for precision agriculture. The health hazards caused by pesticides and the excess use of chemicals fertilizers are major concerns that need to be addressed immediately. Genome editing can be employed to develop climate-resilient crops, to mitigate the adverse impacts on farmers. CRISPR/Cas based precision breeding technology will enable designing of suitable cultivars that can produce more with low inputs, including fertilizers and pesticides. The longstanding issues of Indian agriculture can be resolved through appropriate use of genome editing tools. India will be able to attain self-sufficiency in edible oil and pulses through technology-driven revolution. In addition to meeting the domestic need, export-oriented crops can be produced through smart use of CRISPR/Cas. The biofortification of crops through these technologies can help in achieving the government initiatives of Poshan Abhiyan through alleviating hidden hunger. Genome editing needs to be utilized for enhancing the productivity of several Indian landraces and cultivars with unique properties. Considering the high precision nature of this breeding tool, it can improve yield, without undesired traits and without altering the uniqueness of these landraces. Such interventions will create a new market for these products with medicinal or aesthetic value. For example, black rice of Manipur or Navara rice of Kerala and several such genetic resources can be popularized by enhancing their productivity through genome editing.

In addition to product development, genome editing needs to be utilized for functional genomics to unravel and characterize new genes for high-priority traits, e.g., false smut, sheath blight, and YSB resistance in rice and in high-priority crops including oilseeds and pulses. Considering the great biodiversity, specifically microbial diversity, we need to invest in the discovery of novel sources Cas proteins. The success of CRISPR-based tools in enhancing agriculture productivity will be dependent on development of crops and cultivar-specific efficient transformation and regeneration protocols and also development of indigenous CRISPR/Cas vector systems. Most importantly, the active involvement of government to acquire technology license for agriculture research and development will boost the enthusiasm of researchers and create an enabling environment for wider applications of the technology in crop improvement. Compared to China and USA, India's investment in genome editing research and development has been very insignificant. There is an urgent need to make sufficient investment on human resource development, infrastructure development, and accountable funding toward CRISPR/Cas research. Already a few of the Indian laboratories [ICAR-Indian Agricultural Research Institute (IARI), ICAR-Indian Institute of Rice Research (IIRR), ICAR-National Rice Research Institute (NRRI), Tamil Nadu Agricultural University (TNAU), National Agri-Food Biotechnology Institute (NABI), National Institute of Plant Genome Research (NIPGR), etc.] have demonstrated capabilities of product development using genome editing technologies. Upscaling and equipping of these laboratories along with linkages with promising organizations is necessary to intensify the efforts toward transforming Indian agriculture through genome editing.

Acknowledgments SKM is thankful to ICAR-NASF for funding (Grant numbers: NASF/CRISPR-Cas-7003/2017-18 and F. No. NASF/BGAM-9021/2022-23) the projects on genome editing research in rice.

References

Ainley WM, Sastry-Dent L, Welter ME, Murray MG, Zeitler B, Amora R, Corbin DR, Miles RR, Arnold NL, Strange TL, Simpson MA, Cao Z, Carroll C, Pawelczak KS, Blue R, West K, Rowland LM, Perkins D, Samuel P et al (2013) Trait stacking via targeted genome editing. Plant Biotechnol J 11(9):1126–1134. https://doi.org/10.1111/pbi.12107

Al Amin N, Ahmad N, Wu N, Pu X, Ma T, Du Y, Bo X, Wang N, Sharif R, Wang P (2019) CRISPR-Cas9 mediated targeted disruption of FAD2-2 microsomal omega-6 desaturase in soybean (*Glycine max*. L). BMC Biotechnol 19(1):1–10. https://doi.org/10.1186/s12896-019-0501-2

Allen MR, Dube OP, Solecki WA, Aragón-Durand W, Cramer F, Humphreys S, Kainuma M, Kala J, et al (2018) Framing and context. In: Global warming of 1.5°C. An IPCC special report on the impacts of global warming of 1.5°C above pre-industrial levels and related global greenhouse gas emission pathways, in the context of strengthening the global response to the threat of climate change, sustainable development

Antony G, Zhou J, Huang S, Li T, Liu B, White F, Yang B (2010) Rice xa13 recessive resistance to bacterial blight is defeated by induction of the disease susceptibility gene Os-11N3. Plant Cell 22(11):3864–3876. https://doi.org/10.1105/tpc.110.078964

Anzalone AV, Randolph PB, Davis JR, Sousa AA, Koblan LW, Levy JM, Chen PJ, Wilson C, Newby GA, Raguram A, Liu DR (2019) Search-and-replace genome editing without double-strand breaks or donor DNA. Nature 576(7785):149–157. https://doi.org/10.1038/s41586-019-1711-4

Baltes NJ, Hummel AW, Konecna E, Cegan R, Bruns AN, Bisaro DM, Voytas DF (2015) Conferring resistance to geminiviruses with the CRISPR-Cas prokaryotic immune system. Nat Plants 1(10):4–7. https://doi.org/10.1038/NPLANTS.2015.145

Bandyopadhyay A, Kancharla N, Javalkote VS, Dasgupta S, Brutnell TP (2020) CRISPR-Cas12a (Cpf1): a versatile tool in the plant genome editing tool box for agricultural advancement. Front Plant Sci 11:584151. https://doi.org/10.3389/fpls.2020.584151

Bao A, Chen H, Chen L, Chen S, Hao Q, Guo W, Qiu D, Shan Z, Yang Z, Yuan S, Zhang C, Zhang X, Liu B, Kong F, Li X, Zhou X, Tran LSP, Cao D (2019) CRISPR/Cas9-mediated targeted mutagenesis of GmSPL9 genes alters plant architecture in soybean. BMC Plant Biol 19(1):1–12. https://doi.org/10.1186/s12870-019-1746-6

Bao A, Zhang C, Huang Y, Chen H, Zhou X, Cao D (2020) Genome editing technology and application in soybean improvement. Oil Crop Sci 5(1):31–40. https://doi.org/10.1016/j.ocsci.2020.03.001

Bardhan K, York LM, Hasanuzzaman M, Jena S, Pandya MN (2021) Can smart nutrient applications optimize the plant's hidden half to improve drought resistance? Physiol Plant 172(2):1007–1015. https://doi.org/10.1111/ppl.13332

Barrangou R, Fremaux C, Deveau H, Richards M, Boyaval P, Moineau S, Romero DA, Horvath P (2007) CRISPR provides acquired resistance against viruses in prokaryotes. Science 315(5819):1709–1712. https://doi.org/10.1126/science.1138140

Bhowmik P, Konkin D, Polowick P, Hodgins CL, Subedi M, Xiang D, Yu B, Patterson N, Rajagopalan N, Babic V, Ro D, Tar'an B, Bandara M, Smyth SJ, Cui Y, Kagale S (2021) CRISPR/Cas9 gene editing in legume crops: opportunities and challenges. Legume Sci 3(3):e96. https://doi.org/10.1002/leg3.96

Biswal AK, Mangrauthia SK, Reddy MR, Yugandhar P (2019) CRISPR mediated genome engineering to develop climate smart rice: challenges and opportunities. In: Seminars in cell and developmental biology, vol 96. Elsevier, Amsterdam, pp 100–106. https://doi.org/10.1016/j.semcdb.2019.04.005

Blanvillain-Baufumé S, Reschke M, Solé M, Auguy F, Doucoure H, Szurek B, Meynard D, Portefaix M, Cunnac S, Guiderdoni E, Boch J, Koebnik R (2017) Targeted promoter editing for rice resistance to *Xanthomonas oryzae pv. oryzae* reveals differential activities for SWEET14-inducing TAL effectors. Plant Biotechnol J 15(3):306–317. https://doi.org/10.1111/pbi.12613

Buchholzer M, Frommer WB (2023) An increasing number of countries regulate genome editing in crops. New Phytol 237(1):12–15. https://doi.org/10.1111/nph.18333

Cai L, Cao Y, Xu Z, Ma W, Zakria M, Zou L, Cheng Z, Chen G (2017) A transcription activator-like effector Tal7 of *Xanthomonas oryzae pv. oryzicola* activates rice gene Os09g29100 to suppress rice immunity. Sci Rep 7(1):1–13. https://doi.org/10.1038/s41598-017-04800-8

Cai Y, Chen L, Sun S, Wu C, Yao W, Jiang B, Han T, Hou W (2018) CRISPR/Cas9-mediated deletion of large genomic fragments in soybean. Int J Mol Sci 19(12):3835. https://doi.org/10.3390/ijms19123835

Cai Y, Wang L, Chen L, Wu T, Liu L, Sun S, Wu C, Yao W, Jiang B, Yuan S, Han T, Hou W (2020) Mutagenesis of GmFT2a and GmFT5a mediated by CRISPR/Cas9 contributes for expanding the regional adaptability of soybean. Plant Biotechnol J 18(1):298–309. https://doi.org/10.1111/pbi.13199

Cantos C, Francisco P, Trijatmiko KR, Slamet-Loedin I, Chadha-Mohanty PK (2014) Identification of 'safe harbor' loci in indica rice genome by harnessing the property of zinc-finger nucleases to induce DNA damage and repair. Front Plant Sci 5:1–8. https://doi.org/10.3389/fpls.2014.00302

CAST (Council for Agricultural Science and Technology). 2024. Applications, Benefits, and Challenges of Genome Edited Crops. Issue Paper 74. CAST, Ames, Iowa

Čermák T, Baltes NJ, Čegan R, Zhang Y, Voytas DF (2015) High-frequency, precise modification of the tomato genome. Genome Biol 16(1):232. https://doi.org/10.1186/s13059-015-0796-9

Chandrasekaran J, Brumin M, Wolf D, Leibman D, Klap C, Pearlsman M, Sherman A, Arazi T, Gal-On A (2016) Development of broad virus resistance in non-transgenic cucumber using CRISPR/Cas9 technology. Mol Plant Pathol 17(7):1140–1153. https://doi.org/10.1111/mpp.12375

Chang L, Wu S, Tian L (2019) Effective genome editing and identification of a regiospecific gallic acid 4-O-glycosyltransferase in pomegranate (*Punica granatum* L.). Hortic Res 6:123. https://doi.org/10.1038/s41438-019-0206-7

Char SN, Unger-Wallace E, Frame B, Briggs SA, Main M, Spalding MH, Vollbrecht E, Wang K, Yang B (2015) Heritable site-specific mutagenesis using TALENs in maize. Plant Biotechnol J 13(7):1002–1010. https://doi.org/10.1111/pbi.12344

Chen W, Chen L, Zhang X, Yang N, Guo J, Wang M, Ji S, Zhao X, Yin P, Cai L, Xu J, Zhang L, Han Y, Xiao Y, Xu G, Wang Y, Wang S, Wu S, Yang F et al (2022) Convergent selection of a WD40 protein that enhances grain yield in maize and rice. Science 375(6587):eabg7985. https://doi.org/10.1126/science.abg7985

Clasen BM, Stoddard TJ, Luo S, Demorest ZL, Li J, Cedrone F, Tibebu R, Davison S, Ray EE, Daulhac A, Coffman A, Yabandith A, Retterath A, Haun W, Baltes NJ, Mathis L, Voytas DF, Zhang F (2016) Improving cold storage and processing traits in potato through targeted gene knockout. Plant Biotechnol J 14(1):169–176. https://doi.org/10.1111/pbi.12370

Cohen J, Desai N (2019) With its CRISPR revolution, China becomes a world leader in genome editing. https://www.science.org/content/article/its-crispr-revolution-china-becomes-world-leader-genome-editing

Cong L, Ran FA, Cox D, Lin S, Barretto R, Habib N, Hsu PD, Wu X, Jiang W, Marraffini LA, Zhang F (2013) Multiplex genome engineering using CRISPR/Cas systems. Science 339(6121):819–823. https://doi.org/10.1126/science.1231143

Curtin SJ, Xiong Y, Michno JM, Campbell BW, Stec AO, Čermák T, Starker C, Voytas DF, Eamens AL, Stupar RM (2018) CRISPR/Cas9 and TALENs generate heritable mutations for genes involved in small RNA processing of Glycine max and *Medicago truncatula*. Plant Biotechnol J 16(6):1125–1137. https://doi.org/10.1111/pbi.12857

Dahan-Meir T, Filler-Hayut S, Melamed-Bessudo C, Bocobza S, Czosnek H, Aharoni A, Levy AA (2018) Efficient in planta gene targeting in tomato using geminiviral replicons and the CRISPR/Cas9 system. Plant J 95(1):5–16. https://doi.org/10.1111/tpj.13932

Demorest ZL, Coffman A, Baltes NJ, Stoddard TJ, Clasen BM, Luo S, Retterath A, Yabandith A, Gamo ME, Bissen J, Mathis L, Voytas DF, Zhang F (2016) Direct stacking of sequence-specific nuclease-induced mutations to produce high oleic and low linolenic soybean oil. BMC Plant Biol 16(1):225. https://doi.org/10.1186/s12870-016-0906-1

Deng L, Wang H, Sun C, Li Q, Jiang H, Du M, Li CB, Li C (2018) Efficient generation of pink-fruited tomatoes using CRISPR/Cas9 system. J Genet Genomics 45(1):51–54. https://doi.org/10.1016/j.jgg.2017.10.002

Dey P, Subramaniam SM, Sellamuthu KM (2017) Status of phosphorus and potassium in the Indian soils vis-à-vis world soils ad hoc project on effect of conventional and zero tillage practices on crop production and soil properties under reclaimed sodic soils of north-West India view project AICRP on sub-tropical fruits view project. https://www.researchgate.net/publication/316086948

Do JH, Park SY, Park SH, Kim HM, Ma SH, Mai TD, Shim JS, Joung YH (2022) Development of a genome-edited tomato with high ascorbate content during later stage of fruit ripening through mutation of SlAPX4. Front Plant Sci 13:1–11. https://doi.org/10.3389/fpls.2022.836916

Doench JG, Hartenian E, Graham DB, Tothova Z, Hegde M, Smith I, Sullender M, Ebert BL, Xavier RJ, Root DE (2014) Rational design of highly active sgRNAs for CRISPR-Cas9-mediated gene inactivation. Nat Biotechnol 32(12):1262–1267. https://doi.org/10.1038/nbt.3026

Doudna JA, Charpentier E (2014) Genome editing. The new frontier of genome engineering with CRISPR-Cas9. Science 346(6213):1258096. https://doi.org/10.1126/science.1258096

Du H, Zeng X, Zhao M, Cui X, Wang Q, Yang H, Cheng H, Yu D (2016) Efficient targeted mutagenesis in soybean by TALENs and CRISPR/Cas9. J Biotechnol 217:90–97. https://doi.org/10.1016/j.jbiotec.2015.11.005

El-Mounadi K, Morales-Floriano ML, Garcia-Ruiz H (2020) Principles, applications, and biosafety of plant genome editing using CRISPR-Cas9. Front Plant Sci 11:56. https://doi.org/10.3389/fpls.2020.00056

FAO (2023) Gene editing and food safety. Food and Agriculture Organization, Rome. https://doi.org/10.4060/cc5136en

Fister AS, Landherr L, Maximova SN, Guiltinan MJ (2018) Transient expression of CRISPR/Cas9 machinery targeting TcNPR3 enhances defense response in *Theobroma cacao*. Front Plant Sci 9:1–15. https://doi.org/10.3389/fpls.2018.00268

Fukao T, Bailey-Serres J (2008) Submergence tolerance conferred by Sub1A is mediated by SLR1 and SLRL1 restriction of gibberellin responses in rice. Proc Natl Acad Sci U S A 105(43):16814–16819. https://doi.org/10.1073/pnas.0807821105

Fukao T, Yeung E, Bailey-Serres J (2011) The submergence tolerance regulator SUB1A mediates crosstalk between submergence and drought tolerance in rice. Plant Cell 23(1):412–427. https://doi.org/10.1105/tpc.110.080325

Gago C, Drosou V, Paschalidis K, Guerreiro A, Miguel G, Antunes D, Hilioti Z (2017) Targeted gene disruption coupled with metabolic screen approach to uncover the LEAFY COTYLEDON1-LIKE4 (L1L4) function in tomato fruit metabolism. Plant Cell Rep 36(7):1065–1082. https://doi.org/10.1007/s00299-017-2137-9

Gao C (2021) Genome engineering for crop improvement and future agriculture. Cell 184(6):1621–1635. https://doi.org/10.1016/j.cell.2021.01.005

Gaudelli NM, Komor AC, Rees HA, Packer MS, Badran AH, Bryson DI, Liu DR (2017) Programmable base editing of A•T to G•C in genomic DNA without DNA cleavage. Nature 551(7681):464–471. https://doi.org/10.1038/nature24644

Gomez MA, Lin ZD, Moll T, Chauhan RD, Hayden L, Renninger K, Beyene G, Taylor NJ, Carrington JC, Staskawicz BJ, Bart RS (2019) Simultaneous CRISPR/Cas9-mediated editing of cassava eIF4E isoforms nCBP-1 and nCBP-2 reduces cassava brown streak disease symptom severity and incidence. Plant Biotechnol J 17(2):421–434. https://doi.org/10.1111/pbi.12987

González MN, Massa GA, Andersson M, Turesson H, Olsson N, Fält A, Storani L (2020) Reduced enzymatic browning in potato tubers by specific editing of a polyphenol oxidase gene via ribonucleoprotein complexes delivery of the CRISPR/Cas9 system. Front Plant Sci 10:1–12. https://doi.org/10.3389/fpls.2019.01649

Graham N, Patil GB, Bubeck DM, Dobert RC, Glenn KC, Gutsche AT, Kumar S, Lindbo JA, Maas L, May GD, Vega-Sanchez ME, Stupar RM, Morrell PL (2020) Plant genome editing and the relevance of off-target changes. Plant Physiol 183(4):1453–1471. https://doi.org/10.1104/pp.19.01194

Haun W, Coffman A, Clasen BM, Demorest ZL, Lowy A, Ray E, Retterath A, Stoddard T, Juillerat A, Cedrone F, Mathis L, Voytas DF, Zhang F (2014) Improved soybean oil quality by targeted mutagenesis of the fatty acid desaturase 2 gene family. Plant Biotechnol J 12(7):934–940. https://doi.org/10.1111/pbi.12201

Hellal F, El-Sayed S, Zewainy R, Amer A (2019) Importance of phosphate pock application for sustaining agricultural production in Egypt. Bull Natl Res Cent 43(1):11. https://doi.org/10.1186/s42269-019-0050-9

Henikoff S, Comai L (2003) Single-nucleotide mutations for plant functional genomics. Annu Rev Plant Biol 54:375–401. https://doi.org/10.1146/annurev.arplant.54.031902.135009

Hess GT, Tycko J, Yao D, Bassik MC (2017) Methods and applications of CRISPR-mediated base editing in eukaryotic genomes. Mol Cell 68(1):26–43. https://doi.org/10.1016/j.molcel.2017.09.029

Hill CB, Li C (2022) Genetic improvement of heat stress tolerance in cereal crops. Agronomy 12(5):1205. https://doi.org/10.3390/agronomy12051205

Hoegh-Guldberg O, Jacob D, Taylor M, Bindi M, Brown S, Camilloni I, Diedhiou A, et al (2018) Impacts of 1.5°C global warming on natural and human systems. In: Global warming of 1.5°C. An IPCC special report on the impacts of global warming of 1.5°C above pre-industrial levels and related global greenhouse gas emission pathways. Climate change, sustainable development, and efforts to eradicate poverty

Hummel AW, Chauhan RD, Cermak T, Mutka AM, Boyher A, Starker CG, Bart R, Voytas DF, Taylor NJ (2018) Allele exchange at the EPSPS locus confers glyphosate tolerance in cassava. Plant Biotechnol J 16(7):1275–1282. https://doi.org/10.1111/pbi.12868

Hunziker J, Nishida K, Kondo A, Kishimoto S, Ariizumi T (2020) Multiple gene substitution by target-AID base-editing technology in tomato. Sci Rep 10(1):20471. https://doi.org/10.1038/s41598-020-77379-2

Hwarari D, Radani Y, Ke Y, Chen J, Yang L (2024) CRISPR/Cas genome editing in plants: mechanisms, applications, and overcoming bottlenecks. Funct Integr Genomics 24(2):50. https://doi.org/10.1007/s10142-024-01314-1. PMID: 38441816

IPCC, 2013: Climate Change 2013: The Physical Science Basis. Contribution of Working Group I to the Fifth Assessment Report of the Intergovernmental Panel on Climate Change [Stocker, T.F., D. Qin, G.-K. Plattner, M. Tignor, S.K. Allen, J. Boschung, A. Nauels, Y. Xia, V. Bex and P.M. Midgley (eds.)]. Cambridge University Press, Cambridge, United Kingdom and New York, NY, USA, 1535 pp, https://doi.org/10.1017/CBO9781107415324

Ito Y, Nishizawa-Yokoi A, Endo M, Mikami M, Shima Y, Nakamura N, Kotake-Nara E, Kawasaki S, Toki S (2017) Re-evaluation of the RIN mutation and the role of RIN in the induction of tomato ripening. Nat Plants 3(11):866–874. https://doi.org/10.1038/s41477-017-0041-5

Jacobs TB, LaFayette PR, Schmitz RJ, Parrott WA (2015) Targeted genome modifications in soybean with CRISPR/Cas9. BMC Biotechnol 15(1):1–10. https://doi.org/10.1186/s12896-015-0131-2

Ji X, Zhang H, Zhang Y, Wang Y, Gao C (2015) Establishing a CRISPR–Cas-like immune system conferring DNA virus resistance in plants. Nat Plants 1(10):1–4. https://doi.org/10.1038/NPLANTS.2015.144

Jia H, Zhang Y, Orbović V, Xu J, White FF, Jones JB, Wang N (2017) Genome editing of the disease susceptibility gene CsLOB1 in citrus confers resistance to citrus canker. Plant Biotechnol J 15(7):817–823. https://doi.org/10.1111/pbi.12677

Jiang WZ, Henry IM, Lynagh PG, Comai L, Cahoon EB, Weeks DP (2017) Significant enhancement of fatty acid composition in seeds of the allohexaploid, *Camelina sativa*, using CRISPR/Cas9 gene editing. Plant Biotechnol J 15(5):648–657. https://doi.org/10.1111/pbi.12663

Jinek M, Chylinski K, Fonfara I, Hauer M, Doudna JA, Charpentier E (2012) A programmable dual-RNA-guided DNA endonuclease in adaptive bacterial immunity. Science 337(6096):816–821. https://doi.org/10.1126/science.1225829

Juma BS, Mukami A, Mweu C, Ngugi MP, Mbinda W (2022) Targeted mutagenesis of the CYP79D1 gene via CRISPR/Cas9-mediated genome editing results in lower levels of cyanide in cassava. Front Plant Sci 13:1009860. https://doi.org/10.3389/fpls.2022.1009860

Juranić M, Nagahatenna DSK, Rigel Salinas-Gamboa MLH, Sánchez-León N, Leong WH, Al E (2020) A detached leaf assay for testing transient gene expression and gene editing in cowpea (*Vigna unguiculata* [L.] Walp.). Plant Methods 16:88

Kannan B, Jung JH, Moxley GW, Lee SM, Altpeter F (2018) TALEN-mediated targeted mutagenesis of more than 100 COMT copies/alleles in highly polyploid sugarcane improves saccharification efficiency without compromising biomass yield. Plant Biotechnol J 16(4):856–866. https://doi.org/10.1111/pbi.12833

Karavolias NG, Horner W, Abugu MN, Evanega SN (2021) Application of gene editing for climate change in agriculture. Front Sustain Food Syst 5:685801. https://doi.org/10.3389/fsufs.2021.685801

Karmakar A, Taufiqa S, Baig MJ, Molla KA (2022) Increasing disease resistance in host plants through genome editing. Proc Indian Natl Sci Acad 88(3):417–429. https://doi.org/10.1007/s43538-022-00100-6

Karunarathna NL, Wang H, Harloff H, Jiang L, Jung C (2020) Elevating seed oil content in a polyploid crop by induced mutations in SEED FATTY ACID REDUCER genes. Plant Biotechnol J 18(11):2251–2266. https://doi.org/10.1111/pbi.13381

Kathage J, Qaim M (2012) Economic impacts and impact dynamics of Bt (*Bacillus thuringiensis*) cotton in India. Proc Natl Acad Sci U S A 109(29):11652–11656. https://doi.org/10.1073/pnas.1203647109

Kaur N, Alok A, Kumar P, Kaur N, Awasthi P, Chaturvedi S, Pandey P, Pandey A, Pandey AK, Tiwari S (2020) CRISPR/Cas9 directed editing of lycopene epsilon-cyclase modulates metabolic flux for β-carotene biosynthesis in banana fruit. Metab Eng 59:76–86. https://doi.org/10.1016/j.ymben.2020.01.008

Kelliher T, Starr D, Richbourg L, Chintamanani S, Delzer B, Nuccio ML, Green J, Chen Z, McCuiston J, Wang W, Liebler T, Bullock P, Martin B (2017) MATRILINEAL, a sperm-specific phospholipase, triggers maize haploid induction. Nature 542(7639):105–109. https://doi.org/10.1038/nature20827

Khalil AM (2020) The genome editing revolution: review. J Genet Eng Biotechnol 18(1):68. https://doi.org/10.1186/s43141-020-00078-y
Kis A, Hamar É, Tholt G, Bán R, Havelda Z (2019) Creating highly efficient resistance against wheat dwarf virus in barley by employing CRISPR/Cas9 system. Plant Biotechnol J 17(6):1004–1006. https://doi.org/10.1111/pbi.13077
Klap C, Yeshayahou E, Bolger AM, Arazi T, Gupta SK, Shabtai S, Usadel B, Salts Y, Barg R (2017) Tomato facultative parthenocarpy results from SlAGAMOUS-LIKE 6 loss of function. Plant Biotechnol J 15(5):634–647. https://doi.org/10.1111/pbi.12662
Klimek-Chodacka M, Oleszkiewicz T, Lowder LG, Qi Y, Baranski R (2018) Efficient CRISPR/Cas9-based genome editing in carrot cells Magdalena. Plant Cell Rep 37:575–586
Kola VSR, Renuka P, Padmakumari AP, Mangrauthia SK, Balachandran SM, Ravindra Babu V, Madhav MS (2016) Silencing of CYP6 and APN genes affects the growth and development of rice yellow stem borer, Scirpophagaincertulas. Front Physiol 7:20. https://doi.org/10.3389/fphys.2016.00020
Komor AC, Kim YB, Packer MS, Zuris JA, Liu DR (2016) Programmable editing of a target base in genomic DNA without double-stranded DNA cleavage. Nature 533(7603):420–424. https://doi.org/10.1038/nature17946
Krishnan P, Ramakrishnan B, Reddy KR, Reddy VR (2011) High-temperature effects on rice growth, yield, and grain quality. Adv Agron 111:87–206. https://doi.org/10.1016/B978-0-12-387689-8.00004-7
Kumam Y, Rajadurai G, Kumar KK, Varanavasiappan S, Reddy MK, Krishnaveni D, Mangrauthia SK, Raveendran M, Arul L, Kokiladevi E, Sudhakar D (2022) Genome editing of indica rice ASD16 for imparting resistance against rice Tungro disease. J Plant Biochem Biotechnol 31(4):880–893. https://doi.org/10.1007/s13562-021-00765-y
Kumar D, Yadav A, Ahmad R, Dwivedi UN, Yadav K (2022) CRISPR-based genome editing for nutrient enrichment in crops: a promising approach toward global food security. Front Genet 13:932859. https://doi.org/10.3389/fgene.2022.932859
Kurokawa Y, Nagai K, Huan PD, Shimazaki K, Qu H, Mori Y, Toda Y, Kuroha T, Hayashi N, Aiga S, Itoh JI, Yoshimura A, Sasaki-Sekimoto Y, Ohta H, Shimojima M, Malik AI, Pedersen O, Colmer TD, Ashikari M (2018) Rice leaf hydrophobicity and gas films are conferred by a wax synthesis gene (LGF1) and contribute to flood tolerance. New Phytol 218(4):1558–1569. https://doi.org/10.1111/nph.15070
Li T, Liu B, Spalding MH, Weeks DP, Yang B (2012) High-efficiency TALEN-based gene editing produces disease-resistant rice. Nat Biotechnol 30(5):390–392. https://doi.org/10.1038/nbt.2199
Li X-M, Chao D-Y, Wu Y, Huang X, Chen K, Cui L-G, Su L, Ye W-W, Chen H, Chen H-C, Dong N-Q, Guo T, Shi M, Feng Q, Zhang P, Han B, Shan J-X, Gao J-P, Lin H-X (2015a) Natural alleles of a proteasome α2 subunit gene contribute to thermotolerance and adaptation of African rice. Nat Genet 47(7):827–833. https://doi.org/10.1038/ng.3305
Li Z, Liu ZB, Xing A, Moon BP, Koellhoffer JP, Huang L, Ward RT, Clifton E, Falco SC, Cigan AM (2015b) Cas9-guide RNA directed genome editing in soybean. Plant Physiol 169(2):960–970. https://doi.org/10.1104/pp.15.00783
Li J, Meng X, Zong Y, Chen K, Zhang H, Liu J, Li J, Gao C (2016) Gene replacements and insertions in rice by intron targeting using CRISPR-Cas9. Nat Plants 2(10):1–6. https://doi.org/10.1038/nplants.2016.139
Li S, Wang Q, Chun JA (2017a) Impact assessment of climate change on rice productivity in the Indochinese Peninsula using a regional-scale crop model. Int J Climatol 37:1147–1160. https://doi.org/10.1002/joc.5072
Li X, Zhou W, Ren Y, Tian X, Lv T, Wang Z, Fang J, Chu C, Yang J, Bu Q (2017b) High-efficiency breeding of early-maturing rice cultivars via CRISPR/Cas9-mediated genome editing. J Genet Genomics 44(3):175–178. https://doi.org/10.1016/j.jgg.2017.02.001
Li A, Jia S, Yobi A, Ge Z, Sato SJ, Zhang C, Angelovici R, Clemente TE, Holding DR (2018a) Editing of an alpha-kafirin gene family increases digestibility and protein quality in sorghum. Plant Physiol 177(4):1425–1438. https://doi.org/10.1104/pp.18.00200

Li R, Li R, Li X, Fu D, Zhu B, Tian H, Luo Y, Zhu H (2018b) Multiplexed CRISPR/Cas9-mediated metabolic engineering of γ-aminobutyric acid levels in *Solanum lycopersicum*. Plant Biotechnol J 16(2):415–427. https://doi.org/10.1111/pbi.12781

Li T, Yang X, Yu Y, Si X, Zhai X, Zhang H, Dong W, Gao C, Xu C (2018c) Domestication of wild tomato is accelerated by genome editing. Nat Biotechnol 36:1160–1163. https://doi.org/10.1038/nbt.4273

Li R, Liu C, Zhao R, Wang L, Chen L, Yu W, Zhang S, Sheng J, Shen L (2019) CRISPR/Cas9-mediated SlNPR1 mutagenesis reduces tomato plant drought tolerance. BMC Plant Biol 19(1):38. https://doi.org/10.1186/s12870-018-1627-4

Li J, Scarano A, Gonzalez NM, D'Orso F, Yue Y, Nemeth K, Saalbach G, Hill L, de Oliveira Martins C, Moran R, Santino A, Martin C (2022) Biofortified tomatoes provide a new route to vitamin D sufficiency. Nat Plants 8(6):611–616. https://doi.org/10.1038/s41477-022-01154-6

Li Z, Zhong Z, Wu Z, Pausch P, Al-Shayeb B, Amerasekera J, Doudna JA, Jacobsen SE (2023) Genome editing in plants using the compact editor CasΦ. Proc Natl Acad Sci U S A 120(4):e2216822120. https://doi.org/10.1073/pnas.2216822120

Liu H, Soyars CL, Li J, Fei Q, He G, Peterson BA, Meyers BC, Nimchuk ZL, Wang X (2018) CRISPR/Cas9-mediated resistance to cauliflower mosaic virus. Plant Direct 2(3):1–9. https://doi.org/10.1002/pld3.47

Lou D, Wang H, Liang G, Yu D (2017) OsSAPK2 confers abscisic acid sensitivity and tolerance to drought stress in rice. Front Plant Sci 8:993. https://doi.org/10.3389/fpls.2017.00993

Lu Y, Zhu JK (2017) Precise editing of a target base in the rice genome using a modified CRISPR/Cas9 system. Mol Plant 10(3):523–525. https://doi.org/10.1016/j.molp.2016.11.013

Lu H-P, Luo T, Fu H-W, Wang L, Tan Y-Y, Huang J-Z, Wang Q, Ye G-Y, Gatehouse AMR, Lou Y-G, Shu Q-Y (2018) Resistance of rice to insect pests mediated by suppression of serotonin biosynthesis. Nat Plants 4(6):338–344. https://doi.org/10.1038/s41477-018-0152-7

Ma L, Zhu F, Li Z, Zhang J, Li X, Dong J, Wang T (2015) TALEN-based mutagenesis of lipoxygenase LOX3 enhances the storage tolerance of rice (*Oryza sativa*) seeds. PLoS One 10(12):1–16. https://doi.org/10.1371/journal.pone.0143877

Macovei A, Sevilla NR, Cantos C, Jonson GB, Slamet-Loedin I, Čermák T, Voytas DF, Choi IR, Chadha-Mohanty P (2018) Novel alleles of rice eIF4G generated by CRISPR/Cas9-targeted mutagenesis confer resistance to rice tungro spherical virus. Plant Biotechnol J 16(11):1918–1927. https://doi.org/10.1111/pbi.12927

Maioli A, Gianoglio S, Moglia A, Acquadro A, Valentino D, Milani AM, Prohens J, Orzaez D, Granell A, Lanteri S (2020) Simultaneous CRISPR/Cas9 editing of three PPO genes reduces fruit flesh browning in *Solanum melongena* L. Front Plant Sci 11:607161. https://doi.org/10.3389/fpls.2020.607161

Malnoy M, Viola R, Jung MH, Koo OJ, Kim S, Kim JS, Velasco R, Kanchiswamy CN (2016) DNA-free genetically edited grapevine and apple protoplast using CRISPR/Cas9 ribonucleoproteins. Front Plant Sci 7:1–9. https://doi.org/10.3389/fpls.2016.01904

Mehta D, Stürchler A, Anjanappa RB, Zaidi SSEA, Hirsch-Hoffmann M, Gruissem W, Vanderschuren H (2019) Linking CRISPR-Cas9 interference in cassava to the evolution of editing-resistant geminiviruses. Genome Biol 20(1):1–10. https://doi.org/10.1186/s13059-019-1678-3

Menegat S, Ledo A, Tirado R (2022) Greenhouse gas emissions from global production and use of nitrogen synthetic fertilisers in agriculture. Sci Rep 12(1):14490. https://doi.org/10.1038/s41598-022-18773-w

Miao J, Guo D, Zhang J, Huang Q, Qin G, Zhang X, Wan J, Gu H, Qu LJ (2013) Targeted mutagenesis in rice using CRISPR-Cas system. Cell Res 23(10):1233–1236. https://doi.org/10.1038/cr.2013.123

Miao C, Xiao L, Hua K, Zou C, Zhao Y, Bressan RA, Zhu JK (2018) Mutations in a subfamily of abscisic acid receptor genes promote rice growth and productivity. Proc Natl Acad Sci U S A 115(23):6058–6063. https://doi.org/10.1073/pnas.1804774115

Michno JM, Wang X, Liu J, Curtin SJ, Kono TJ, Stupar RM (2015) CRISPR/Cas mutagenesis of soybean and *Medicago truncatula* using a new web-tool and a modified Cas9 enzyme. GM Crops Food 6(4):243–252. https://doi.org/10.1080/21645698.2015.1106063

Nagamine A, Ezura H (2022) Genome editing for improving crop nutrition. Front Genome Editing 4:850104. https://doi.org/10.3389/fgeed.2022.850104

Nakayasu M, Akiyama R, Jae H, Osakabe K, Osakabe Y (2018) Plant physiology and biochemistry generation of α-solanine-free hairy roots of potato by CRISPR/Cas9 mediated genome editing of the St16DOX gene. Plant Physiol Biochem 131:70–77. https://doi.org/10.1016/j.plaphy.2018.04.026

Nekrasov V, Wang C, Win J, Lanz C, Weigel D, Kamoun S (2017) Rapid generation of a transgene-free powdery mildew resistant tomato by genome deletion. Sci Rep 7(1):1–6. https://doi.org/10.1038/s41598-017-00578-x

Nieves-Cordones M, Mohamed S, Tanoi K, Kobayashi NI, Takagi K, Vernet A, Guiderdoni E, Christophe P (2017) Production of low-Cs+ rice plants by inactivation of the K+ transporter OsHAK1 with the CRISPR-Cas system. Plant J 92(1):43–56. https://doi.org/10.1111/tpj.13632

Nonaka S, Arai C, Takayama M, Matsukura C, Ezura H (2017) Efficient increase of ɣ-aminobutyric acid (GABA) content in tomato fruits by targeted mutagenesis. Sci Rep 7(1):7057. https://doi.org/10.1038/s41598-017-06400-y

Okuzaki A, Ogawa T, Koizuka C, Kaneko K, Inaba M, Imamura J (2018) CRISPR/Cas9-mediated genome editing of the fatty acid desaturase 2 gene in *Brassica napus*. Plant Physiol Biochem 131:63–69. https://doi.org/10.1016/j.plaphy.2018.04.025

Oliva R, Ji C, Atienza-Grande G, Huguet-Tapia JC, Perez-Quintero A, Li T, Eom J-S, Li C, Nguyen H, Liu B, Auguy F, Sciallano C, Luu VT, Dossa GS, Cunnac S, Schmidt SM, Slamet-Loedin IH, Vera Cruz C, Szurek B et al (2019) Broad-spectrum resistance to bacterial blight in rice using genome editing. Nat Biotechnol 37(11):1344–1350. https://doi.org/10.1038/s41587-019-0267-z

Ortigosa A, Gimenez-Ibanez S, Leonhardt N, Solano R (2019) Design of a bacterial speck resistant tomato by CRISPR/Cas9-mediated editing of SlJAZ2. Plant Biotechnol J 17(3):665–673. https://doi.org/10.1111/pbi.13006

Park M-E, Kim HU (2022) Applications and prospects of genome editing in plant fatty acid and triacylglycerol biosynthesis. Front Plant Sci 13:969844. https://doi.org/10.3389/fpls.2022.969844

Pausch P, Al-Shayeb B, Bisom-Rapp E, Tsuchida CA, Li Z, Cress BF, Knott GJ, Jacobsen SE, Banfield JF, Doudna JA (2020) CRISPR-CasΦ from huge phages is a hypercompact genome editor. Science 369(6501):333–337. https://doi.org/10.1126/science.abb1400

Peng S, Huang J, Sheehy JE, Laza RC, Visperas RM, Zhong X, Centeno GS, Khush GS, Cassman KG (2004) Rice yields decline with higher night temperature from global warming. Proc Natl Acad Sci U S A 101(27):9971–9975. https://doi.org/10.1073/pnas.0403720101

Peng A, Chen S, Lei T, Xu L, He Y, Wu L, Yao L, Zou X (2017) Engineering canker-resistant plants through CRISPR/Cas9-targeted editing of the susceptibility gene CsLOB1 promoter in citrus. Plant Biotechnol J 15(12):1509–1519. https://doi.org/10.1111/pbi.12733

Pixley KV, Falck-Zepeda JB, Paarlberg RL, Phillips PWB, Slamet-Loedin IH, Dhugga KS, Campos H, Gutterson N (2022) Genome-edited crops for improved food security of smallholder farmers. Nat Genet 54(4):364–367. https://doi.org/10.1038/s41588-022-01046-7

Pyott DE, Sheehan E, Molnar A (2016) Engineering of CRISPR/Cas9-mediated potyvirus resistance in transgene-free Arabidopsis plants. Mol Plant Pathol 17(8):1276–1288. https://doi.org/10.1111/mpp.12417

Qaim M (2020) Role of new plant breeding technologies for food security and sustainable agricultural development. Appl Econ Perspect Policy 42(2):129–150. https://doi.org/10.1002/aepp.13044

Sanchez-Leon S, Gil-humanes J, Ozuna CV, Sousa C, Voytas DF, Barro F (2018) Low-gluten, nontransgenic wheat engineered with CRISPR/Cas9. Plant Biotechnol J 16(4):902–910. https://doi.org/10.1111/pbi.12837

Santosh Kumar VV, Verma RK, Yadav SK, Yadav P, Watts A, Rao MV, Chinnusamy V (2020) CRISPR-Cas9 mediated genome editing of drought and salt tolerance (OsDST) gene in indica mega rice cultivar MTU1010. Physiol Mol Biol Plants 26(6):1099–1110. https://doi.org/10.1007/s12298-020-00819-w

Sathee L, Jagadhesan B, Pandesha PH, Barman D, Sandeep Adavi B, Nagar S, Krishna GK, Tripathi S, Jha SK, Chinnusamy V (2022) Genome editing targets for improving nutrient use

efficiency and nutrient stress adaptation. Front Genet 13:900897. https://doi.org/10.3389/fgene.2022.900897
Sauer NJ, Narváez-vásquez J, Mozoruk J, Miller RB, Warburg ZJ, Woodward MJ, Mihiret YA, Lincoln TA, Segami RE, Sanders SL, Walker KA, Beetham PR, Schöpke CR, Gocal GFW, Diego S (2016) Oligonucleotide-mediated genome editing provides precision and function to engineered nucleases and antibiotics in plants. Plant Physiol 170:1917–1928. https://doi.org/10.1104/pp.15.01696
Shan Q, Zhang Y, Chen K, Zhang K, Gao C (2015) Creation of fragrant rice by targeted knockout of the OsBADH2 gene using TALEN technology. Plant Biotechnol J 13(6):791–800. https://doi.org/10.1111/pbi.12312
Shen H, Zhong X, Zhao F, Wang Y, Yan B, Li Q, Chen G, Mao B, Wang J, Li Y, Xiao G, He Y, Xiao H, Li J, He Z (2015) Overexpression of receptor-like kinase ERECTA improves thermotolerance in rice and tomato. Nat Biotechnol 33(9):996–1003. https://doi.org/10.1038/nbt.3321
Shi J, Gao H, Wang H, Lafitte HR, Archibald RL, Yang M, Hakimi SM, Mo H, Habben JE (2017) ARGOS8 variants generated by CRISPR-Cas9 improve maize grain yield under field drought stress conditions. Plant Biotechnol J 15(2):207–216. https://doi.org/10.1111/pbi.12603
Shimatani Z, Kashojiya S, Takayama M, Terada R, Arazoe T, Ishii H, Teramura H, Yamamoto T, Komatsu H, Miura K, Ezura H, Nishida K, Ariizumi T, Kondo A (2017) Targeted base editing in rice and tomato using a CRISPR-Cas9 cytidine deaminase fusion. Nat Biotechnol 35(5):441–443. https://doi.org/10.1038/nbt.3833
Shukla VK, Doyon Y, Miller JC, Dekelver RC, Moehle EA, Worden SE, Mitchell JC, Arnold NL, Gopalan S, Meng X, Choi VM, Rock JM, Wu YY, Katibah GE, Zhifang G, McCaskill D, Simpson MA, Blakeslee B, Greenwalt SA et al (2009) Precise genome modification in the crop species *Zea mays* using zinc-finger nucleases. Nature 459(7245):437–441. https://doi.org/10.1038/nature07992
Singh P, Sinha AK (2016) A positive feedback loop governed by SUB1A1 interaction with MITOGEN-ACTIVATED PROTEIN KINASE3 imparts submergence tolerance in rice. Plant Cell 28(5):1127–1143. https://doi.org/10.1105/tpc.15.01001
Soyk S, Müller NA, Park SJ, Schmalenbach I, Jiang K, Hayama R, Zhang L, Van Eck J, Jiménez-gómez JM, Lippman ZB (2017) Variation in the flowering gene SELF PRUNING 5G promotes day-neutrality and early yield in tomato. Nat Genet 49(1):162–168. https://doi.org/10.1038/ng.3733
Srinivas KR (2023) CISGENESIS AND CISGENIC CROPS: need for a paradigm shift in harnessing and governance. In: Cisgenic crops: safety, legal and social issues. Concepts and strategies in plant sciences. Springer, Cham, pp 255–268. https://doi.org/10.1007/978-3-031-10721-4_11
Subedi U, Ozga JA, Chen G, Foroud NA, Singer SD (2020) CRISPR/Cas-mediated genome editing for the improvement of oilseed crop productivity. Crit Rev Plant Sci 39(3):195–221. https://doi.org/10.1080/07352689.2020.1782568
Sun Z, Li N, Huang G, Xu J, Pan Y, Wang Z, Tang Q, Song M, Wang X (2013) Site-specific gene targeting using transcription activator-like effector (TALE)-based nuclease in *Brassica oleracea*. J Integr Plant Biol 55(11):1092–1103. https://doi.org/10.1111/jipb.12091
Sun X, Hu Z, Chen R, Jiang Q, Song G, Zhang H, Xi Y (2015) Targeted mutagenesis in soybean using the CRISPR-Cas9 system. Sci Rep 5:1–10. https://doi.org/10.1038/srep10342
Sun Y, Jiao G, Liu Z, Zhang X, Li J, Guo X, Du W, Du J, Francis F, Zhao Y, Xia L (2017) Generation of high-amylose rice through CRISPR/Cas9-mediated targeted mutagenesis of starch branching enzymes. Front Plant Sci 8:1–15. https://doi.org/10.3389/fpls.2017.00298
Svitashev S, Young JK, Schwartz C, Gao H, Falco SC, Cigan AM (2015) Targeted mutagenesis, precise gene editing, and site-specific gene insertion in maize using Cas9 and Guide RNA. Plant Physiol 169:931–945. https://doi.org/10.1104/pp.15.00793
Symington LS, Gautier J (2011) Double-strand break end resection and repair pathway choice. Annu Rev Genet 45:247–271. https://doi.org/10.1146/annurev-genet-110410-132435
Takagi H, Tamiru M, Abe A, Yoshida K, Uemura A, Yaegashi H, Obara T, Oikawa K, Utsushi H, Kanzaki E, Mitsuoka C, Natsume S, Kosugi S, Kanzaki H, Matsumura H, Urasaki N,

Kamoun S, Terauchi R (2015) MutMap accelerates breeding of a salt-tolerant rice cultivar. Nat Biotechnol 33(5):445–449. https://doi.org/10.1038/nbt.3188

Tang L, Mao B, Li Y, Lv Q, Zhang L, Chen C, He H (2017) Knockout of OsNramp5 using the CRISPR/Cas9 system produces low Cd-accumulating indica rice without compromising yield. Sci Rep 7(1):14438. https://doi.org/10.1038/s41598-017-14832-9

Tashkandi M, Ali Z, Aljedaani F, Shami A, Mahfouz MM (2018) Engineering resistance against tomato yellow leaf curl virus via the CRISPR/Cas9 system in tomato. Plant Signal Behav 13(10):e1525996. https://doi.org/10.1080/15592324.2018.1525996

Tikunov YM, Roohanitaziani R, Meijer-Dekens F, Molthoff J, Paulo J, Finkers R, Capel I, Carvajal Moreno F, Maliepaard C, Nijenhuis-de Vries M, Labrie CW, Verkerke W, van Heusden AW, van Eeuwijk F, Visser RGF, Bovy AG (2020) The genetic and functional analysis of flavor in commercial tomato: the FLORAL4 gene underlies a QTL for floral aroma volatiles in tomato fruit. Plant J 103(3):1189–1204. https://doi.org/10.1111/tpj.14795

Tripathi JN, Ntui VO, Ron M, Muiruri SK, Britt A, Tripathi L (2019) CRISPR/Cas9 editing of endogenous banana streak virus in the B genome of *Musa spp.* overcomes a major challenge in banana breeding. Commun Biol 2(1):46. https://doi.org/10.1038/s42003-019-0288-7

Tyagi S, Kesiraju K, Saakre M, Rathinam M, Raman V, Pattanayak D, Sreevathsa R (2020) Genome editing for resistance to insect pests: an emerging tool for crop improvement. ACS Omega 5(33):20674–20683. https://doi.org/10.1021/acsomega.0c01435

Vu TV, Nguyen NT, Kim J, Hong JC, Kim J-Y (2024) Prime editing: Mechanism insight and recent applications in plants. Plant Biotechnol J 22:19–36. https://doi.org/10.1111/pbi.14188

Wada N, Ueta R, Osakabe Y, Osakabe K (2020) Precision genome editing in plants: state-of-the-art in CRISPR/Cas9-based genome engineering. BMC Plant Biol 20(1):234. https://doi.org/10.1186/s12870-020-02385-5

Waltz E (2016) Gene-edited CRISPR mushroom escapes US regulation, September 2015

Wang Y, Cheng X, Shan Q, Zhang Y, Liu J, Gao C, Qiu J (2014) Simultaneous editing of three homoeoalleles in hexaploid bread wheat confers heritable resistance to powdery mildew. Nat Biotechnol 32(9):947–951. https://doi.org/10.1038/nbt.2969

Wang F, Wang C, Liu P, Lei C, Hao W, Gao Y, Liu Y-G, Zhao K (2016) Enhanced rice blast resistance by CRISPR/Cas9-targeted mutagenesis of the ERF transcription factor gene OsERF922. PLoS One 11(4):e0154027. https://doi.org/10.1371/journal.pone.0154027

Wang M, Mao Y, Lu Y, Tao X, Zhu JK (2017) Multiplex gene editing in rice using the CRISPR-Cpf1 system. Mol Plant 10(7):1011–1013. https://doi.org/10.1016/j.molp.2017.03.001

Wang L, Rubio MC, Xin X, Zhang B, Fan Q, Wang Q, Ning G, Becana M, Duanmu D (2019) CRISPR/Cas9 knockout of leghemoglobin genes in *Lotus japonicus* uncovers their synergistic roles in symbiotic nitrogen fixation. New Phytol 224(2):818–832. https://doi.org/10.1111/nph.16077

Wang W, Wang W, Pan Y, Tan C, Li H, Chen Y, Liu X, Wei J, Xu N, Han Y, Gu H, Ye R, Ding Q, Ma C (2022) A new gain-of-function OsGS2/GRF4 allele generated by CRISPR/Cas9 genome editing increases rice grain size and yield. Crop J 10(4):1207–1212. https://doi.org/10.1016/j.cj.2022.01.004

Wei H, Liu J, Wang Y, Huang N, Zhang X, Wang L, Zhang J, Tu J, Zhong X (2013) A dominant major locus in chromosome 9 of rice (*Oryza sativa L.*) confers tolerance to 48°C high temperature at seedling stage. J Hered 104(2):287–294. https://doi.org/10.1093/jhered/ess103

Wu K, Wang S, Song W, Zhang J, Wang Y, Liu Q, Yu J, Ye Y, Li S, Chen J, Zhao Y, Wang J, Wu X, Wang M, Zhang Y, Liu B, Wu Y, Harberd NP, Fu X (2020) Enhanced sustainable green revolution yield via nitrogen-responsive chromatin modulation in rice. Science 367(6478):eaaz2046. https://doi.org/10.1126/science.aaz2046

Xia X, Cheng X, Li R, Yao J, Li Z, Cheng Y (2021) Advances in application of genome editing in tomato and recent development of genome editing technology. Theor Appl Genet 134(9):2727–2747. https://doi.org/10.1007/s00122-021-03874-3

Xu K, Xu X, Fukao T, Canlas P, Maghirang-Rodriguez R, Heuer S, Ismail AM, Bailey-Serres J, Ronald PC, Mackill DJ (2006) Sub1A is an ethylene-response-factor-like gene that confers submergence tolerance to rice. Nature 442(7103):705–708. https://doi.org/10.1038/nature04920

Xu R, Li H, Qin R, Wang L, Li L, Wei P, Yang J (2014) Gene targeting using the *Agrobacterium tumefaciens*-mediated CRISPR-Cas system in rice. Rice 7(1):5. https://doi.org/10.1186/s12284-014-0005-6

Xu R, Yang Y, Qin R, Li H, Qiu C, Li L, Wei P, Yang J (2016) Rapid improvement of grain weight via highly efficient CRISPR/Cas9-mediated multiplex genome editing in rice. J Genet Genomics 43(8):529–532. https://doi.org/10.1016/j.jgg.2016.07.003

Yan S, Chen N, Huang Z, Li D, Zhi J, Yu B, Liu X, Cao B, Qiu Z (2020) Anthocyanin fruit encodes an R2R3-MYB transcription factor, SlAN2-like, activating the transcription of SlMYBATV to fine-tune anthocyanin content in tomato fruit. New Phytol 225(5):2048–2063. https://doi.org/10.1111/nph.16272

Yang L, Machin F, Wang S, Saplaoura E, Kragler F (2023) Heritable transgene-free genome editing in plants by grafting of wild-type shoots to transgenic donor rootstocks. Nat Biotechnol 41(7):958–967. https://doi.org/10.1038/s41587-022-01585-8

Yao L, Zhang Y, Liu C, Liu Y, Wang Y, Liang D, Liu J, Sahoo G, Kelliher T (2018) OsMATL mutation induces haploid seed formation in indica rice. Nat Plants 4(8):530–533. https://doi.org/10.1038/s41477-018-0193-y

Yin K, Qiu J-L (2019) Genome editing for plant disease resistance: applications and perspectives. Philos Trans R Soc Lond Ser B Biol Sci 374(1767):20180322. https://doi.org/10.1098/rstb.2018.0322

Yin X, Biswal AK, Dionora J, Perdigon KM, Balahadia CP, Mazumdar S, Chater C, Lin HC, Coe RA, Kretzschmar T, Gray JE, Quick PW, Bandyopadhyay A (2017) CRISPR-Cas9 and CRISPR-Cpf1 mediated targeting of a stomatal developmental gene EPFL9 in rice. Plant Cell Rep 36(5):745–757. https://doi.org/10.1007/s00299-017-2118-z

Yuan M, Zhu J, Gong L, He L, Lee C, Han S, Chen C (2019) Mutagenesis of FAD2 genes in peanut with CRISPR/Cas9 based gene editing. BMC Biotechnol 19:24

Zabel F, Müller C, Elliott J, Minoli S, Jägermeyr J, Schneider JM, Franke JA, Moyer E, Dury M, Francois L, Folberth C, Liu W, Pugh TAM, Olin S, Rabin SS, Mauser W, Hank T, Ruane AC, Asseng S (2021) Large potential for crop production adaptation depends on available future varieties. Glob Chang Biol 27(16):3870–3882. https://doi.org/10.1111/gcb.15649

Zetsche B, Gootenberg JS, Abudayyeh OO, Slaymaker IM, Makarova KS, Essletzbichler P, Volz SE, Joung J, van der Oost J, Regev A, Koonin EV, Zhang F (2015) Cpf1 is a single RNA-guided endonuclease of a class 2 CRISPR-Cas system. Cell 163(3):759–771. https://doi.org/10.1016/j.cell.2015.09.038

Zetsche B, Heidenreich M, Mohanraju P, Fedorova I, Kneppers J, DeGennaro EM, Winblad N, Choudhury SR, Abudayyeh OO, Gootenberg JS, Wu WY, Scott DA, Severinov K, van der Oost J, Zhang F (2017) Multiplex gene editing by CRISPR-Cpf1 using a single crRNA array. Nat Biotechnol 35(1):31–34. https://doi.org/10.1038/nbt.3737

Zhang Y, Bai Y, Wu G, Zou S, Chen Y, Gao C, Tang D (2017a) Simultaneous modification of three homoeologs of TaEDR1 by genome editing enhances powdery mildew resistance in wheat. Plant J 91(4):714–724. https://doi.org/10.1111/tpj.13599

Zhang Z, Chen Y, Wang C, Wang P, Tao F (2017b) Future extreme temperature and its impact on rice yield in China. Int J Climatol 37(14):4814–4827. https://doi.org/10.1002/joc.5125

Zhang Y, Li D, Zhang D, Zhao X, Cao X, Dong L, Liu J, Chen K, Zhang H, Gao C, Wang D (2018) Analysis of the functions of TaGW2 homoeologs in wheat grain weight and protein content traits. Plant J 94(5):857–866. https://doi.org/10.1111/tpj.13903

Zhang A, Liu Y, Wang F, Li T, Chen Z, Kong D, Bi J, Zhang F, Luo X, Wang J, Tang J, Yu X, Liu G, Luo L (2019) Enhanced rice salinity tolerance via CRISPR/Cas9-targeted mutagenesis of the OsRR22 gene. Mol Breed 39:47. https://doi.org/10.1007/s11032-019-0954-y

Zhao P, You Q, Lei M (2019) A CRISPR/Cas9 deletion into the phosphate transporter SlPHO1;1 reveals its role in phosphate nutrition of tomato seedlings. Physiol Plant 167(4):556–563. https://doi.org/10.1111/ppl.12897

Zhi J, Liu X, Li D, Huang Y, Yan S, Cao B, Qiu Z (2020) CRISPR/Cas9-mediated SlAN2 mutants reveal various regulatory models of anthocyanin biosynthesis in tomato plant. Plant Cell Rep 39(6):799–809. https://doi.org/10.1007/s00299-020-02531-1

Zhou J, Peng Z, Long J, Sosso D, Liu B, Eom JS, Huang S, Liu S, Vera Cruz C, Frommer WB, White FF, Yang B (2015) Gene targeting by the TAL effector PthXo2 reveals cryptic resistance gene for bacterial blight of rice. Plant J 82(4):632–643. https://doi.org/10.1111/tpj.12838

Zhu Z, Sun B, Cai W, Zhou X, Mao Y, Chen C, Wei J, Cao B, Chen C, Chen G, Lei J (2019) Natural variations in the MYB transcription factor MYB31 determine the evolution of extremely pungent peppers. New Phytol 223(2):922–938. https://doi.org/10.1111/nph.15853

Zong Y, Wang Y, Li C, Zhang R, Chen K, Ran Y, Qiu JL, Wang D, Gao C (2017) Precise base editing in rice, wheat and maize with a Cas9-cytidine deaminase fusion. Nat Biotechnol 35(5):438–440. https://doi.org/10.1038/nbt.3811

Zsögön A, Čermák T, Naves ER, Notini MM, Edel KH, Weinl S, Freschi L, Voytas DF, Kudla J, Eustáquio L, Peres P (2018) De novo domestication of wild tomato using genome editing. Nat Biotechnol. https://doi.org/10.1038/nbt.4272

Genetically Modified Crops for Sustainable Agriculture and Food Security in Indian Scenario

2

Ruby Tiwari, Sambhavana Chauhan, Krishan Kumar, Vanga S. Reddy, and Manchikatla V. Rajam

Abstract

Genetic engineering (GE) is a powerful tool that allows the introduction of foreign DNA sequence(s) into plants resulting in the development of improved crops. Adoption of such genetically modified (GM) crops in the last two decades has been shown to benefit agriculture immensely, viz., higher crop yields with reduced input costs, development of crops resistant to a wide range of biotic and abiotic stresses, crops with improved nutritional quality, better shelf-life, and hence greater food security. GM crops developed using transgenic technology are currently grown in 29 countries spread across the globe, covering more than 190 million hectares of land. Among various commercialized GM crops, herbicide-tolerant and insect-resistant GM cotton, soybean, and maize have been in cultivation for more than 20 years in some parts of the world, proving the safety and benefits of such crops. However, in India, insect-resistant cotton is the only GM/transgenic crop approved so far for commercial cultivation, while *Bt* Brinjal developed in India is cultivated in Bangladesh. More recently, the Indian government allowed the environmental release of transgenic mustard, Dhara Mustard Hybrid-11, developed by a public sector institution. Further, several other crops are in the pipeline or awaiting regulatory approval. Considering the numerous challenges posed by agriculture to produce enough

R. Tiwari · S. Chauhan · M. V. Rajam (✉)
Department of Genetics, University of Delhi, South Campus, New Delhi, India

K. Kumar
ICAR-Indian Institute of Maize Research, Delhi Unit, Pusa Campus, New Delhi, India

V. S. Reddy (✉)
Plant Transformation Group, International Centre for Genetic Engineering and Biotechnology, New Delhi, India
e-mail: vsreddy@gmail.com

G. John, A. K. Panda (eds.), *Biotechnology in India - Reworking A Strategy*,
https://doi.org/10.1007/978-981-97-0896-3_2

quality food to feed the growing population with depleting land availability, reduced water and other resources, GM crops have a huge potential to contribute to the food security of India and address global warming and other environmental issues. However, the adoption of GM crops having foreign gene(s) raises public concerns related to the potential adverse effect on the environment and human health, etc. To address such concerns, the adoption of alternative tools like genome editing (SDN-1 and SDN-II type) may be a viable approach since the transgene-free crop plants with the desirable traits can be generated using this novel tool. In this chapter, we briefly discuss the importance of GM crops in the context of agricultural sustainability and food as well as nutritional security in the Indian scenario.

2.1 Introduction

Presently, global food production systems are under immense pressure to enhance productivity significantly in order to feed the increasing population. Climate change further aggravates the challenge that the global food production system is facing, as crop yields are expected to reduce in response to every degree Celsius rise in global mean temperature (Varshney et al. 2020). The prevalence of extreme weather conditions is anticipated to affect pests/pathogens dynamics and compromise the plant defense response (Atlin et al. 2017). Recently, it has been reported that annual yield gains for major crops, such as wheat, rice, maize, and soybean remain less than what was projected to achieve the target of doubling global production (Ray et al. 2013). Most importantly, a tremendous increase in food production has to be achieved even with depleting land and water resources while simultaneously meeting the demand for ecosystem preservation (Ronald 2014). Thus, ensuring food availability to every human being and eliminating malnutrition by 2030 will be a major challenge for the scientific community globally (including India), which can only be achieved by the collective efforts of all the stakeholders (FAO 2017). To address these challenges, the use of biotechnological approaches like genetic engineering (transgenics, cisgenic, and genome editing) is of utmost importance. Integrating these biotechnological tools with conventional breeding, speed breeding, and double haploid (DH) technology for rapid genetic gains would be beneficial to achieve food and nutritional security.

Genetic engineering (GE) refers to the artificial modification and manipulation of an organism's genetic element/genome to produce improved or novel traits. The advent of recombinant DNA and genetic transformation methods allows the transfer of beneficial gene(s) (known as a transgene) from one organism (plants, animals, or microorganisms) into a completely unrelated and sexually incompatible organism. The improved organisms/crops developed using the GE process are known as genetically modified organisms/crops (GMOs/GM crops). GM plants having foreign/transgene are termed as transgenic plants, and foods produced from such GM plants are referred to as GM foods. The "*Flavr Savr*" tomato was the first transgenic crop commercialized successfully in the USA in 1994. This modified tomato contains reduced levels of the cell wall softening enzyme, polygalacturonase resulting in delayed fruit ripening (Smith et al. 1988; Bruening and Lyons 2000). Subsequently, several other GM/transgenic food and nonfood crops have been developed and commercialized globally, including pest-resistant *Bt* cotton, *Bt* canola, herbicide glyphosate-resistant soybean and cotton, viral-disease resistant potatoes, papaya, and squash, etc. (Kumar et al. 2018, 2020). Besides, many other GM crops with the traits of phytoremediation, biofortification, and production of pharmaceuticals are either under development or in field trials and awaiting commercial release (Shahid et al. 2016). The details of GM crop events commercialized in different parts of the world are summarized in Table 2.1. In the past 23 years, the global area of transgenic crops has increased significantly from 1.7 million hectares in 1996 to 190.4 million hectares in 2019, i.e., ~112-fold increase (ISAAA 2019). In 2019, a total of 72 countries (29 countries planted while 43 imported) adopted GM crops globally.

In India, only *Bt* cotton was allowed for commercial cultivation, and no transgenic food crops have been granted approval for cultivation due to the persistent concerns associated with the use of transgenic technology, thereby delaying the wider acceptability of transgenic technology in India. India has nearly 17.5% of the world's population, which depends on merely 2.3% of the world's land. Thus, food security for its 141 crore population is the major concern for the Government of India (GOI). Recently, 109 Nobel laureates signed a letter stating, "Scientific and regulatory agencies around the world have repeatedly and consistently found crops and foods improved through biotechnology to be as safe as, if not safer, than those derived from any other method of production." The experts urged the environmental NGO, Greenpeace, and various governments across the globe to end their opposition to GMOs. Undoubtedly, this appeal is necessary for a country like India, where about 60% of the population is engaged in agriculture and there is a constant need to increase crop productivity.

Table 2.1 A list of commercially released/approved GM crops globally

Transgenic/GM crops	Trade name	Trait improved	Countries granted approval	Developers
Alfalfa (*Medicago sativa*)	Roundup Ready™ Alfalfa, HarvXtra™	Altered lignin production, glyphosate herbicide tolerance	Japan, Mexico, New Zealand, Philippines, Singapore, South Korea, United States	Monsanto Company (including fully and partly owned companies), Monsanto Company and Forage Genetics International (United States)
Apple (*Malus domestica*)	Arctic™, Arctic™ Fuji Apple, Arctic™ "Golden Delicious" Apple	Non-browning	United States	Okanagan Specialty Fruits Incorporated
Argentine Canola (*Brassica napus*)	Laurical™ Canola, Optimum® Gly canola, Roundup Ready™ Canola, Liberty Link™ Independence™, InVigor™ × TruFlex™ Roundup Ready™ Canola, Liberty Link™ Innovator™, Liberty Link™ Innovator™ × TruFlex™ Roundup Ready™ Canola, TruFlex™ Roundup Ready™ Canola, Phytaseed™ Canola, InVigor™ × TruFlex™ Roundup Ready™ Canola, Navigator™ Canola	Dicamba herbicide tolerance, fertility restoration, glufosinate herbicide tolerance, glyphosate herbicide tolerance, imazamox herbicide tolerance, male sterility, modified oil/fatty acid, oxynil herbicide tolerance, phytase production	Japan, Malaysia, Mexico, New Zealand, Philippines, Singapore, South Africa, South Korea, Taiwan, United States	BASF, Bayer Crop Science (including fully and partly owned companies), DuPont (Pioneer Hi-Bred International Inc.), Monsanto Company (including fully and partly owned companies), Nuseed Pty Ltd
Bean (*Phaseolus vulgaris*)	BRS FC401 RMD	Viral disease resistance		EMBRAPA (Brazil)
Carnation (*Dianthus caryophyllus*)	Moondust™, Moonshadow™, Moonlite™, Moonaqua™, Moonvista™, Moondust™, Moonique™, Moonpearl™, Moonberry™, Moonvelvet™	Delayed ripening/senescence, modified flower color, sulfonylurea herbicide tolerance	Japan, Malaysia, Norway	Florigene Pty Ltd. (Australia), Suntory Limited (Japan)
Chicory (*Cichorium intybus*)	Seed Link™	Glufosinate herbicide tolerance, male sterility	United States	BejoZaden BV (Netherlands)
Cotton (*Gossypium hirsutum* L.)	WideStrike™ Cotton, WideStrike™ Roundup Ready™ Cotton, Widestrike™ Roundup Ready Flex™ Cotton, Widestrike™ × Roundup Ready Flex™ × VIPCOT™ Cotton, BXN™ Plus Bollgard™ Cotton, BXN™ Cotton, VIPCOT™ Cotton, VIPCOT™ Roundup Ready Flex™ Cotton, Bollgard® III, Bollgard® III × Roundup Ready™ Flex™, JK 1, GlyTol™, GlyTol™ Liberty Link™, Glytol™ × Twinlink™ × VIPCOT™ Cotton, Fibermax™ Liberty Link™, Fibermax™ Liberty Link™ Bollgard II™, Bollgard™ Cotton, Roundup Ready™ Cotton, Bollgard II™ Cotton, Roundup Ready™ Bollgard II™ Cotton, Bollgard™ Cotton, Ingard™, Roundup Ready™ Bollgard™ Cotton, Bollgard™ Cotton, Roundup Ready™ Flex™ Cotton, Roundup Ready™ Flex™ Bollgard II™ Cotton, Ngwe Chi 6 Bt, TwinLink™ Cotton	2,4-D herbicide tolerance, dicamba herbicide tolerance, glufosinate herbicide tolerance, glyphosate herbicide tolerance, hemipteran insect resistance, isoxaflutole herbicide tolerance, lepidopteran insect resistance, low gossypol, multiple insect resistance, oxynil herbicide tolerance, sulfonylurea herbicide tolerance	India, Japan, Malaysia, Mexico, Myanmar, New Zealand, Nigeria, Pakistan, Paraguay, Philippines, Singapore, South Africa, South Korea, Sudan, Taiwan, United States	BASF, BASF and Bayer Crop Science (including fully and partly owned companies), Bayer Crop Science (including fully and partly owned companies), Cotton Research and University of Agricultural Sciences Dharwad (India), Chinese Academy of Agricultural Sciences, Cotton and Sericulture Department (Myanmar), Dow AgroSciences LLC, DuPont (Pioneer Hi-Bred International Inc.), JK Agri Genetics Ltd (India), Metahelix Life Sciences Pvt. Ltd (India), Monsanto Company (including fully and partly owned companies), Monsanto Company and Dow AgroSciences LLC, Nath Seeds/Global Transgenes Ltd (India), Syngenta, Syngenta and Monsanto Company, Texas A&M AgriLife Research University

Cowpea (*Vigna unguiculata*)	–	Lepidopteran insect resistance	Ghana, Nigeria	African Agricultural Technology Foundation (AATF)
Creeping Bentgrass (*Agrostis stolonifera*)	Roundup Ready™ Creeping Bentgrass	Glyphosate herbicide tolerance	United States	Monsanto Company and Scotts Seeds
Eggplant (*Solanum melongena*)	BARI Bt Begun-1, -2, -3 and -4	Lepidopteran insect resistance	Philippines	Maharashtra Hybrid Seed Company (MAHYCO)
Eucalyptus (*Eucalyptus* sp.)	GM Eucalyptus	Volumetric wood increase		FuturaGene Group
Flax (*Linumusitatissimum* L.)	CDC Triffid Flax	Nopaline synthesis, sulfonylurea herbicide tolerance	United States	University of Saskatchewan
Maize (*Zea mays* L.)	32138 SPT maintainer, Enogen™, Agrisure® Duracade™, Agrisure® Duracade™ 5122, Herculex™ RW, Herculex™ RW Roundup Ready™ 2, Optimum™ GAT™, Bt10, Agrisure™ CB/LL, Agrisure® 3122, Agrisure™ GT/CB/LL, Agrisure® Viptera™ 2100, Agrisure® Viptera™ 3110, Agrisure® Viptera™ 3111, Agrisure® Viptera™ 4, Agrisure™ Viptera 3220, Agrisure™ CB/LL/RW, Agrisure™ 3000GT, NaturGardKnockOut™, Maximizer™, Starlink™ Maize, Enlist™ Maize, BtXtra™ Maize, Roundup Ready™ Maize, Agrisure™GT, Roundup Ready™ YieldGard™ maize, Mavera™ Maize, Mavera™ YieldGard™ Maize, Agrisure™ Viptera, Agrisure™ RW, Agrisure™ GT/RW, YieldGard™, MaizeGard™, YieldGard™ VT Triple, Roundup Ready™ Maize, YieldGard™ Rootworm RW, MaxGard™, YieldGard™ Plus, YieldGard™ Plus with RR, YieldGard™ RW + RR, Roundup Ready™ Maize, SmartStax™ Pro × Enlist™, Genuity® DroughtGard™, YieldGard™ VT™ Rootworm™ RR2, YieldGard™ VT Pro™, Genuity® VT Triple Pro™, Genuity® VT Double Pro™, Genuity® SmartStax™, Power Core™ × MIR162 × Enlist™, InVigor™ Maize, Roundup Ready™ 2 Maize, YieldGard™ CB + RR, Roundup Ready™ Liberty Link™ Maize, GralNzyme Phytase, Liberty Link™ Maize, Liberty Link™ Yieldgard™ Maize, Herculex™ I, Herculex™ CB, Optimum™ Intrasect Xtreme, Herculex XTRA™, Optimum™ Intrasect XTRA, Herculex XTRA™ RR, Optimum™ TRIsect, Optimum™ Intrasect, Herculex™ I RR	2,4-D herbicide tolerance, coleopteran insect resistance, dicamba herbicide tolerance, drought stress tolerance, enhanced photosynthesis/yield, fertility restoration, glufosinate herbicide tolerance, glyphosate herbicide tolerance, increased ear biomass, lepidopteran insect resistance, male sterility, mannose metabolism, modified alpha amylase, modified amino acid, multiple insect resistance, phytase production, sulfonylurea herbicide tolerance	Hondurus, Indonesia, Iran, Japan, Malaysia, Mexico, New Zealand, Nigeria, Pakistan, Panama, Paraguay, Philippines, Russia, Singapore, South Africa, South Korea, Switzerland, Taiwan, Thailand, Turkey, United States, Uruguay, Vietnam, Zambia	Agrivida, Inc., Bayer Australia, Bayer Crop Science (including fully and partly owned companies), Beijing DaBeiNong Biotechnology Co. Ltd. (DBNBC), Dow AgroSciences LLC, Dow AgroSciences LLC and DuPont (Pioneer Hi-Bred International Inc.), DuPont (Pioneer Hi-Bred International Inc.), Genective S.A., Monsanto Company (including fully and partly owned companies), Monsanto Company and BASF, Monsanto Company and Bayer CropScience, Monsanto Company and Dow AgroSciences LLC, Monsanto Company and Dupont (Pioneer Hi-Bred International Inc.), Origin Agritech (China), Renessen LLC (Netherlands), Renessen LLC (Netherlands) and Monsanto Company, Stine Seed Farm, Inc. (USA), Syngenta, Syngenta and Monsanto Company

(continued)

Table 2.1 (continued)

Transgenic/GM crops	Trade name	Trait improved	Countries granted approval	Developers
Melon (*Cucumis melo*)	–	Delayed ripening/senescence	United States	Agritope Inc. (USA)
Papaya (*Carica papaya*)	Rainbow, SunUp, Huanong No. 1	Viral disease resistance	United States	Cornell University and University of Hawaii, South China Agricultural University, University of Florida
Petunia (*Petunia hybrida*)	–	Modified flower color	United States	Beijing University, WesthoffVertriebsgesellschaftmbH
Pineapple (*Ananas comosus*)	Rosé	Delayed ripening/senescenc, modified fruit color	United States	Del Monte Fresh Produce Company
Plum (*Prunus domestica*)	–	Viral disease resistance	United States	United States Department of Agriculture - Agricultural Research Service
Polish canola (*Brassica rapa*)	Hysyn 101 RR Roundup-Ready™	Glufosinate herbicide tolerance, glyphosate herbicide tolerance		Bayer Crop Science (including fully and partly owned companies), University of Florida
Poplar (*Populus* sp.)	–	Lepidopteran insect resistance, multiple insect resistance		Research Institute of Forestry (China)
Potato (*Solanum tuberosum* L.)	Lugovskoi plus, Elizaveta plus, Starch Potato, Atlantic NewLeaf™ potato, New Leaf™ Russet Burbank potato, Innate® Cultivate, Innate® Generate, Simplot Innate, Hi-Lite NewLeaf™ Y potato, New Leaf™ Y Russet Burbank potato, ShepodyNewLeaf™ Y potato, Superior NewLeaf™ potato, Innate® Invigorate, Innate® Acclimate, Innate® Hibernate, Innate	Coleopteran insect resistance, foliar late blight resistance, glyphosate herbicide tolerance, late blight disease resistance, lowered free asparagine, lowered reducing sugars, modified starch/carbohydrate, reduced black spot	Japan, Indonesia, Malaysia, Mexico, New Zealand, Philippines, Russia, Singapore, South Korea, United States	BASF, Centre Bioengineering, Russian Academy of Sciences, ICABIOGRAD, J.R. Simplot Co., Monsanto Company (including fully and partly owned companies), Technoplant Argentina
Rice (*Oryza sativa* L.)	BT Shanyou 63, Golden Rice, Liberty Link™ rice	Anti-allergy, enhanced provitamin A content, glufosinate herbicide tolerance, lepidopteran insect resistance, mannose metabolism	Mexico, New Zealand, Honduras, Iran, Philippines, Russia, South Africa, United States	Agricultural Biotech Research Institute (Iran), Bayer Crop Science (including fully and partly owned companies), Huazhong Agricultural University (China), International Rice Research Institute, National Institute of Agrobiological Sciences (Japan)
Rose (*Rosa hybrida*)	–	Modified flower color	Japan, United States	Suntory Limited (Japan)
Safflower (*Carthamus tinctorius* L.)	–	Modified oil/fatty acid		Go Resources Pvt Ltd

Soybean (*Glycine max* L.)	Liberty Link® soybean, Cultivance, Conkesta Enlist E3™ Soybean, Treus™, Plenish™, Optimum GAT™, Liberty Link® GT27™, Roundup Ready™ soybean, Liberty Link™ soybean, Verdeca HB4 Soybean, Intacta™ Roundup Ready™ 2 Pro, Vistive Gold™, Genuity® Roundup Ready™ 2 Xtend™, Genuity® Roundup Ready 2 Yield™, Herbicide-tolerant Soybean line, Liberty Link™ soybean	2,4-D herbicide tolerance, dicamba herbicide tolerance, drought stress tolerance, enhanced photosynthesis/yield, glufosinate herbicide tolerance, glyphosate herbicide tolerance, isoxaflutole herbicide tolerance, lepidopteran insect resistance, mesotrione herbicide tolerance, modified oil/fatty acid, nematode resistance, sulfonylurea herbicide tolerance, tolerance to HPPD inhibiting herbicides	Japan, Malaysia, Mexico, New Zealand, India, Indonesia, Iran, Nigeria, Paraguay, Philippines, Russia, Singapore, South Africa, South Korea, Switzerland, Taiwan, Thailand, Turkey, United States, Uruguay, Vietnam	BASF, Bayer Crop Science (including fully and partly owned companies), Dow AgroSciences LLC, DuPont (Pioneer Hi-Bred International Inc.), INDEAR, Monsanto Company (including fully and partly owned companies), Verdeca
Squash (*Cucurbita pepo*)	–	Viral disease resistance	United States	Seminis Vegetable Seeds (Canada) and Monsanto Company (Asgrow)
Sugar Beet (*Beta vulgaris*)	InVigor™ sugarbeet, Roundup Ready™ sugar beet, Liberty Link™ sugarbeet	Glufosinate herbicide tolerance, glyphosate herbicide tolerance	Japan, Mexico, New Zealand, Philippines, Russia, Singapore, South Korea, Taiwan, United States	Bayer Crop Science (including fully and partly owned companies), Monsanto Company (including fully and partly owned companies), Novartis Seeds and Monsanto Company
Sugarcane (*Saccharum* sp.)	–	Drought stress tolerance, lepidopteran insect resistance	United States, Indonesia	Centro de TecnologiaCanavieira (CTC), PT Perkebunan Nusantara XI (Persero)
Sweet pepper (*Capsicum annuum*)	–	Viral disease resistance		Beijing University
Tobacco (*Nicotiana tabacum* L.)	–	Nicotine reduction, oxynil herbicide tolerance	United States	SEITA S.A. (France), Vector Tobacco Inc. (USA)
Tomato (*Lycopersicon esculentum*)	FLAVR SAVR™	Delayed fruit softening, delayed ripening/senescenc, lepidopteran insect resistance, viral disease resistance	Mexico, United States	Agritope Inc. (USA), Beijing University, DNA Plant Technology Corporation (USA), Huazhong Agricultural University (China), Institute of Microbiology, CAS (China), Monsanto Company (including fully and partly owned companies), Zeneca Plant Science and Petoseed Company
Wheat (*Triticum aestivum*)	HB4 Wheat, Roundup Ready™ wheat	Drought stress tolerance, glyphosate herbicide tolerance	New Zealand, Nigeria, United States	Bioceres S.A., Monsanto Company (including fully and partly owned companies)

Source: ISAAA, GM Approval Database retrieved on 1 February 2023 (http://www.isaaa.org/)

2.2 Regulatory Framework for Research and Development of GM Crops in India

In India, living modified organisms (LMOs) regulation was initiated in 1982, when the National Biotechnology Board which became the Department of Biotechnology (DBT) implemented guidelines on biotechnology research (Chaturvedi 2004; Sharma et al. 2003). At present, all activities pertaining to LMOs/GMOs, cells, or hazardous microorganisms and products thereof are regulated as per the "Rules for the manufacture, use/import/export, and storage of hazardous micro-organisms/GMOs or cells 1989" notified by the Ministry of Environment, Forest and Climate Change (MoEF&CC), Government of India, under the Environment (Protection) Act, 1986 (www.dbtindia.nic.in). The EPA not only covered the regulations on the use of hazardous chemicals and waste management but also laid down strict guidelines for the use of GM organisms raised using biotechnology, under the section hazardous substances (Shukla et al. 2018). There are six competent authorities/bodies to implement biosafety regulations under the Rules 1989: RDAC (Recombinant DNA Advisory Committee), IBSC (Institutional Biosafety Committees), RCGM (Review Committee on Genetic Manipulation), GEAC (Genetic Engineering Appraisal Committee), SBCC (State Biotechnology Coordination Committees), and DLC (District Level Committees). All these authorities have very specific roles. The RDAC is advisory in function, while IBSC, RCGM, and GEAC are responsible for the regulation and approval of GMOs. SBCC and DLC are involved in monitoring the activities related to GMOs at the state and district levels, respectively (www.dbtindia.nic.in; Shukla et al. 2018).

RDAC is a DBT-constituted committee to review developments in biotechnology within the country and also provide suggestions for the successful application and appropriate regulation of recombinant DNA (rDNA) technology. RCGM is involved in formulating guidelines for regulatory processes related to the activities involving GE organisms and updating them from time to time for the regulation of recombinant DNA technology (Shukla et al. 2018). It monitors the safety-related aspects of all ongoing research projects and activities involving GE organisms/hazardous microorganisms. RCGM approves research on transgenics, provides clearance on the use of vectors, transgenic germplasm, and other related research material, and also directs the generation of associated relevant scientific data on other aspects such as allergenicity and toxicity (Shukla et al. 2018; Kanaujia and Bhattacharya 2021). IBSC is formulated by individual institutions to implement guidelines and keep a regular check on research involving genetic manipulations of organisms such as microorganisms (bacteria, bacteriophages, fungus, etc.), plants, and animals, as per EPA 1986 rules. GEAC is the apex regulatory body governed by MoEF & CC for the regulation of GE organisms/hazardous microorganisms as per the rules of EPA 1989. GEAC is the governing body, which has the overall responsibility for appraisal of actions related to large-scale use or production of genetically manipulated or hazardous microorganisms in research or industry, for ensuring environmental safety. It is also responsible for the appraisal of proposals pertaining to the release of GE organisms and products into the environment including experimental field trials. After GEAC approval, a "no objection certificate" from the respective state is a must for the field trial of any GM crop. However, such NOC is not mandatory for Event Selection Trials (EST) when conducted in the applicant's own premises or on long-term leased land. Nevertheless, ESTs shall follow SOPs similar to BRL trials related to handling, storage, and disposal of seed and/or waste generated during the trial. SBCC and DLC are executed by the chief secretary of state and district administrative officer, respectively. The role of these committees is to monitor and ensure the safety regulations related to the use and application of LMOs/GMOs in institutions/installations at the state and district levels (Shukla et al. 2018). Apart from these six major bodies, other subcommittees are constituted by RCGM and GEAC based on cases. These sub-committees comprise experts from different areas concerning ethical rights, environmental rights, and research, and they formulate the various guidelines under EPA and examine the biosafety data with precision (Ahuja 2018). Thus, a comprehensive biosafety regulatory system is in place to review the ongoing research on GE crops in India.

2.3 Regulatory Process for Commercialization of GM Crops in India

The regulatory steps taken by RCGM and GEAC in India ensure the stringent screening of GM crops to be certain about the biosafety of humans and the environment, before their commercialization. The applicant has to first file for approval of the GMOs with the RCGM through IBSC, where RCGM reviews the proposal and recommends generating all the necessary biosafety data by conducting field trials and producing the necessary allergenicity, toxicity, and off-targeting data (Craig et al. 2012). If the risk is minimal, the RCGM grants further approval for conducting field trials to generate data related to the agronomic impacts of GM crops (Craig et al. 2012). These trials are either conducted by ICAR or any other organization/institute, and the reports are directly submitted to RCGM or GEAC. Based on field trial data and the performance of GMOs, the RCGM refers the case to GEAC for further investigation of rightful implementation of all the rules and regulations as per EPA (Craig et al. 2012). From here, the GEAC supervises the case and may ask the third party, i.e., industry or government/private institutions to conduct trials at different locations. At this step, GEAC informs every person related or unrelated to the case and associated with the trials to inform the committee immediately, if any adverse effects are observed. GEAC reviews the field trial data and makes recommendations to MoEF & CC about the commercialization of GMOs, and this forms the basis for further evaluation of the GMOs and decides upon the liability of the product (Craig et al. 2012). The environmentally safe and viable GM product shall be considered for commercialization under specific terms and conditions with a limited period, which has to be renewed after evaluation every time (Craig et al. 2012).

2.4 Milestones Related to GM Crops in India

The timeline of controversies and moratoriums associated with GM crops in India is as follows:

- 2002: Introduction of insect-resistant transgenic *Bt* cotton in India (Whitfield 2003).
- 2005: Public interest litigation filed by activists against GM crops in the Supreme Court of India (Padma 2014).
- October 2009: GEAC cleared transgenic *Bt* brinjal/eggplant for environmental release (http://www.geacindia.gov.in/Uploads/MoMPublished/2009-geac-97.pdf).
- February 2010: India puts an indefinite moratorium on the approval of *Bt* eggplant to conduct long-term studies to ensure human and environmental safety (https://www.isaaa.org/kc/cropbiotechupdate/article/default.asp?ID=5430).
- May 2017: The locally developed transgenic mustard hybrid was cleared by GEAC, however, the government deferred the approval.
- August 2017: The Supreme Court issued directives for the Food Safety and Standards Authority of India (FSSAI) to regulate the approval of GE food products (https://apps.fas.usda.gov/newgainapi/api/Report/DownloadReportByFileName?fileName=Agricultural%20Biotechnology%20Annual_New%20Delhi_India_10-20-2019#:~:text=On%20August%2011%2C%202017%2C%20the,products%20to%20FSSAI%20for%20approval).
- 2020: The FSSAI ordered that every food product consignment requires an accompanying "non-GM origin cum GM free" certificate issued by the competent authority in the country of export, effective from 1st March 2021 (https://www.fssai.gov.in/upload/advisories/2020/10/5f8583d9bac5eClarification_GM_Food_13_10_2020.pdf).
- March 2022: The government of India exempted genome-edited plants falling under the categories of SDN1 and SDN2 from the provisions of Rules 7–11 (both inclusive) of the Rules, 1989, i.e., exemption from stringent GM regulations (https://ibkp.dbtindia.gov.in/Content/Rules).
- May 2022: RCGM, Department of Biotechnology (DBT) notifies the guidelines for safety assessment of genome-edited plants (https://ibkp.dbtindia.gov.in/Content/Rules).
- October 2022: RCGM, DBT prepared the standard operating procedures for regulatory review of genome-edited plants under SDN-I and SDN-II categories (https://ibkp.dbtindia.gov.in/Content/Rules).

2.5 Current Scenario of GM Crops in India

In India, several public research institutions and universities funded by the Indian Council of Agricultural Research (ICAR) and DBT are working on many research projects related to the development of GM crops with improved trait(s) such as banana, cabbage, cassava, cauliflower, chickpea, cotton, eggplant, papaya, peanut, pigeon pea, potatoes, rapeseed/mustard, rice, sorghum, sugarcane, tomato, watermelon wheat, etc. (Singh 2020; Singh and Beillard 2020). Apart from government institutions, many private companies have also invested in research and development of GM crops. Many GM food and non-food crops are in different stages of development or limited field trials or are ready for commercialization amid strong opposition from non-government organizations (NGOs) and activists.

Owing to the moratorium on transgenic crops, GM (except *Bt* cotton) crops are not allowed for cultivation in India. However, vegetable oils derived from GM soybean (five events) and canola are the only products approved for import in India. Further, a lot of transgenic germplasm have been imported previously into India (mostly by private companies) for breeding purposes, by following established norms. Further, the multi-location field trials [Biosafety Research Level-1 (BRL-1) and/or BRL-II trials] of these imported transgenic germplasm have also been done, mainly for insect resistance and herbicide tolerance traits. The moratorium and ban on transgenic food crops constrained investment in the research and development of GE crops by the private sector in the last decade. Also, companies working on the development of GE crops have explored options to commercialize their products in other countries, for instance, *Bt* eggplant event EE1 developed by Maharashtra Hybrid Seed Company (MAHYCO) was commercialized in Bangladesh and the Philippines in 2013 and 2022, respectively (Singh 2020; Singh and Beillard 2020: ISAAA database).

2.5.1 *Bt* Cotton

Bt cotton, an insect-resistant transgenic crop that contains the *cry* gene(s) from *Bacillus thuringiensis* (*Bt*) bacteria to control the bollworms by selectively damaging insect midgut lining, is the only GM crop approved for commercial cultivation in 2002 by GEAC and MoEF & CC in India. To date, six events of *Bt* cotton have been approved for commercial cultivation (Table 2.2) and nearly 95% (11.9 million hectares) of India's total cotton acreage are under *Bt* cotton hybrids.

2.5.2 *Bt* Brinjal

Fruit and shoot borer (*Leucinodes orbonalis*) resistant transgenic *Bt* brinjal was developed by MAHYCO in collaboration with Tamil Nadu Agricultural University, Coimbatore, University of Agricultural Sciences, Dharwad, and ICAR-Indian Institute of Vegetable Research (IIVR), Varanasi, using the *cry1Ac* gene. In 2009, after reviewing the *Bt* brinjal dossier encompassing data on safety and potential risks, event "EE-1" was declared safe for environmental release by GEAC and recommended its commercial approval to the MoEF & CC, Government of India (GOI) (Millstone 2014). However, in 2010, due to the unwillingness of many state

Table 2.2 GM cotton events approved in India

Event name	Gene used for insect resistance	Developer	Year of approval in India
MON531	*cry1Ac*	Monsanto	2002
MON15985	*cry1Ac* and *cry2Ab2* (stacked gene)	Monsanto	2006
GFM Cry1A	*cry1Ab-Ac* (synthetic fusion gene)	Nath Seeds/Global Transgenes Ltd	2006
BNLA-601	*cry1Ac*	ICAR-Central Institute for Cotton Research (CICR) and University of Agricultural Sciences Dharwad	2008
MLS 9124	*cry1C*	Metahelix Life Sciences Pvt. Ltd	2009
Event1	*cry1Ac*	JK Agri Genetics Ltd	2018

Source: ISAAA, GM Approval Database (http://www.isaaa.org/)

governments to permit commercial cultivation of *Bt* brinjal in their states, and concerns raised by a few nongovernment organizations (NGOs)/activists, farmers, and scientists regarding the inadequacy of tests related to the toxicity and safety of *Bt* brinjal, MoEF & CC imposed a 10-year moratorium on the commercial release of "EE-1" event (Srinivas 2017). Recently in 2020, the GEAC approved confined field trials (BRL-II in 8 states for 3 years, i.e., 2020–2023) of 2 *Bt* brinjal varieties, namely "Janak" and "BSS-793" developed by ICAR-National Institute for Plant Biotechnology (NIPB), New Delhi, using the indigenous *cryFa1* gene (event 142).

2.5.3 GM Mustard

GM mustard (*Brassica juncea*), Dhara Mustard Hybrid-11 (DMH 11), and its parental lines, namely bn 3.6 and modbs 2.99 were developed in 2002 by Prof. Deepak Pental's team at the University of Delhi, South Campus, to improve mustard yield in the country, which is typically lower than the global average (Rani et al. 2018; Padma 2022). Hybrids are known for higher yield due to hybrid vigor or heterosis. Since mustard is a self-pollinating crop, natural hybridization does not take place. To prevent self-pollination, the Delhi University team used *barnase/barstar* transgene system for hybrid seed production (Kumar et al. 2018) in DMH 11. The male sterile line was obtained by expressing *barnase* gene derived from the soil bacterium *Bacillus amyloliquefaciens* in an Indian mustard variety and the resulting transgenic plants were producing sterile pollen grains and hence used as the female line. The male parent line (restorer) was obtained by expressing the *barstar* gene derived from the same soil bacteria in an East European variety to restore fertility in the F_1 hybrid, i.e., DMH 11. Thus, DMH 11 hybrid contains both *barnase* and *barstar* genes, and they produce fertile pollen grains so that the farmer gets fully fertile seeds (Jayaraman 2017). To facilitate hybrid seed production, a third gene, *bar,* derived from another soil bacterium, governing glufosinate herbicide tolerance was introduced for selecting transgenic plants/events. In October

2022, GEAC and MoEF & CC approved the conditional environmental release of GM mustard for 4 years for its seed production and testing as per existing ICAR guidelines and other extant rules/regulations. GEAC also recommended that GM mustard can be used for developing new parental lines and hybrids, and the field demonstration studies on the effect of transgenic mustard on honeybees and other pollinators shall also be conducted post-environmental release by the applicant within 2 years under the supervision of ICAR. Further, the usage of herbicide was recommended only under controlled and specified conditions exclusively for hybrid seed production and not for cultivation in the farmer's field (147th GEAC meeting, 2022). However, following the environmental release, an NGO named Gene Campaign and some activists filed petitions in the Supreme Court of India seeking a moratorium/ban on GM mustard approval based on the potential threat to biodiversity, biosafety, and the environment. In the apex court, the GOI advocated environmental release stating that conditional approval covering every aspect of risk is given only after following stringent regulatory procedures and rigorous scrutiny over 12 years. Undoubtedly, GM technology has the potential to reduce forex drain on agricultural imports as India meets about 60% of vegetable oil requirements by importing oil derived from GM crops like canola and soybean. All three transgenes have nearly 20 years of history of safe use in GM rapeseed, which was approved for cultivation in Canada, the USA, and Australia in 1995, 2002, and 2003, respectively (ISAAA database).

2.6 Concerns Associated with Transgenic GM Crops

It has been shown in many studies that the global adoption of transgenic technology has increased agricultural productivity, diminished pesticide and insecticide use, decreased CO_2 emission, reduced crop production cost, and thereby benefited farmers economically worldwide (Brookes and Barfoot 2013; Brookes 2022a, b, c; ISAAA 2019; Klümper and Qaim 2014). Despite the potential benefits of GM crops, transgenic crops have been an unresolved issue, especially in India. A section of people believe that transgenic GM crops may address hunger issues, bringing food and commodities prices down, while the other section argues the potential risk of such crops to the environment and human health. The major concerns against transgenic crops are described below.

2.6.1 Human and Animal Health Concerns

The issue of the potential human and animal health risks of toxicity and allergenicity associated with transgenic food crops has always been controversial. The potential implications of such crops on human/animal health might be due to (a) the inherent toxicity of transgene, (b) the potential to provoke allergenicity, (c) the potential to alter the level of existing allergens, and (d) the potential for insertional mutagenesis and pleiotropic effect of the products of transgene expression (Chekol

2021). A foreign gene encoding for a protein may interact with the neighboring proteins and regulate differential expression under different external environment influences (epigenetically modified). Despite public concerns, transgenic crops have been consumed by people in more than 20 countries for many years, which proves their safety. Furthermore, with a few exceptions, the majority of research conducted in the last decade on the potential health risks of transgenic crops has not documented the negative consequences of consuming transgenic food on animal health.

2.6.2 Ecological Concerns

Cross-pollination of transgenic plants with wild relatives or non-GM cultivars may result in pollen-mediated transgene flow and thereby pose a risk to biodiversity (Chen et al. 2004; Han et al. 2015; Yan et al. 2015). In addition, if the resistant traits are transferred to weeds or unwanted species, their control would become even more difficult (McHughen 2012; Bawa and Anilakumar 2013; Kumar 2014). Potential unintended effects on pollinators and honeybees are also possible as foreign protein(s) ingestion may lead to chemical reactions, which are not fit for their survival. This may sharply decline the numbers of these beneficial organisms. However, all these possibilities are taken into account while following the established regulatory process before the final release of transgenic crops for commercial cultivation.

2.6.3 Environmental Concerns

GM plants pose considerable risk of contaminating the preexisting gene pool of cultivated or wild-growing plants, which may disturb the natural biodiversity and create new variations, and give rise to a new species. Subsequently, this can increase the competition, which might affect the growth of natural varieties and hence may lead to the loss of cultivated crops (McHughen 2012; Bawa and Anilakumar 2013). Similarly, a lack of knowledge about the modified traits in the dynamic population may result in the transfer of insect or pest-resistant transgene(s) in other species. This could result in the development of new resistant species of insects and pests, which will further lead to insensitivity toward various insecticide and pesticide dosages (McHughen 2012; Bawa and Anilakumar 2013). Further, the extensive cultivation of herbicide-tolerant and insect-resistant transgenic crops may increase the possibility of resistance development in targeted weeds and insect populations, respectively, due to high selection pressure. For instance, field-evolved pest resistance to *Bt* crops such as *Bt* cotton and *Bt* maize has been reported previously (Bagla 2010; Tabashnik and Carrière 2010; Tabashnik et al. 2013). Development of resistance against pesticides over a period of time is also prevalent. The build-up of resistance against long-term use of any control mechanism is a natural phenomenon in insects. Despite these few occurrences, the overall experience with *Bt* crops has shown that resistance in transgenic *Bt* crops can continue to be effective against the majority of pests even after 10 years (Tabashnik et al. 2008; Carrière et al. 2010). Additionally, the pyramiding of two or more insect-resistant genes is an important approach that has been employed recently to successfully delay the resistance breakdown.

In India, each transgenic crop is approved for release only after a thorough examination of health, environment, food, and feed safety assessment studies undertaken systematically and scientifically as per prescribed guidelines and standard operating procedures stipulated by the various regulatory agencies under the Rules, 1989 from time to time. Despite this, less public/consumer acceptance of transgenic food crops due to misinformation and confusion about such crops is a major issue, as the development and commercialization of any GM product requires a huge investment (Wunderlich and Gatto 2015; Savadori et al. 2013). This is because most people are unaware of transgenic technology and its benefits. There is a lack of knowledge and understanding in public on this topic. The development of GM crops and approval by GEAC for commercialization is only fruitful if people accept GM crops for consumption (Goyal and Gurtoo 2011; Wunderlich and Gatto 2015). Thus in India, there is an urgent need for evidence-based educational programs to raise public awareness about transgenic crops through social media, scientists, politicians, administration and individuals, etc. (Craig et al. 2012; Shukla et al. 2018).

2.7 CRISPR/Cas Genome-Editing Tool

Public concern and less consumer acceptance of transgenic crops prompted researchers to develop alternative technologies such as genome editing in the last decade. Genome-editing tools make use of the engineered site-specific nucleases (SSNs) that recognize and cleave the genomic DNA in a sequence-specific manner. There are three main genome-editing tools, namely zinc-finger nucleases (ZFNs), transcription activator-like effector nucleases (TALENs), and recently developed clustered regularly interspaced short palindromic repeats (CRISPR)/CRISPR-associated protein-9 nuclease (Cas9). Among these tools, CRISPR/Cas9 is the most precise, cheap, highly efficient, robust, and hence widely used, which has enabled scientists to modify crop genomes with unprecedented ease and accuracy. The genome-editing approach such as CRISPR/Cas utilizes a genetic modification step, but the resulting end products/plants mostly (but not always) do not contain any transgene (Kumar et al. 2020). Therefore, such transgene-free genome-edited (GEd) crop varieties have the potential to address most of the public concerns raised against transgenic GM crops. The CRISPR/Cas genome-editing system is discussed below briefly.

In the last decade, CRISPR/Cas genome-editing tool has been successfully employed to create beneficial mutations/variations in the pre-existing germplasm or gene pool of different crop plants (Jiang et al. 2013; Li et al. 2013; Shan et al. 2013; Globus and Qimron 2018; Agarwal et al. 2018; Kumar et al. 2020). This versatile and user-friendly technology can edit the targeted sequences such as gene/promoter/

loci in the genome of any plant (Li et al. 2013; Shan et al. 2013). The CRISPR/Cas system has a Cas endonuclease derived from bacteria and a single guide RNA (sgRNA)—containing a 20 base pair (bp) target site sequence and a hairpin structure (Agarwal et al. 2018). The RNA-guided endonuclease, i.e., Cas, recognizes and cleaves the target genomic DNA complementary to sgRNA and thereby generates double-strand breaks (DSBs) in homologous pairs of chromosomes. These DSBs are repaired by either homology-directed repair (HDR) or error-prone nonhomologous end joining method (NHEJ), leading to gene modification (insertion/deletion/substitution of a few nucleotides or targeted insertion of new DNA sequence/gene, etc.) at the target sites (Kumar et al. 2020). CRISPR/Cas-based editing results in three different types of output, namely site-directed nuclease (SDN)-I, SDN-II, and SDN-III. Transgene-free genetically engineered genotypes with improved or novel traits can be developed using SDN-I and II-type editing approaches. Mutations induced via SDN-I and II approaches may either disturb the reading frame of the targeted endogenous gene or alter the regulatory element sequence resulting in gene knockouts or altered expression of the target gene, respectively.

2.8 Regulation of Genome-Edited Crops in India

As we know, transgenic crops with any foreign gene have to pass stringent, time-consuming, and expensive regulatory processes before commercialization in India. However, in 2022 GOI exempted SDN-I and SDN-II type CRISPR edited crops from the provisions of Rules 7–11 (both inclusive) of the Rules, 1989, i.e., cumbersome GM regulations, which in turn would result in easy and fast commercialization of such GEd crops. More recently, RCGM notified the standard operating procedures for regulatory review of SDN-I and SDN-II type GEd plants (https://ibkp.dbtindia.gov.in/).

Many transgene-free SDN-I and SDN-II type crops are under different stages of development in India. For instance, transgene-free drought and salt-tolerant rice developed at IARI, New Delhi, via CRISPR/Cas9-based editing of *dst* gene in mega rice variety MTU1010 (Kumar et al. 2020) is in the advanced stage of field trials and may be commercialized in near future.

Every technology has limitations of its own. For instance, for developing non-transgenic GEd crops with novel or improved traits, the target gene or genetic elements to be edited must be present within the same crop genome, i.e., endogenously. Further, the target gene should also be well-characterized functionally. This means that if the gene governing the desired trait is present in other species, i.e., sexually incompatible species/gene pool and not present in the same genome, in such a situation, one must use transgenic/GM technology for introducing the desired trait (for example, *cry* gene from *Bacillus thuringiensis* bacterium utilized for imparting insect resistance trait in crop plants). In a nutshell, despite the advent of genome-editing techniques, transgenic technology will probably continue to be important in the future.

2.9 Conclusions and Way Forward

In India, the regulation of transgenic GM crops is very stringent as they have to pass through various committees. Basic biotechnology research on any organism, be it microorganisms, plants, or animals has to be approved by the IBSC and RCGM committees, as per EPA 1989 rules. For commercialization, the applicant has to apply to RCGM through IBSC and follow the various guidelines and SOPs issued by RCGM/GEAC from time to time. After the RCGM recommendation which is based on BRL-I trials(s) data, the GEAC is the final appraisal committee that recommends to MoEF & CC for environmental release of GM crops, after a complete risk assessment that includes multilocation BRL-II trial data. The entire regulatory process to bring out any GM crop/product successfully to the market is very lengthy, requires huge investments during R&D as well as regulatory appraisal phase, and has to face many challenges including the uncertainty of getting NOC from the relevant state government to conduct field trials for obtaining the final approval by the government. Instead of issuing authorization on a case-by-case and state-by-state basis, it would be preferable if the Government of India grant ICAR institutions site-specific, one-time permission to undertake contained field trials of any transgenic crops. Since the relevant state government's permit is not required separately each time, this could save some time. Public concerns related to foreign/transgene insertion could be addressed by employing alternative technologies like cis-genesis and CRISPR/Cas-based genome editing. Using the CRISPR/Cas editing system, precise variations at the desired locus can be created, without any transgene in the final product/plant in case of SDN-I or SDN-II events. Such events not only reduce the cost involved and more significantly make the regulatory approval process simplified for environmental release. However, SDN-III events require RCGM/GEAC approvals for environmental release which are similar to the one followed for GM crops as they do contain foreign DNA. Therefore, despite the advancement of genome-editing technologies, one cannot completely overlook the significance of transgenic technology and transgenic crops in the future. In order to bring out new varieties of crop plants with better yields using less water, withstanding biotic and abiotic stress, and nutritional quality for human and animal consumption, it is desirable to have a more transparent and predictable regulatory approval system for GM crops. Over the years, DBT has played a significant role in promoting biotechnological interventions to improve crop plants through its R&D support. Also, DBT has played a proactive role to streamline the process, revised various biosafety guidelines that are now at par with internationally well-established guidelines for R&D at the institutional level, as well as for product development leading to commercialization. More significantly, DBT through BIRAC has supported a number of Indian public as well as private sector institutions to generate biosafety data which is crucial for regulatory appraisal. Another new and positive development is the invitation of comments from the general public and other stakeholders on the safety assessment dossier prepared by the GEAC before recommending to the government by making the entire safety assessment report of GM crop available and making the decision-making process more inclusive and transparent. In addition, DBT/RCGM has established an Indian Biosafety Knowledge Portal

(IBKP) for the submission of all applications seeking regulatory approvals for R&D and product development. However, a major hurdle in the Indian regulatory appraisal system is the uncertainty and/or inability to get a mandatory NOC from the concerned state governments to conduct biosafety research at field level to generate data that is necessary for the safety assessment by the RCGM and GEAC for decision-making. To overcome such state NOC requirement for each BRL trial, GEAC is making efforts to establish notified field trial sites (NFTs) across the country where such trials can be conducted to generate biosafety data. By using cisgenic crops where a naturally occurring gene sequence is taken from the same species or sexually compatible plant species and introduced into cultivated variety using recombinant DNA technology, there is a reduction of 5–7 years in developing new varieties as compared to conventional breeding. Therefore, cisgenic crops free from any maker gene may be considered for the exemption from the provision of Rules 7–11 under the Rule 20 of Rules, 1989 of the EPA, 1986. Streamlining the regulatory pathway by addressing all the bottlenecks is key to bringing in additional investments required to develop environmentally resilient crops that will benefit Indian farmers, agri-industry, and most importantly the public at large and make India highly self-reliant in terms of food security and nutrition security.

Author Contributions Statements VSR and MVR have conceived the concept. RT, SC, and KK wrote the manuscript. MVR, VSR, and KK revised the manuscript. All the authors approved the final manuscript for submission.

Conflict of Interest The authors declare that they have no conflicts of interest.

References

Agarwal A, Yadava P, Kumar K, Singh I, Kaul T, Pattanayak A, Agrawal PA (2018) Insights into maize genome editing via CRISPR/Cas9. Physiol Mol Biol Plants 24:175–183

Ahuja V (2018) Regulation of emerging gene technologies in India. BMC Proc 12:5–11

Atlin GN, Cairns JE, Das B (2017) Rapid breeding and varietal replacement are critical to adaptation of cropping systems in the developing world to climate change. Glob Food Sec 12:31–37

Bagla P (2010) Hardy cotton-munching pests are latest blow to GM crops. Science 327:1439

Bawa AS, Anilakumar KR (2013) Genetically modified foods: safety risks and public concerns—a review. J Food Sci Technol 50:1035–1046

Brookes G (2022a) Genetically modified (GM) crop use 1996–2020: environmental impacts associated with pesticide use change. GM Crops Food 13:262–289

Brookes G (2022b) Farm income and production impacts from the use of genetically modified (GM) crop technology 1996-2020. GM Crops Food 13:171–195

Brookes G (2022c) Genetically modified (GM) crop use 1996–2020: impacts on carbon emissions. GM Crops Food 13(1):242–261

Brookes G, Barfoot P (2013) The global income and production effects of genetically modified (GM) crops 1996-2011. GM Crops Food 4:74–83

Bruening G, Lyons JM (2000) The case of the FLAVR SAVR tomato. Calif Agric 54:6–7

Carrière Y, Crowder DW, Tabashnik BE (2010) Evolutionary ecology of insect adaptation to Bt crops. Evol Appl 3:561–573

Chaturvedi S (2004) Biosafety regulation: need for fine balancing. Econ Polit Wkly 39(33):3693–3697

Chekol C (2021) The health effects of genetically modified foods: a brief review. Int J Nutr Sci 6(1):1047

Chen LJ, Lee DS, Song ZP, Suh HS, Lu B (2004) Gene flow from cultivated rice (*Oryza sativa*) to its weedy and wild relatives. Ann Bot 93:67–73

Craig W, Vanga SR, Medaglia JC (2012) Commercialisation of GM crops: comparison of regulatory frameworks. VGB Kraftwerkstechnik 77(10):802–806

FAO, IFAD, UNICEF, WFP, WHO (2017) The state of food security and nutrition in the world. Building resilience for peace and food security. FAO, Rome

Globus R, Qimron U (2018) A technological and regulatory outlook on CRISPR crop editing. J Cell Biochem 119(2):1291–1298

Goyal P, Gurtoo S (2011) Factors influencing public perception: genetically modified organisms. GMO Biosaf Res 2

Han SM, Lee B, Won OJ, Hwang KS, Suh SJ, Kim C, Park KW (2015) Gene flow from herbicide resistant genetically modified rice to conventional rice (*Oryza sativa* L.) cultivars. J Ecol Environ 38:397–403

ISAAA (2019) Global status of commercialized biotech/GM crops in 2019: biotech crops drive socio-economic development and sustainable environment in the new frontier continue to help meet the challenges of increased population and climate change. *ISAAA* brief no. 55. ISAAA, Ithaca

Jayaraman K (2017) Activists bury India's GM mustard hopes. Nat Biotechnol 35:1124. https://doi.org/10.1038/nbt1217-1124a

Jiang W, Zhou H, Bi H, Fromm M, Yang B, Weeks DP (2013) Demonstration of CRISPR/Cas9/sgRNA-mediated targeted gene modification in Arabidopsis, tobacco, sorghum and rice. Nucleic Acids Res 41(20):e188

Kanaujia A, Bhattacharya S (2021) Genetically Modified Crops and Indian Agriculture: Issues Relating to Governance and Regulation. In: Sudesh Ratna R, Sharma SK, Kumar R, Dobhal A. (eds) Indian Agriculture Under the Shadows of WTO and FTAs. India Studies in Business and Economics. Springer, Singapore. https://doi.org/10.1007/978-981-33-6854-5_11

Klümper W, Qaim M (2014) A meta-analysis of the impacts of genetically modified crops. PLoS One 9:e111629

Kumar S (2014) Biosafety issues of genetically modified organisms. Biosafety 3:e150

Kumar K, Aggarwal C, Sapna B, Singh I, Yadava P (2018) Microbial genes in crop improvement. Crop improvement through microbial biotechnology. Elsevier, Amsterdam, pp 39–56

Kumar SVV, Verma RK, Yadav SK, Yadav P, Watts A, Rao MV, Chinnusamy V (2020) CRISPR-Cas9 mediated genome editing of drought and salt tolerance (OsDST) gene in *indica* mega rice cultivar MTU1010. Physiol Mol Biol Plants 26:1099–1110

Li JF, Norville JE, Aach J, McCormack M, Zhang D, Bush J, Church GM, Sheen J (2013) Multiplex and homologous recombination–mediated genome editing in Arabidopsis and *Nicotiana benthamiana* using guide RNA and Cas9. Nat Biotechnol 31:688–691

McHughen A (2012) Introduction to the GM crops special issue on biosafety, food and GM regulation. GM Crops Food 3(1):6–8

Millstone E (2014) Science and politics in Indian GM crop regulation: a u-turn down a blind alley. In: Moor R, Gowda MVR (eds) India's risks: democratizing the management of threats to environment, health and values. Oxford University Press, New Delhi, pp 205–226

Padma, T (2014) India stalls on GM crops. Nature https://doi.org/10.1038/nature.2014.15677

Padma TV (2022) When will India approve its first GM crop? Nature 611:648–649

Rani MS, Satish Y, Rani C, Prasad NVVSDBS, Sri Lakshmi B, Kumari RS (2018) History, status and impact of genetically modified crops in India. Int J Chem Stud 6(5):2735–2739

Ray DK, Mueller ND, West PC, Foley JA (2013) Yield trends are insufficient to double global crop production by 2050. PLoS One 8:e66428

Ronald PC (2014) Lab to farm: applying research on plant genetics and genomics to crop improvement. PLoS Biol 126:e1001878

Savadori L, Savio S, Nicotra E, Rumiati R, Finucane ML, Slovic P (2013) Expert and public perception of risk from biotechnology. In: Slovic P (ed) The feeling of risk. Routledge, London, pp 273–288

Shahid R, Sahrish F, Hira M (2016) GMO's: applications, uses and future prospects. Eur J Pharm Med Res 3(1):53–55

Shan Q, Wang Y, Li J, Zhang Y, Chen K, Liang Z, Zhang K, Liu J, Xi JJ, Qiu JL, Gao C (2013) Targeted genome modification of crop plants using a CRISPR-Cas system. Nat Biotechnol 31(8):686–688

Sharma M, Charak KS, Ramanaiah TV (2003) Agricultural biotechnology research in India: status and policies. Curr Sci 84:297–302

Shukla M, Al-Busaidi KT, Trivedi M, Tiwari RK (2018) Status of research, regulations and challenges for genetically modified crops in India. GM Crops Food 9:173–188

Singh SK (2020) Agricultural biotechnology annual report. In: Wallace M (ed) United State Department of Agriculture. New Delhi, IN2019-0109

Singh SK, Beillard MJ (2020) Agricultural biotechnology annual—2020. In: Beillard M (ed) United State Department of Agriculture, New Delhi, IN2020-0148

Smith CJS, Watson CF, Ray J, Bird CR, Schuch W, Grierson D (1988) Antisense inhibition of polygalacturonase expression in transgenic tomatoes. Nature 334:724–726

Srinivas KR (2017) The regulatory regime for genetically modified crops in India. In: Adenle AA et al (eds) Genetically modified organisms in developing countries: risk analysis and governance. Cambridge University Press, Cambridge, pp 236–246

Tabashnik BE, Carrière Y (2010) Field-evolved resistance to Bt cotton: bollworm in the US and pink bollworm in India. Southwest Entomol 35:417–424

Tabashnik BE, Gassman AJ, Crowder DW, Carriere Y (2008) Insect resistance to Bt crops: evidence versus theory. Nat Biotechnol 26:199–202

Tabashnik BE, Brévault T, Carrière Y (2013) Insect resistance to Bt crops: lessons from the first billion acres. Nat Biotechnol 31:510–521

Varshney RK, Sinha P, Singh VK, Kumar A, Zhang Q, Bennetzen JL (2020) 5Gs for crop genetic improvement. Curr Opin Plant Biol 56:190–196

Whitfield, J (2003) Transgenic cotton a winner in India. Nature https://doi.org/10.1038/news030203-12

Wunderlich S, Gatto KA (2015) Consumer perception of genetically modified organisms and sources of information. Adv Nutr 6:842–851

Yan S, Zhu J, Zhu W, Li Z, Shelton AM, Luo J, Cui J, Zhang Q, Liu X (2015) Pollen-mediated gene flow from transgenic cotton under greenhouse conditions is dependent on different pollinators. Sci Rep 5:15917

Part II

Animal Biotechnology

Reaping the Benefits of Gene Modulations in Animals in the Era of Genomics

3

Subeer S. Majumdar, Neerja Wadhwa, Souvik Sen Sharma, and Nirmalya Ganguli

Abstract

Since the last two decades, animal science has been transformed by rapid advances in biotechnology—a field of biology that encompasses a variety of technologies, including transgenics, recombinant DNA techniques and cloning, among others. Transgenic technology has attracted widespread attention of Indian researchers during this period. Research in animal biotechnology is driven by the application of this technology to unravel functions of genes, improve productivity, consistency, and quality of food, producing biopharmaceuticals and generating organs and tissues for xenotransplantation.

3.1 Transgenic Technology

Understanding the physiological function of identified genes has been a fundamental objective in the field of genomics research. One of the fundamental goals of biological sciences is to decipher the physiological function of any identified gene. In this context, transgenic technology has completely revolutionized biological

S. S. Majumdar (✉)
National Institute of Immunology, New Delhi, India

National Institute of Animal Biotechnology, Hyderabad, India

Gujarat Biotechnology University, Gandhinagar, India
e-mail: dg@gbu.edu.in

N. Wadhwa
National Institute of Immunology, New Delhi, India

S. S. Sharma · N. Ganguli
National Institute of Immunology, New Delhi, India

National Institute of Animal Biotechnology, Hyderabad, India

G. John, A. K. Panda (eds.), *Biotechnology in India - Reworking A Strategy*, https://doi.org/10.1007/978-981-97-0896-3_3

sciences by enabling the analysis of gene function in vivo when an animal bearing the added gene is born. Transgenic animals are genetically modified animals generated by a procedure called "transgenesis" in which a foreign gene (transgene) is introduced into an animal so that the gene is expressed and inherited by the offspring. Several transgenic techniques have been developed over the past few decades for the generation of transgenic animals. Each transgenic technique has its own efficiency and associated advantages and disadvantages. A prominent area of contemporary animal biotechnology research involves the development of transgenic animals through genetic modification. Since the generation of the first transgenic animal by Gordon and Ruddle (1981, 1984), several methods and procedures have been developed for manipulation of the genome. Over the past few decades, several techniques have been developed to achieve successful transgenesis, such as pronuclear microinjection, embryonic stem cell-mediated gene transfer, viral vectors, and gene editing technologies like CRISPR-Cas9 to generate unique animal models. Each technique exhibits varying levels of efficiency and gene expression, accompanied by distinct advantages and disadvantages. Consequently, researchers have successfully utilized these transgenic technologies to create diverse animal models for facilitating the study of human development, diseases, and disorders.

3.2 Historical Background

The concept of genetic modification and the potential to alter the genetic makeup of organisms was first proposed by Hermann J. Muller in 1927 (Muller 1927). However, it was not until the 1970s that the development of recombinant DNA technology by Paul Berg and his colleagues laid the foundation for the creation of transgenic animals (Berg 1971). This breakthrough allowed scientists to manipulate DNA sequences and transfer genes between different organisms.

In 1974, Rudolf Jaenisch and Beatrice Mintz successfully generated the first transgenic mice by introducing foreign DNA into early mouse embryos using retroviral vectors (Jaenisch and Mintz 1974). This groundbreaking achievement opened new doors for the creation of genetically modified animal models. In 1981, Gordon and Ruddle generated a transgenic mouse in which specified foreign genes were inserted into its genome and transmitted to the offspring, indicating stable inheritance (Gordon and Ruddle 1981). The development of pronuclear microinjection, a technique pioneered by Ralph Brinster and Richard Palmiter in the early 1980s, further accelerated the generation of transgenic animals (Palmiter and Brinster 1986). This technique involved injecting DNA directly into the pronuclei of fertilized eggs, allowing for efficient gene transfer and subsequent generation of transgenic offspring. Pronuclear microinjection became a widely used method for creating transgenic mice and laid the foundation for the generation of transgenic animals in other species.

In the following years, researchers expanded the scope of transgenic animal generation to various organisms. In 1985, first transgenic rabbit produced by Hammer and his colleagues marked a significant milestone beyond mice (Hammer et al.

1985). Subsequently, transgenic animals were generated in diverse species, including rats, pigs, sheep, cattle, fish, and insects. Each species presented unique challenges and required the development of species-specific transgenic techniques.

In the 1990s, the advent of embryonic stem cell-mediated transgenesis revolutionized the field of transgenic animal generation. In 1992, Mario Capecchi, Martin Evans, and Oliver Smithies demonstrated the potential of embryonic stem (ES) cells in creating targeted gene modifications in mice, a technique known as gene targeting or knockout mice (Capecchi et al. 2001). This breakthrough allowed researchers to study the effects of specific gene alterations and gain valuable insights into gene function.

The advent of the new millennium brought about revolutionary advancements in transgenic technology. In the early 2000s, the introduction of zinc finger nucleases (ZFNs), transcription activator-like effector nucleases (TALENs), and later, the CRISPR-Cas9 system provided powerful tools for targeted genome editing in a wide range of organisms (Boch et al. 2009; Cong et al. 2013; Kim et al. 1996). In recent years, CRISPR-Cas9 system has further propelled the field of transgenic animal generation. CRISPR-Cas9 offers unprecedented precision and efficiency in genome editing, allowing researchers to create transgenic animals with precise modifications and investigate the functions of specific genes. This breakthrough technology has accelerated the generation of transgenic animals and expanded the scope of genetic research. Since the advent of transgenesis, numerous transgenic animals have been generated in India and worldwide, employing various technologies.

This chapter highlights several technical advancements in animal transgenesis during the last few decades in India that has enabled researchers to decipher the function of several genes in vivo. The Department of Biotechnology (DBT) has made tremendous efforts since late 1990s in promoting transgenic animal technology facilitating Indian researchers to generate transgenic animals and animal models to study human development, diseases, and disorders.

3.3 DBT Supported Institutions having State-of-the-Art Transgenic Facility

3.3.1 National Institute of Immunology (NII)

NII of the Department of Biotechnology (DBT) was the first institute in the country to initiate transgenesis-related work and standardize the technique of transgenesis in the year 1997. Dr. Satish Totey and his team at the institute pioneered the transgenic technique of microinjection of DNA into the pronucleus of mouse embryos. In 1997, NII generated India's first neomycin-resistant transgenic mice, the results of which were published in 1998 (Kaur et al. 1998). A team of Indian scientists and technologists at NII showed the capability of germline transmission of neomycin-resistance gene and established transgenic technology in India. The first disease model of hepatitis, using HBVx gene provided by ICGEB, was also produced at NII creating animals having liver cancer (Singh and Kumar 2003). Following that, NII organized several national and international training workshops in the early years to spread the technology across the country.

In 2001, the program was taken over by Dr. Subeer Majumdar who was with the same team of technologists that generated numerous transgenic lines.

Transgenic Facility of NII offered its collaborative assistance to other institutions in the country, by generating transgenic mice for them which were utilized as animal models for their biomedical research. NII began offering free services for generating transgenic animals to various institutions across the country, generating animal models for genes such as *Fmr1, Aif, Smar1, Dkk3, Hox10a*, etc. (Alam et al. 2010; Banerjee et al. 2012; Baskaran et al. 2002; Chemmannur et al. 2015; Das et al. 2013; Datta et al. 2011; Kaul-Ghanekar et al. 2005; Mehta et al. 2009; Mishra et al. 2022; Pradhan and Majumdar 2016; Yadav et al. 2015).

Despite the success of microinjection technique at NII, the complexity of the technique, expensive infrastructure, and the requirement of skilled technologists was restricting the adaptability of pronuclear microinjection in other laboratories of the country. These limitations generated an urgent need to develop an alternate, cost-effective, user friendly, and rapid approach for obtaining transgenic animals, preferably with minimum or no loss of animal life. NII continued working toward the development of new easier techniques, finding solutions to these limitations. In 2008, for the first time, NII developed a technique for generating transgenic mice by in vivo electroporation of the desired transgene in the testicular germ cells of mice. This death-less transgenesis technique enabled the generation of transgenic mice with no loss of animal life (Dhup and Majumdar 2008). The new easier technique increased the probability of generating transgenic large animals (cattle and non-human primates) also (Wadhwa et al. 2021).

Furthermore, keeping pace with the changing needs and working on the caveats of this newly developed technique of testicular electroporation which required surgical exposure of the testes, the NII team fine-tuned it further and developed a non-invasive procedure for integration of the foreign DNA in repopulating spermatogonial cells of the testes (Usmani et al. 2013). The relative ease of using this non-surgical method enabled the rapid generation of transgenic mice without the help of specialized labs, at a low cost and without the need for highly skilled personnel.

Testis-mediated gene transfer (TMGT) involves the in vivo transfer of genes into the testes to deliver exogenous DNA into eggs via the spermatozoa. To generate transgenic animals, different techniques such as chemical-based (DMSO, DMA) and liposome-based delivery system were used to transfer foreign DNA molecules into the testicular germ cells (Yonezawa et al. 2001). In some studies, DNA was injected into caput epididymis also. Furthermore, using the non-liposomal lipid–DNA complexes, DNA was successfully transmitted even into F2 offspring (Sato et al. 1999). Our own team developed a unique method of in vivo testicular electroporation-based technique, in which the transfection of DNA construct into the germ cells was performed by electric pulses after the delivery of DNA in the testis (Dhup and Majumdar 2008). The electric pulses helped in opening pores in the cell membranes of germ cells, thereby allowing the construct to internalize and incorporate into the testicular germ cells. This method of generating transgenic animals is well established by the NII team in mice and rats. To ensure that this technique will not cause substantial damage to the testis and spermatogenesis, several factors are required to be analyzed for a particular species, e.g., age of the animal, volume or concentration of DNA to be injected, injection site, surgical procedure, and the voltage and pulse duration of the electroporation itself, and thereafter optimize these conditions to avoid tissue damage while using electroporation for gene transfer. To overcome these difficulties, non-nsurgical external electroporation of testis was established using a modified protocol, which did not hamper the reproductive performance of the animal post-electroporation (Pradhan and Majumdar 2016; Usmani et al. 2013) (Fig. 3.1).

Utilizing these multiple techniques for generating transgenic mice, NII continued to generate transgenic mice and other animal models for biomedical research. Collaborative work from across the country was undertaken, and transgenic lines were generated to address their respective scientific goals. NII collaborated with numerous investigators/institutions across the country—Dr. Vijay Kumar (ICGEB), Dr. Vani Brahmachari (ACBR), Dr. Panicker (NCBS), Dr. Shyamal Goswami (JNU, New Delhi), Dr. Rakesh Tyagi, JNU, Dr. Satyajit Rath (NII, New Delhi), Dr. Vineeta Bal (NII, New Delhi), Dr. Satish Adiga (Manipal Academy of higher Education, Manipal), Dr. Suresh B. Gokhale (BAIF, Pune), Dr. Sagar Sengupta (NII, New Delhi), Dr. Deepak Modi (NIRRH, Mumbai), Dr. Samit Chattopadhyay (NCCS, Pune), Dr. Milind Vaidya (ACTREC, Mumbai), Dr. Rajesh Gokhale (IGIB), and many others. Transgenic overexpression and knockdown lines were generated for genes like neomycin-resistance gene, interleukin 2, hepatitis B virus gene, serotonin receptors, fragile X syndrome, SMAR 1, Dickkopf homolog 3 (DKK3), keratin 8, HOXA10, ROR alpha, Smok1, Gas5, NOR 1, tetraspanins, YY1, Sostdc1, etc. The list of transgenic rodent models generated at NII is detailed in Table 3.1.

Working on the progressive transgenic trend, NII started exploring alternate approaches for generating transgenic large animals. Researchers at NII started working on somatic cell nuclear transfer, i.e., somatic cell cloning. Somatic cloning involves reconstruction of an embryo by transferring genetic material from a donor somatic cell to a recipient egg whose own genetic material has been removed. A transgene could be introduced into the genome of somatic cell nucleus in vitro, which may then be used as a source of donor nucleus for somatic cell nuclear transfer (SCNT). Unfortunately, the success rate of somatic cell nuclear transfer in mammals is low. Hence, the work initiated at NII was designed to determine if SCNT efficiency could be augmented in vitro by modifying culture conditions and to explore if transgenesis is possible via cloning. In 2009, NII established a procedure of somatic cloning by nuclear transfer and produced transgenic cloned embryos of mice and buffalo carrying the targeted genes (Fig. 3.2).

This was the first report from an Indian institution regarding the production of transgenic cloned buffalo embryos (Wadhwa et al. 2009). Embryo transfer of such transgenic cloned buffalo embryos was performed at BAIF, Pune. Due to lack of a sufficient number of buffalo recipients for embryo transfer, live cloned buffalo were

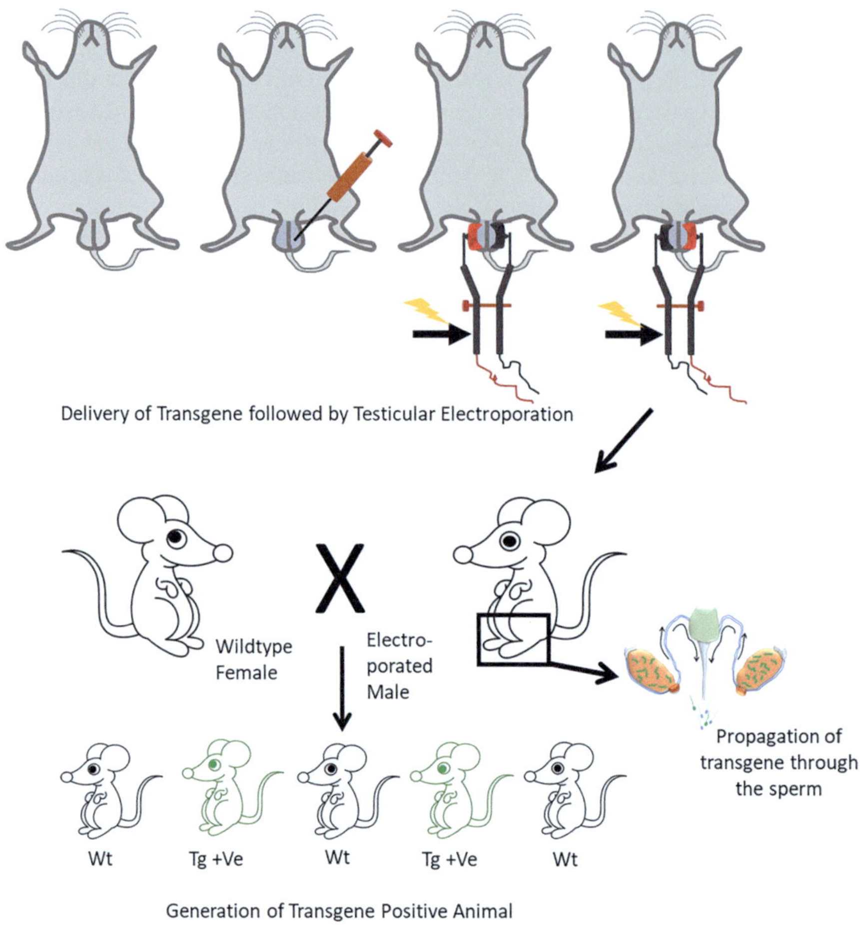

Fig. 3.1 Schematic diagram showing the process of testicular external electroporation-based transgenesis in mice

not obtained in that study. However, research groups at National Dairy Research Institute (NDRI), Karnal, led by Dr. M. S. Chauhan and Dr. S. K. Singla generated cloned buffalo female calf in 2010 using embryonic stem cells through "hand-guided cloning technique" (George et al. 2011; Mohapatra et al. 2015; Shah et al. 2008).

Besides these transgenic techniques, NII also worked on several related technologies like in vitro fertilization (Minhas et al. 2016), stem cell renewal and differentiation, autologous germ cell transplantation (Ganguli et al. 2016), etc. NII has made tremendous contribution in the field of transgenics and kept pace with the emerging technologies while refining existing methods.

In parallel to establishment of transgenic facility at NII, the Department of Biotechnology made tremendous efforts in promoting transgenic animal science.

Table 3.1 List of transgenic rodent lines generated at the National Institute of Immunology

SI no.	Gene	Species (mouse/ rat)	Method of transgenesis	Type of genetic modification	References
1	Neomycin-resistance gene	Mouse	Embryo micro-injection	Overexpression	Kaur et al. (1998)
2	SMAR1	Mouse	Embryo microinjection/ testicular electroporation	Overexpression	Chemmannur et al. (2015)
3	AIF	Mouse	Testicular electroporation	Overexpression	Banerjee et al. (2012)
4	HOX10A	Mouse	Testicular electroporation	Knockdown	Mishra et al. (2022)
5	Meis 1	Mouse	Embryo microinjection	Knockdown	Sarkar et al. (2021)
6	Sostdc1	Rat	Testicular electroporation	Knockdown	Pradhan et al. (2019)
7	YY1	Rat	Testicular electroporation	Knockdown	Mandal et al. (2017)
8	ROR-α	Rat	Testicular electroporation	Knockdown	Mandal et al. (2018)
9	Nor1	Mouse	Testicular electroporation	Knockdown	Shukla et al. (2018)
10	Wnt3	Mouse	Testicular electroporation	Knockdown	Basu et al. (2018)
11	Tetraspanin8	Rat	Testicular electroporation	Overexpression	Pradhan et al. (2020)
12	Mir-382-3p	Mouse	Testicular electroporation	Overexpression	Gupta et al. (2021)
13	Mir-92a-5p	Mouse	Testicular electroporation	Overexpression	Gupta et al. (2022)
14	Dkk3	Mouse	Testicular electroporation	Overexpression	Das et al. (2013)
15	EGFP under mice PEPCK1 promoter	Mouse	Testicular electroporation	Overexpression	Usmani et al. (2013)
16	EGFP under human K14 promoter	Mouse	Testicular electroporation	Overexpression	Usmani et al. (2013)
17	Fetuin-A	Mouse	Testicular electroporation	Knockdown	Usmani et al. (2016)
18	K8 under K14 promoter	Mouse	Testicular electroporation	Overexpression	Tiwari et al. (2021)
19	EGFP under buffalo β-casein promoter	Mouse	Testicular electroporation	Overexpression	Nirmalya Ganguli et al. (2015)
20	PXR	Mouse	Testicular electroporation	Overexpression	Kotiya et al. (2016)

(continued)

Table 3.1 (continued)

SI no.	Gene	Species (mouse/ rat)	Method of transgenesis	Type of genetic modification	References
21	EGFP under CMV promoter	Mouse	Testicular electroporation	Overexpression	Ganguli et al. (2016)

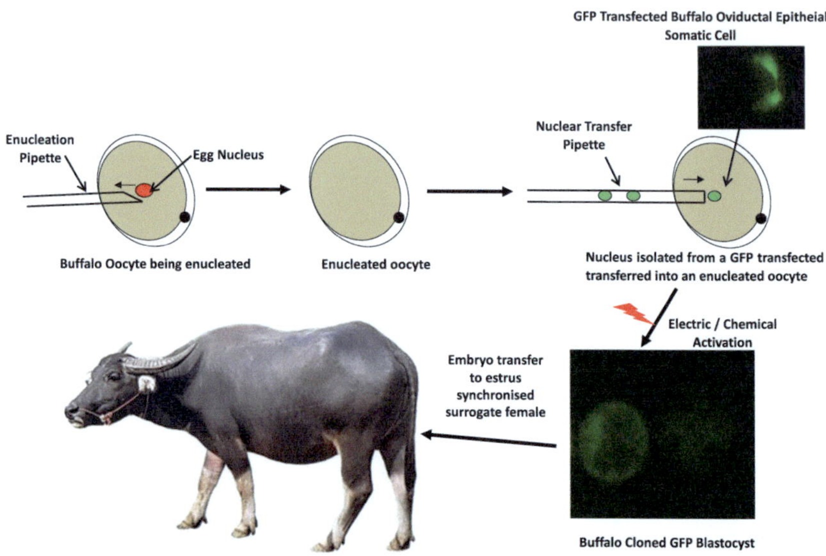

Fig. 3.2 A flow chart of the various steps involved in generation of transgenic embryo by somatic cell nuclear transfer

The department laid emphasis on the generation of transgenic animals for enhanced efficiency, productivity in the field of functional genomics, drug discovery, agriculture, and animal health. Some of the other transgenic centers supported by DBT in the country are detailed below.

3.3.2 Indian Institute of Science, Bangalore (IISc)

In the year 1995, Indian Institute of Science (IISc) embarked on establishing a transgenic animal facility. The impetus, including the financial support, was provided by the Department of Biotechnology (DBT), Govt of India. The overall objective of the project was to establish a Transgenic Animal (mouse model) Facility at the IISc in order to carry out modern biology research for studying regulation of gene expression, employing transgenic technology. This project was implemented by Prof. Polani Seshagiri (Dept of MRDG) and Prof. MRS Rao (Dept of Biochemistry).

The specific objectives were: (a) To create a modern transgenic animal (mouse model) infrastructure at IISc to facilitate gene expression studies in the mouse model; (b) To generate EGF-expressing transgenic mouse and to improve the transgenic technology by selective choice of EGFP expressing-transgenic embryos for uterine transfer; (c) To generate rat transition protein (rTP-2) transgenic animal to study tissue-specific and developmental stage-specific expression of rTP-2 gene and characterize its promoter. IISc established a de novo transgenic facility for the purpose of generating transgenic mouse models having ubiquitous or tissue-specific or inducible promoter constructs. These have wide applications, in addition to transgenic animal generation, ranging from generation of knockout mice and developing animal models for developmental studies and for cancer and genetic disorders.

After establishing de novo infrastructure for mouse transgenesis, attempts were made to generate two types of transgenic animals, harboring the transgenes, viz. enhanced green fluorescence protein (EGFP) or rat transition protein (rTP)-2 (Devgan et al. 2003; Devgan and Seshagiri 2003; Singh et al. 2012).

The EGFP gene was used as a model gene for studying developmental regulation of gene expression and to improve the transgenic technology involving embryo culture/uterine transfer. The rTP-2 transgene was used to understand its tissue-specific and developmental stage-specific expression, in terms of promoter characterization and to study its function during mammalian spermiogenesis. In addition to establishing the facility, the system provided huge human resource development opportunities and derivation of embryonic stem cell (GS-2 ESC) and induced pluripotent stem cell (N9-iPSC) lines which were extensively used for lineage-specific differentiation toward cardiac and neural cell types (Abbey et al. 2019; Abbey and Seshagiri 2013; Verma and Seshagiri 2018). The long-term goal of the project was to have a fully functional, State-of-the-Art Mouse Transgenic Facility at IISc, which could act as a core facility to carry out modern biology research on the in vivo regulation of gene expression in developmental systems, employing transgenic technology. Unfortunately, this facility has had the same fate as that of NII after the retirement of Prof. Sheshagiri. However, recently, IISc scientists have revived the transgenic work and reported transgenic mouse model related studies of diarrheal diseases (Mishra et al. 2021). Currently, a modern dedicated facility is being established.

3.3.3 Centre for Cellular and Molecular Biology (CCMB), Hyderabad

In 2002, the country's first National Facility for Gene Knock-Out Mice was set up at Centre for Cellular and Molecular Biology (CCMB). CCMB created infrastructure to generate transgenic and knockout mice models. In 2006, Dr. Satish Kumar and his team was credited with the establishment of the first Gene Knockout Technology in India at CCMB. Some of the important contributions of CCMB include deciphering the role of κ-casein in lactation (Shekar et al. 2006), role of WDR13 as a potential drug target for ameliorating hyperglycemia and inflammation in diabetic conditions (Singh et al. 2015) and published research on few knockout models. Dr. Chandrashekhar is now running this facility for CCMB work.

3.3.4 Advanced Centre for Treatment, Research and Education in Cancer (ACTREC), Tata Memorial Centre, Navi Mumbai

ACTREC has also published several research articles related to transgenic and reproductive biology work. Dr. Sorab Dalal from ACTREC reported several novel techniques for generation of transgenic mice by in vivo manipulation of spermatogonial stem cells with a high rate of success (Sehgal et al. 2011). He published on sperm-mediated gene transfer technique for generating transgenic mice (Sehgal et al. 2012, 2014; Vishal et al. 2017).

3.3.5 Mouse Genome Engineering Facility (MGEF) of Bangalore Life Science Cluster (A Unit of National Centre for Biological Sciences, Institute for Stem Cell Science and Regenerative Medicine, Centre for Cellular and Molecular Platforms, and Tata Institute for Genetics and Society)

MGEF is a multifunctional state-of-the-art advanced animal facility at the Bangalore Life Science Cluster (BLiSC) setup with substantial support from the DBT. Some of the services provided by MGEF include mouse stock cryopreservation/vitrification, in vitro fertilization, rederivation, and generation of novel genome engineered mice. The facility created several CRISPR/Cas Knock-IN, CRISPR/Cas Knock-OUT, Conditional cKO, cKI Point mutations and Deletions Humanized Knock-IN. Most recently, the facility created India's First Humanized ACE2 CRISPR/Cas ACE2 Knock-OUT Mouse Models for studying COVID-19. As part of its mission to enhance access to these resources and expertise nationwide, MGEF conducts regular hands-on workshops and individual training of scientists all over India and neighboring countries. Their training program has given an impetus to work involving transgenesis.

3.3.6 National Facility for Gene Function in Health and Disease, IISER, Pune

Realizing that the availability of a small animal research facility is critical to address basic biological as well as biomedical questions relevant to disease biology, IISER initiated this center with the help of DBT's support. The facility to generate transgenic and gene knockouts tailored to meet the research requirements was created at IISER Pune which was well equipped to generate in-house transgenic/knockout mice and act as a repository of such animal models. Availability of such animal models would facilitate research focused on a variety of biologically important phenomena and diseases. Since 2019, this facility at IISER Pune is also a member of the International Mouse Phenotyping Consortium (IMPC). They are also involved in creating human resources with international collaboration.

3.4 Application of Transgenic Technology for Improving Human Health

The ability to generate transgenic animals using methods like embryo manipulation and testicular electroporation has enabled scientists to decipher the function(s) of a large number of genes in development and disease. Transgenic mice with cell-specific knockout/knockdown and overexpression of specific genes have been used for deciphering the etiology of various diseases. Furthermore, transgenic animal expressing gene coding for therapeutic proteins can help in the large-scale production of costly therapeutic proteins, in a cheap manner through milk. In the following section, some of the applications of transgenic and high-throughput technologies relevant to studies in India are discussed.

3.4.1 Transgenic Animals Created for Identifying Novel Genes Involved in Regulation of Male Fertility

The development of modern biotechnological tools and high-throughput technologies like next-generation sequencing, proteomics, and metabolomics has revolutionized biological sciences and medicine, providing differential expression of many genes and protein in normal vs. disease conditions (like normal and cancerous tissues, fertile and infertile situation, normal and abnormal metabolic condition, etc.). Such tools have enabled the study of biological systems at molecular levels to decipher the mechanisms underlying biological processes and pathological conditions. As a working example, the use of high-throughput techniques and transgenic technology for identifying genes essential for male fertility is discussed here.

Infertility/subfertility is a global health concern. It is estimated that nearly 15% of couples suffer from fertility-related issues. Sperm count is constantly falling, globally. A large body of scientific evidence has established the indispensable role of testicular Sertoli cells in regulating spermatogenic progression (França et al. 2016; Griswold 1998). It is important to note that infantile Sertoli cells (Sc) are incapable of supporting spermatogenesis in spite of adequate levels of circulating gonadotropins, which is probably due to low responsiveness of Sertoli cells toward FSH and testosterone (Sharpe et al. 2003). Sertoli cells undergo functional maturation at the onset of puberty which is characterized by enhanced responsiveness of these cells toward FSH and testosterone and elevated expression of genes that are essential for germ cell proliferation and differentiation. The functional maturation of Sc, which occurs during puberty, leads to a robust initiation of germ cell differentiation, highlighting the importance of Sc in initiation and maintenance of spermatogenesis post-puberty.

The mRNAs, micro-RNAs, and transcription factors regulated differentially in the testicular Sc were compared between infertile-like (infant) and fertile (pubertal) conditions. The level of gene expression at a specific age was reversed using transgenic technology in mice and rats, and importance of those genes in induction or prevention of spermatogenesis was established. We provided knowledge about eight new genes which regulate male fertility and can be considered important for diagnosis and treatment of idiopathic infertility, creating transgenic mice models. A study was also carried out in cancer biology using testicular transgenesis in collaboration with ACTREC, Mumbai, and also one about endometritis with ICMR-NIRRH, Mumbai (Mishra et al. 2022; Tiwari et al. 2021). To be specific, using the techniques described above, our laboratory has identified genes like *Wnt3*, *Nor1*, *YY1*, *Ror-α*, *Dkk3*, *Tspn8,* etc. and small RNAs like mir-92a and mir-382-3p as crucial regulators of Sc function and that the defective expression of such genes may be an underlying cause of idiopathic infertility in males (Basu et al. 2018; Das et al. 2013; Gupta et al. 2021, 2022; Mandal et al. 2017, 2018; Pradhan et al. 2020; Shukla et al. 2018).

3.4.2 Production of Biotherapeutics in Transgenic Animal Bioreactor

In the era of high-throughput sequencing technology driven genomics, information about genes and their products (proteins) in remediation of diseases is accumulating rapidly. These proteins may easily be supplemented to patients who lack them or in whom disrupted production of such proteins is noticed as a result or cause of the disease. For replenishment, these proteins need to be produced in large amounts as biotherapeutics (Echelard 2006). In vitro production of therapeutic proteins in cultured cells is a common choice for production with several drawbacks like tedious recovery process, improper folding and modification of expressed proteins, etc. Transgenic animals provide a very good replacement system for production and isolation of therapeutic proteins. Mammary gland, being exocrine in nature, produces large amounts of milk at the time of lactation. Expression of therapeutic protein in milk, utilizing mammary gland as a bioreactor is an ideal choice for producing these at a cheaper and affordable cost. The obvious way to transform a mammary gland into a bioreactor is by generating a transgenic animal through presently available methods of genetic modifications. Before such modifications are introduced in the animals in the form of a transgene, the generation of a transgene cassette by assembling various regulatory and signaling components is the major step. The major factor in this transgene cassette is the promoter elements. Promoter elements are the ones that drive the expression of the downstream cDNA (of any therapeutic protein) in tissue/cell-specific manner, for example, β-casein promoter would do so only in the udder gland. The most widely used promoter for driving mammary gland/milk specific expression of the therapeutic protein gene is goat β-casein gene promoter, which is already patented. To initiate such work in India, we isolated and characterized the promoter element of the β-casein gene from the genome of Indian river buffalo *Bubalus bubalis*. This promoter has proven its ability to drive expression of tagged genes in vitro in MCF7 cells and cultured mammary epithelial cells of buffalo as well as in vivo, in transgenic mice (Ganguli et al. 2015). The promoter was found to be specifically active in the mammary gland of transgenic mice at the time of lactation. We designed a transgene cassette using this promoter harboring cDNA of different human therapeutic proteins like gamma interferon, BMP2, and also bovine FSH as well as LH. Methods of transgenesis in large animals are very cumbersome, time-consuming, and a costly affair. Using the newly developed testicular transgenesis method (explained in detail in the preceding section of this chapter), we developed transgenic mice and rabbits that express human interferon gamma in the milk. We are continuously working on extrapolating this method for other farm animals like goat and cow.

At present, a concern is raised about the generation and use of genetically modified organisms (GMO) as they bear the risk of unrestrained germline transfer of transgene in the normal herd. Even nonspecific integration of transgene in genome may add unknown anomalies and give rise to disease-susceptible animals. To address this concern, we have developed an easy technique to directly transfect the epithelial cells in the mammary glands for production of therapeutic proteins. This involved generation of virosomes from purified membrane of Sendai virus for delivery of transgene (human source) entrapped within it. The functional cassette of β-casein promoter driven human interferon gamma (hIFNg) was entrapped in the virosomes and delivered in the mammary glands through the milk duct. We found expression of hIFNg in the mammary gland as well as in the secreted milk of virosome delivered mice at the time of lactation (Ganguli et al. 2018). Along with this, we developed another method for direct transfection of mammary epithelial cells based on the 25 kDa branched polyethyleneimine that harbors a 6-carbon fatty acid chain (FA6PEI) modification that can be used as a new generation transfection agent for gene insertion. This transfection reagent is inexpensive and has also been very promising in transfecting the mammary gland (Ulgekar et al. 2022). These methods of non-germline modification (Fig. 3.3) will help in the generation of safe mammary bioreactors devoid of environmental risk posed by the GMOs, in a very cost-effective manner.

3.5 Transgenic Technology in India: Way Forward and Limitations

Genetic engineering has evolved to a great extent due to the advent of techniques like the CrispR-Cas system and has provided immense scope for the generation of transgenic animals for various applications. Furthermore, recent techniques like testicular transgenesis and direct udder gland gene transfection developed in India offer cost-effective alternatives for transgenic animal generation.

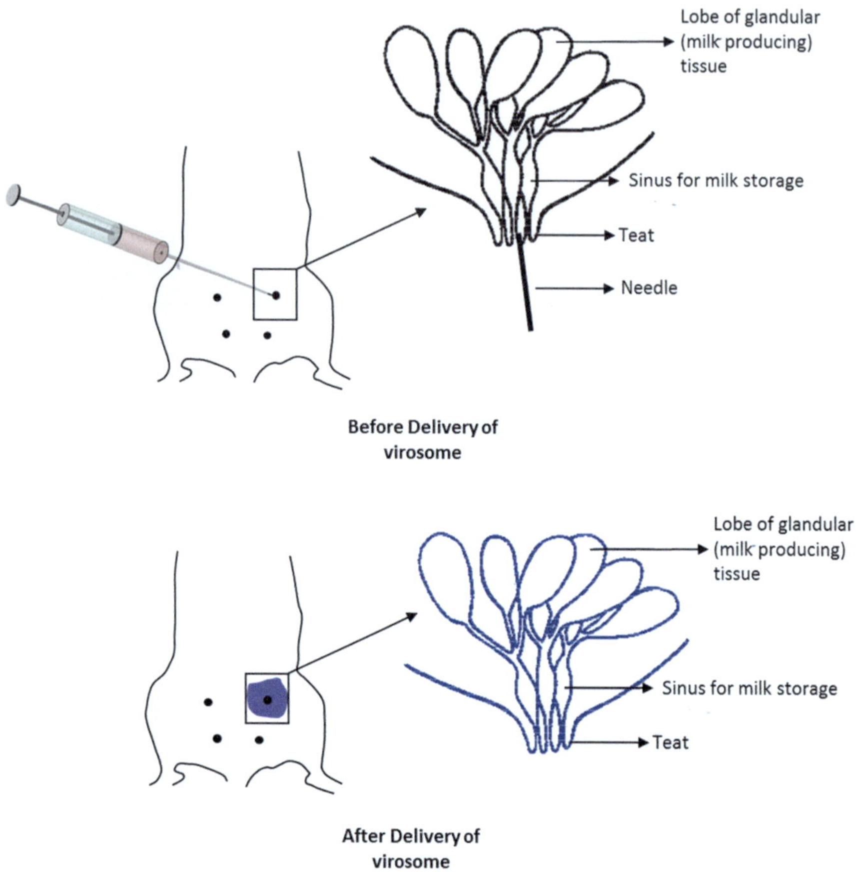

Fig. 3.3 Schematic diagram showing the process of direct mammary gland transfection in mice

Introducing disease resistance in farm animals is a very important aspect of the livestock sector as livestock losses due to bacterial and viral pathogens are immense. Traditional methods like vaccination, culling of the infected animal, and antibiotic treatment have shown success, but the efficacy of these methods is very limited. Furthermore, excessive use of antibiotics causes pathogens to develop resistance which is a major issue. Transgenic technology can serve as an effective alternative to introduce disease resistance in animals. The development of herds with resistance to diseases prevalent in the Indian subcontinent is of utmost importance, and testicular transgenesis may make that possible, avoiding limitations of embryo-mediated transgenesis for large animals. India also faces many outbreaks of transboundary zoonotic diseases. Transgenic livestock can help in developing herd resistance to those diseases and also as serve as models to study the pathophysiology of those diseases.

Transgenic technology has allowed us to produce proteins of therapeutic use in the mammary gland. A mammary gland bioreactor could be a cheaper alternative than cell line-based production system for generating recombinant proteins in farm animals. It does not require a costly industrial infrastructural setup and hence can produce therapeutic proteins during every lactation cycle at much cheaper cost. For example, Atryn®, which is recombinant antithrombin III, is the commercially approved therapeutic protein produced in the mammary gland of goats.

There have been reports of production of transgenic chicken, sheep, and bull with the myostatin gene knocked down/out which allows these transgenic animals to grow more muscle mass (He et al. 2018; Kim et al. 2020; McPherron and Lee 1997). India with its growing population needs to utilize transgenic technology for generating more protein mass per animal or reducing animal mortality rates through disease-resistant herds.

Transgenic technology has also found immense applications in transplantation medicine. The ever-growing need for transplantable organs for replacing diseased and/or nonfunctional organs in humans has been recognized (Ekser et al. 2012). Although xenotransplantation of organs from large animals like pigs, chimpanzees, and monkeys has been carried out, transplantation of such organs carries with it the risk of severe immunological reactions and rejection of the transplanted tissue (Xi et al. 2023). In recent years, transgenic technology has been used for genetically modifying donor animals in order to prevent immune rejection of the graft upon transplantation (Xi et al. 2023).

Although benefits of transgenic technology are well known in the scientific community, a major step to reap its benefits in India is still lacking. It is a fact that well-established transgenic facilities of NII and IISc, the first Indian facilities to standardize transgenic work ceased to work after retirement of the relevant scientists. Due to this lack of leadership and scientific support and advancement ability, the importance of the program took a back seat, in terms of priority. The leaders who started most recent projects at IISER and InStem involving transgenic technology have now moved on with new ventures. The new leaders of these facilities require to gear up supporting human resource development and start programs which bring benefits of this technology in terms of better and healthy animals which are more productive. They must take steps to regain ability to knock out genes in a targeted fashion to stay afloat at par with international biomedical research and should also develop techniques of turning the genes on and off to study functional genomics and disease progression, making extraordinary, humanized models of diseases.

It is surprising that while a set of leaders in biotechnology feel the importance of this technology in supporting human health and accomplishing feed related needs, as the era of expertise changes, the requirement loses priority and developed technologies remain unused. A constant support in important technologies and revamping them from time to time must become part of policy to ensure we do not lose the race and become dependent on other countries for the technology. It is matter of concern and worry that even now, scientists in India go outside for outsourcing their transgenic or knockout animal work. Although the government has recently supported two institutions with good grants, including funding for training and human resource development, outcomes are not phenomenal. We need to revisit and strategize the achievements, as well as services with a pert chart with a long-term (say 10 years) goal. Human resource development and their employment should be designed within the framework, recognizing the importance of this branch of biotechnology which gives (1) knowledge about function of genes, (2) helps in deciding candidates of gene therapy, (3) provides humanized disease models to develop drugs and vaccines, and (4) provides scope for cheaper source of human proteins produced in animal bioreactors. This also has a huge potential to increase productivity of meat, milk, and eggs. The technology is well deployed in some countries. This calls for establishing a product-oriented center for transgenesis that may enhance efforts of the country toward health supplementation and address the food and feed requirements of the country in long term.

Acknowledgments The authors are thankful to the Department of Biotechnology, India, National Institute of Immunology, National Institute of Animal Biotechnology, and DBT-TATA innovation fellowship (BT/HRD/35/01/01/2010) and SERB-JC Bose fellowships (JCB/2017/000027) for providing generous funds to undertake transgenic research work. We are also thankful to Gujarat Biotechnology University for its support.

References

Abbey D, Seshagiri PB (2013) Aza-induced cardiomyocyte differentiation of P19 EC-cells by epigenetic co-regulation and ERK signaling. Gene 526(2):364–373. https://doi.org/10.1016/J.GENE.2013.05.044

Abbey D, Singh G, Verma I, Derebail S, Kolkundkar U, Chandrashekar DS, Acharya KK, Vemuri MC, Seshagiri PB (2019) Successful derivation of an induced pluripotent stem cell line from a genetically nonpermissive enhanced green fluorescent protein-transgenic FVB/N mouse strain. Cell Reprogram 21(5):270–284. https://doi.org/10.1089/cell.2019.0019

Alam MP, Datta S, Majumdar S, Mehta AK, Baskaran S, Gulati N, Brahmachari V (2010) Comparative analysis of DNA methylation in transgenic mice with unstable CGG repeats from FMR1 gene. Epigenetics 5(3):241–248. https://doi.org/10.4161/epi.5.3.11417

Banerjee H, Das A, Srivastava S, Mattoo HR, Thyagarajan K, Khalsa JK, Tanwar S, Das DS, Majumdar SS, George A, Bal V, Durdik JM, Rath S (2012) A role for apoptosis-inducing factor in T cell development. J Exp Med 209(9):1641–1653. https://doi.org/10.1084/jem.20110306

Baskaran S, Datta S, Mandal A, Gulati N, Totey SM, Anand RR, Brahmachari V (2002) Instability of CGG repeats in transgenic mice. Genomics 80(2):151–157. https://doi.org/10.1006/geno.2002.6813

Basu S, Arya SP, Usmani A, Pradhan BS, Sarkar RK, Ganguli N, Shukla M, Mandal K, Singh S, Sarda K, Majumdar SS (2018) Defective Wnt3 expression by testicular Sertoli cells compromise male fertility. Cell Tissue Res 371(2):351–363. https://doi.org/10.1007/s00441-017-2698-5

Berg P (1971) The viral genome in transformed cells. Proc R Soc Lond B Biol Sci 177(46):65–76. https://doi.org/10.1098/rspb.1971.0016

Boch J, Scholze H, Schornack S, Landgraf A, Hahn S, Kay S, Lahaye T, Nickstadt A, Bonas U (2009) Breaking the code of DNA binding specificity of TAL-type III effectors. Science 326(5959):1509–1512. https://doi.org/10.1126/science.1178811

Capecchi MR, Evans MJ, Smithies O (2001) Generating mice with targeted mutations. Nat Med 7(10):1086–1090. https://doi.org/10.1038/nm1001-1086

Chemmannur SV, Badhwar AJ, Mirlekar B, Malonia SK, Gupta M, Wadhwa N, Bopanna R, Mabalirajan U, Majumdar S, Ghosh B, Chattopadhyay S (2015) Nuclear matrix binding protein SMAR1 regulates T-cell differentiation and allergic airway disease. Mucosal Immunol 8(6):1201–1211. https://doi.org/10.1038/mi.2015.11

Cong L, Ran FA, Cox D, Lin S, Barretto R, Habib N, Hsu PD, Wu X, Jiang W, Marraffini LA, Zhang F (2013) Multiplex genome engineering using CRISPR/Cas systems. Science 339(6121):819–823. https://doi.org/10.1126/science.1231143

Das DS, Wadhwa N, Kunj N, Sarda K, Pradhan BS, Majumdar SS (2013) Dickkopf homolog 3 (DKK3) plays a crucial role upstream of WNT/b-CATENIN signaling for Sertoli cell mediated regulation of spermatogenesis. PLoS One 8(5):e63603. https://doi.org/10.1371/journal.pone.0063603

Datta S, Alam MP, Majumdar SS, Mehta AK, Maiti S, Wadhwa N, Brahmachari V (2011) Nucleosomal occupancy and CGG repeat expansion: a comparative analysis of triplet repeat region from mouse and human fragile X mental retardation gene 1. Chromosom Res 19(4):445–455

Devgan V, Seshagiri PB (2003) Successful development of viable blastocysts from enhanced green fluorescent protein transgene-microinjected mouse embryos: comparison of culture media. Mol Reprod Dev 65(3):269–277. https://doi.org/10.1002/mrd.10306

Devgan V, Thomas M, Ullas KS, Rao MRS, Seshagiri PB (2003) Embryo culture-based generation of enhanced green fluorescent protein-transgenic mice. Biochem Biophys Res Commun 303(4):994–1001. https://doi.org/10.1016/S0006-291X(03)00483-2

Dhup S, Majumdar SS (2008) Transgenesis via permanent integration of genes in repopulating spermatogonial cells in vivo. Nat Methods 5(7):601–603. https://doi.org/10.1038/nmeth.1225

Echelard Y (2006) Production of recombinant therapeutic proteins in the milk of transgenic animals. BioPharm Int 19(8)

Ekser B, Ezzelarab M, Hara H, Van Der Windt DJ, Wijkstrom M, Bottino R, Trucco M, Cooper DKC (2012) Clinical xenotransplantation: the next medical revolution? Lancet 379(9816):672–683. https://doi.org/10.1016/S0140-6736(11)61091-X

França LR, Hess RA, Dufour JM, Hofmann MC, Griswold MD (2016) The Sertoli cell: one hundred fifty years of beauty and plasticity. Andrology 4(2):189–212. https://doi.org/10.1111/andr.12165

Ganguli N, Ganguli N, Usmani A, Majumdar SS (2015) Isolation and functional characterization of buffalo (*Bubalus bubalis*) β-casein promoter for driving mammary epithelial cell-specific gene expression. J Biotechnol 198:53–59. https://doi.org/10.1016/j.jbiotec.2015.02.001

Ganguli N, Wadhwa N, Usmani A, Kunj N, Ganguli N, Sarkar RK, Ghorai SM, Majumdar SS (2016) An efficient method for generating a germ cell depleted animal model for studies related to spermatogonial stem cell transplantation. Stem Cell Res Ther 7(1):142. https://doi.org/10.1186/s13287-016-0405-1

Ganguli N, Ganguli N, Chandra S, Choubey M, Sarkar DP, Majumdar SS (2018) A combinatorial approach for robust transgene delivery and targeted expression in mammary gland for generating biotherapeutics in milk, bypassing germline gene integration. Appl Microbiol Biotechnol 102(14):6221–6234. https://doi.org/10.1007/s00253-018-9094-2

George A, Sharma R, Singh KP, Panda SK, Singla SK, Palta P, Manik R, Chauhan MS (2011) Production of cloned and transgenic embryos using buffalo (Bubalus bubalis) embryonic stem cell-like cells isolated from in vitro fertilized and cloned blastocysts. Cell Reprogram 13(3):263–272. https://doi.org/10.1089/cell.2010.0094

Gordon JW, Ruddle FH (1981) Integration and stable germ line transmission of genes injected into mouse pronuclei. Science 214(4526):1244–1246. https://doi.org/10.1126/science.6272397

Gordon JW, Ruddle FH (1984) The production of transgenic mice. In: Kumar A (ed) Eukaryotic gene expression. GWUMC department of biochemistry annual spring symposia. Springer, Boston, MA

Griswold MD (1998) The central role of Sertoli cells in spermatogenesis. Semin Cell Dev Biol 9:411

Gupta A, Mandal K, Singh P, Sarkar R, Majumdar SS (2021) Declining levels of miR-382-3p at puberty trigger the onset of spermatogenesis. Mol Ther Nucleic Acids 26:192–207. https://doi.org/10.1016/j.omtn.2021.07.001

Gupta A, Vats A, Ghosal A, Mandal K, Sarkar R, Bhattacharya I, Das S, Pal R, Majumdar SS (2022) Follicle-stimulating hormone-mediated decline in miR-92a-3p expression in pubertal mice Sertoli cells is crucial for germ cell differentiation and fertility. Cell Mol Life Sci 79(3):1–21. https://doi.org/10.1007/S00018-022-04174-9

Hammer RE, Pursel VG, Rexroad CE, Wall RJ, Bolt DJ, Ebert KM, Palmiter RD, Brinster RL (1985) Production of transgenic rabbits, sheep and pigs by microinjection. Nature 315(6021):680–683. https://doi.org/10.1038/315680a0

He Z, Zhang T, Jiang L, Zhou M, Wu D, Mei J, Cheng Y (2018) Use of CRISPR/Cas9 technology efficiently targeted goat myostatin through zygotes microinjection resulting in double-muscled phenotype in goats. Biosci Rep 38(6):BSR20180742. https://doi.org/10.1042/BSR20180742

Jaenisch R, Mintz B (1974) Simian virus 40 DNA sequences in DNA of healthy adult mice derived from preimplantation blastocysts injected with viral DNA. Proc Natl Acad Sci U S A 71(4):1250–1254. https://doi.org/10.1073/pnas.71.4.1250

Kaul-Ghanekar R, Majumdar S, Jalota A, Gulati N, Dubey N, Saha B, Chattopadhyay S (2005) Abnormal V(D)J recombination of T cell receptor β locus in SMAR1 transgenic mice. J Biol Chem 280(10):9450–9459. https://doi.org/10.1074/jbc.M412206200

Kaur G, Gulati N, Mandal A, Babu DS, Anand R, Garg L, Totey SM (1998) Expression and stable germline transmission of neomycin-resistance gene in transgenic mice. Indian J Exp Biol 36(12):1209–1215

Kim Y-G, Cha J, Chandrasegaran S (1996) Hybrid restriction enzymes: zinc finger fusions to Fok I cleavage domain (Flavobacterium okeanokoites/chimeric restriction endonuclease/protein engineering/recognition and cleavage domains). Proc Natl Acad Sci U S A 93:1156–1160

Kim GD, Lee JH, Song S, Kim SW, Han JS, Shin SP, Park BC, Park TS (2020) Generation of myostatin-knockout chickens mediated by D10A-Cas9 nickase. FASEB J 34(4):5688–5696. https://doi.org/10.1096/FJ.201903035R

Kotiya D, Jaiswal B, Ghose S, Kaul R, Datta K, Tyagi RK (2016) Role of PXR in hepatic cancer: its influences on liver detoxification capacity and cancer progression. PLoS One 11(10):e0164087. https://doi.org/10.1371/JOURNAL.PONE.0164087

Mandal K, Bader SL, Kumar P, Malakar D, Campbell DS, Pradhan BS, Sarkar RK, Wadhwa N, Sensharma S, Jain V, Moritz RL, Majumdar SS (2017) An integrated transcriptomics-guided genome-wide promoter analysis and next-generation proteomics approach to mine factor(s) regulating cellular differentiation. DNA Res 24(2):143–157. https://doi.org/10.1093/dnares/dsw057

Mandal K, Sarkar RK, Sen Sharma S, Jain A, Majumdar SS (2018) Sertoli cell specific knockdown of RAR-related orphan receptor (ROR) alpha at puberty reduces sperm count in rats. Gene 641:18–24. https://doi.org/10.1016/j.gene.2017.10.032

McPherron AC, Lee SJ (1997) Double muscling in cattle due to mutations in the myostatin gene. Proc Natl Acad Sci U S A 94(23):12457–12461. https://doi.org/10.1073/PNAS.94.23.12457/ASSET/2E9900E3-01E6-458A-A672-37F44C160D42/ASSETS/GRAPHIC/PQ2272802004.JPEG

Mehta AK, Majumdar SS, Alam P, Gulati N, Brahmachari V (2009) Epigenetic regulation of cytomegalovirus major immediate-early promoter activity in transgenic mice. Gene 428(1–2):20–24. https://doi.org/10.1016/j.gene.2008.09.033

Minhas V, Shrestha A, Wadhwa N, Singh R, Gupta SK (2016) Novel sperm and gonadotropin-releasing hormone-based recombinant fusion protein: achievement of 100% contraceptive efficacy by co-immunization of male and female mice. Mol Reprod Dev 83(12):1048–1059. https://doi.org/10.1002/mrd.22743

Mishra V, Bose A, Kiran S, Banerjee S, Shah IA, Chaukimath P, Reshi MM, Srinivas S, Barman A, Visweswariah SS (2021) Gut-associated cGMP mediates colitis and dysbiosis in a mouse model of an activating mutation in GUCY2C. J Exp Med 218(11):e20210479. https://doi.org/10.1084/jem.20210479

Mishra A, Ganguli N, Majumdar SS, Modi D (2022) Loss of HOXA10 causes endometrial hyperplasia progressing to endometrial cancer. J Mol Endocrinol 69(3):431–444. https://doi.org/10.1530/JME-22-0051

Mohapatra SK, Sandhu A, Neerukattu VS, Singh KP, Selokar NL, Singla SK, Chauhan MS, Manik RS, Palta P (2015) Buffalo embryos produced by handmade cloning from oocytes selected using brilliant cresyl blue staining have better developmental competence and quality and are closer to embryos produced by in vitro fertilization in terms of their epigenetic status and gene expression pattern. Cell Reprogram 17(2):141–150. https://doi.org/10.1089/cell.2014.0077

Muller HJ (1927) Artificial transmutation of the gene. Science 66(1699):84–87. https://doi.org/10.1126/science.66.1699.84

Palmiter RD, Brinster RL (1986) Germ-line transformation of mice. Annu Rev Genet 20(1):465–499. https://doi.org/10.1146/annurev.ge.20.120186.002341

Pradhan BS, Majumdar SS (2016) An efficient method for generation of transgenic rats avoiding embryo manipulation. Mol Ther Nucleic Acids 5:e293. https://doi.org/10.1038/mtna.2016.9

Pradhan BS, Bhattacharya I, Sarkar R, Majumdar SS (2019) Downregulation of Sostdc1 in testicular Sertoli cells is prerequisite for onset of robust spermatogenesis at puberty. Sci Rep 9(1):1–11. https://doi.org/10.1038/s41598-019-47930-x

Pradhan BS, Bhattacharya I, Sarkar R, Majumdar SS (2020) Pubertal down-regulation of Tetraspanin 8 in testicular Sertoli cells is crucial for male fertility. Mol Hum Reprod 26(10):760–772. https://doi.org/10.1093/molehr/gaaa055

Sarkar RK, Sen Sharma S, Mandal K, Wadhwa N, Kunj N, Gupta A, Pal R, Rai U, Majumdar SS (2021) Homeobox transcription factor Meis1 is crucial to Sertoli cell mediated regulation of male fertility. Andrology 9(2):689–699. https://doi.org/10.1111/andr.12941

Sato M, Yabuki K, Watanabe T, Kimura M (1999) Testis-mediated gene transfer (TMGT) in mice: successful transmission of introduced DNA from F0 to F2 generations. Transgenics 3(1):11–22

Sehgal L, Thorat R, Khapare N, Mukhopadhaya A, Sawant M, Dalal SN (2011) Lentiviral mediated transgenesis by in vivo manipulation of spermatogonial stem cells. PLoS One 6(7):e21975. https://doi.org/10.1371/journal.pone.0021975

Sehgal L, Thorat R, Khapare N, Mukhopadhaya A, Sawant M, Dalal S (2012) A protocol for generation of transgenic mice by manipulating spermatogonial stem cells in vivo. Protoc Exchange https://doi.org/10.1038/protex.2012.020

Sehgal L, Usmani A, Dalal SN, Majumdar SS (2014) Generation of transgenic mice by exploiting spermatogonial stem cells in vivo. Methods Mol Biol 1194:327–337. https://doi.org/10.1007/978-1-4939-1215-5_18

Shah RA, George A, Singh MK, Kumar D, Chauhan MS, Manik R, Palta P, Singla SK (2008) Hand-made cloned buffalo (Bubalus bubalis) embryos: comparison of different media and culture systems. Cloning Stem Cells 10(4):435–442. https://doi.org/10.1089/clo.2008.0033

Sharpe RM, Mckinnell C, Kivlin C, Fisher JS (2003) Proliferation and functional maturation of Sertoli cells, and their relevance to disorders of testis function in adulthood. Reproduction 125:769

Shekar PC, Goel S, Rani SDS, Sarathi DP, Alex JL, Singh S, Kumar S (2006) κ-Casein-deficient mice fail to lactate. Proc Natl Acad Sci U S A 103(21):8000–8005. https://doi.org/10.1073/pnas.0601611103

Shukla M, Ganguli N, Sen Sharma S, Majumdar SS (2018) Sertoli cell specific decline in NOR-1 leads to germ cell apoptosis and reduced fertility. J Cell Biochem 119(8):6514–6526. https://doi.org/10.1002/jcb.26698

Singh M, Kumar V (2003) Transgenic mouse models of hepatitis B virus-associated hepatocellular carcinoma. Rev Med Virol 13(4):243–253. https://doi.org/10.1002/rmv.392

Singh G, Totiger TM, Seshagiri PB (2012) Successful derivation of EGFP-transgenic embryonic stem cell line from a genetically non-permissive FVB/N mouse. Am J Stem Cells 1(2):163–173

Singh VP, Gurunathan C, Singh S, Singh B, Lakshmi BJ, Mishra AP, Kumar S (2015) Genetic deletion of Wdr13 improves the metabolic phenotype of Lepr db/db mice by modulating AP1 and PPARγ target genes. Diabetologia 58(2):384–392. https://doi.org/10.1007/s00125-014-3438-y

Tiwari R, Ganguli N, Alam H, Sahu I, Vadivel CK, Sinha S, Patel S, Jamghare SN, Bane S, Thorat R, Majumdar SS, Vaidya MM (2021) Generation of a tissue-specific transgenic model for K8 phosphomutants: a tool to investigate the role of K8 phosphorylation during skin carcinogenesis in vivo. Cell Biol Int 45(8):1720–1732. https://doi.org/10.1002/CBIN.11611

Ulgekar G, Kaur D, Ganesan V, Sen Sharma S, Ganguli N, Majumdar SS (2022) Anhydride chemistry-based hexanoylation of polyethylenimine increases transfection efficiency and expression of tagged DNA for therapeutic proteins in cultured cells. Biotechnol Bioeng 119:3275. https://doi.org/10.1002/BIT.28196

Usmani A, Ganguli N, Sarkar H, Dhup S, Batta SR, Vimal M, Ganguli N, Basu S, Nagarajan P, Majumdar SS (2013) A non-surgical approach for male germ cell mediated gene transmission through transgenesis. Sci Rep 3:3430. https://doi.org/10.1038/srep03430

Usmani A, Ganguli N, Jain SK, Ganguli N, Sarkar RK, Choubey M, Shukla M, Sarkar H, Majumdar SS (2016) Robust generation of transgenic mice by simple hypotonic solution mediated delivery of transgene in testicular germ cells. Mol Ther Methods Clin Dev 3:16076. https://doi.org/10.1038/MTM.2016.76

Verma I, Seshagiri PB (2018) Directed differentiation of mouse P19 embryonal carcinoma cells to neural cells in a serum- and retinoic acid-free culture medium. In Vitro Cell Dev Biol Anim 54(8):567–579. https://doi.org/10.1007/s11626-018-0275-1

Vishal SS, Tilwani S, Thorat R, Dalal SN (2017) Generation of mice with tissue specific transgene expression using sperm mediated gene transfer. Int J Pharm Bio Sci 8(2):324–329. https://doi.org/10.22376/ijpbs.2017.8.2.b324-329

Wadhwa N, Kunj N, Tiwari S, Saraiya M, Majumdar SS (2009) Optimization of embryo culture conditions for increasing efficiency of cloning in buffalo (Bubalus bubalis) and generation of transgenic embryos via cloning. Cloning Stem Cells 11(3):387–395

Wadhwa N, Ganguli N, Majumdar SS (2021) Generating transgenic animal models: recent technological advancements. In: Essentials of laboratory animal science: principles and practices. Springer, Berlin, pp 709–724. https://doi.org/10.1007/978-981-16-0987-9_29

Xi J, Zheng W, Chen M, Zou Q, Tang C, Zhou X (2023) Genetically engineered pigs for xenotransplantation: hopes and challenges. Front Cell Dev Biol 10:1093534. https://doi.org/10.3389/FCELL.2022.1093534/BIBTEX

Yadav B, Malonia SK, Majumdar SS, Gupta P, Wadhwa N, Badhwar A, Gupta UD, Katoch VM, Chattopadhyay S (2015) Constitutive expression of SMAR1 confers susceptibility to mycobacterium tuberculosis infection in a transgenic mouse model. Indian J Med Res 142(6):732–741. https://doi.org/10.4103/0971-5916.174566

Yonezawa T, Furuhata Y, Hirabayashi K, Suzuki M, Takahashi M, Nishihara M (2001) Detection of transgene in progeny at different developmental stages following testis-mediated gene transfer. Mol Reprod Dev 60:196

Marine Biotechnology in India: Status and Unfolding Opportunities

4

George John, A. Gopalakrishnan, Kajal Chakraborty, and V. V. R. Suresh

Abstract

There is a need to increase focus on the oceans for producing more food and generating technologies in the non-food sector. India's Blue Economy is the centrepiece where different disciplines are converging to generate food and wealth.

The phenomenal potential of the non-food sector of marine biotechnology unfolds through the promise of new drugs and bioactive compounds. Enzymes from various types of extremophilic marine organisms can lead to industrial applications. Marine nanotechnology applications are becoming increasingly important as they diffuse into many sectors. The development of marine biocomposites should be a priority. High-value products and nutraceuticals present many technological options.

A shift towards biofuel is compelling and holds promise through new technological options that are available. Algal biofuel which is being increasingly looked upon as a substitute for conventional fuel will be a focus for deploying new techniques. The production of marine biomaterial including nanomaterial has shown immense promise. With the increasing pollution of marine ecosystems, it is imperative to develop technoeconomically feasible protocols for bioremediation.

Biotechnological interventions in the food sector will involve farming large swathes of the oceans using new culture techniques, genetically improved strains, and molecular tools, including those for modifying traits and gene editing. Advances in genomics, especially the use of genetic markers and CRISPR tech-

G. John (✉)
Department of Biotechnology, Ministry of Science and Technology, Government of India (Formerly), New Delhi, India
e-mail: gjohn@nic.in

A. Gopalakrishnan · K. Chakraborty · V. V. R. Suresh
Central Marine Fisheries Research Institute, Kochi, India

G. John, A. K. Panda (eds.), *Biotechnology in India - Reworking A Strategy*, https://doi.org/10.1007/978-981-97-0896-3_4

niques, will redraw the contours of mariculture by improving economically important traits. In vitro fish or factory fish meat produced through tissue engineering will be a humane, safe, and environmentally friendly alternative to slaughtered fish flesh.

At the infrastructural level, there is a need to set up a Mariculture Authority and a National Institute of Marine Biotechnology to tap the food and the non-food sectors. The policy framework and the strategic direction needed to capture the emerging potential are discussed. This review is not exhaustive but presents an overview of key sectors for providing the contextual setting and looking at the strategy required.

4.1 Introduction

Marine biotechnology is an overarching field that embraces several sectors, ecosystems, and governance models shaped by international laws of the sea and the commercial interests of coastal countries. It continues to entice the ardent researcher with the promise of multibillion-dollar drugs that offer cures from minor ailments to cancer. The fascination with the seas has been intertwined with Indian mythology as portrayed by the 'Sagar Manthan' or churning of the sea to extract the elixir of life.

The United Nations Decade of Ocean Science for Sustainable Development (2021–2030) will give impetus to marine biotechnology applications. Marine biotechnology, a commercial spin-off from marine biology presents an array of opportunities for new drugs, cosmeceuticals, and biomaterials including nanomaterial, biofuel, enzymes, and designer aquatic foods (Fig. 4.1). Since the oceans are regarded to be the cradle of life, its myriad forms like extremophiles, archaea, and nanobacteria with their novel biochemical pathways hold secrets to industrial enzymes for a wide range of applications.

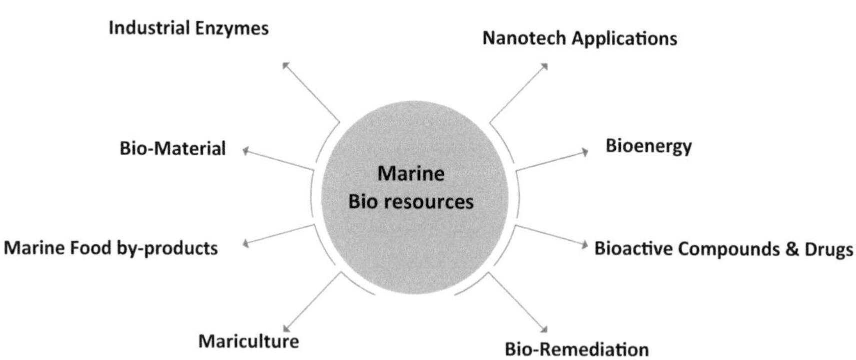

Fig. 4.1 Key drivers of marine biotechnology

Though oceans cover 70% of the Earth's surface, they are not fully explored. While setting the context it must be noted that the vast marine resources of India present endless opportunities for marine drug development (Indap 2009). Townsed et al. (2018) have described the marine environment as one offering a plethora of ecosystem services with societal benefits. The fascination deepens because microorganisms, the invisible majority account for almost 90% of the living biomass in the oceans and hold the key to the health of the ocean systems.

4.2 Resources

Marine biotechnology is governed by CBD, the Nagoya Protocol, UN Convention on the Law of the Sea (Lallier et al. 2014; Broggiato et al. 2018). The interplay between these provisions and the rights of coastal countries in the EEZ needs to be carefully managed. Access and benefit sharing of resources beyond a country's jurisdiction needs to be negotiated.

While looking at key resources, mention must be made of the 'The World Registry of Marine Species', which holds more than 24,100 records, https://marinespecies.org/index.php. Ocean microbiome research is defining important components of marine biotechnology. About 50 million prokaryotes and 100 million eukaryote genes have been described by Project Tara Oceans (Abreu et al. 2022).

Several resources that facilitate and support marine biotechnology research in India are listed in Tables 4.1 and 4.2. In the absence of a single network or resource base, it is prudent to keep track of the information available in these tables. Marine organisms of potential application in marine biotechnology are given in Table 4.3. The National Repository of Fish Cell Lines (NRFC) was established in 2010 at ICAR-NBFGR, Lucknow, to accelerate research. With 63 fish cell line accessions, it is one of the largest collections in the world. Out of this, 28 cell lines were developed and deposited by ICAR-NBFGR and the rest came from other researchers in India (Kumar et al. 2020d), http://mail.nbfgr.res.in/nrfc/. A mitochondrial DNA database of fish has been set up at the National Bureau of Fish Genetic Resources (NBFGR). The Fish Mitogenome Resource (FMiR) has curated mitochondrial genomes from 1302 species from 297 families and 47 orders. The database is useful in locating SSR motifs, repeat loci, and relative abundance of microsatellite repeats (Nagpure et al. 2015).

India is playing a key role in protecting the IPR of traditional medicines. The Traditional Knowledge Library of the Council of Scientific and Industrial Research has details of about 250,000 formulations (Patwardhan and Mashelkar 2009). The Indian Council of Medical Research has assessed the value of traditional medicines, and a more rigorous evaluation covering the marine sector also will be helpful as many products are used globally (Parveen et al. 2015).

Table 4.1 Important resources for marine biotechnology

No.	Resource	Web link
1	Marine Biotechnology ERA-NET	http://www.marinebiotech.eu/
2	Marine Biotechnology in India-ERA-NET	http://www.marinebiotech.eu/wiki/Marine_Biotechnology_in_India
3	Marine Biotechnology-country survey-ERA NET	http://www.marinebiotech.eu/wiki/Category:Marine_Biotechnology_country_survey
4	Marine Pharmacology	https://www.marinepharmacology.org/
5	Marin Lit. On marine natural products	https://marinlit.rsc.org/
6	Chem spider	http://www.chemspider.com/
7	Comprehensive database of secondary metabolites from cyanobacteria	https://www.biorxiv.org/content/10.1101/2020.04.16.038703v1
8	StreptomeDB. Streptomycetes natural products	http://132.230.56.4/streptomedb/
9	An online resource for marine fungi	https://link.springer.com/article/10.1007/s13225-019-00426-5#article-info
10	Marine Fungi	https://www.marinefungi.org/
11	Marine Biodiversity	http://www.imbd.ie/
12	Submariner Network	https://www.submariner-network.eu/
13	Blue Med	http://www.bluemed-initiative.eu/
14	Interreg Mediterranean	https://b-blue.interreg-med.eu/
15	Ocean4Biotech	https://www.ocean4biotech.eu/
16	Sagar Manthan or churning of the seas for the elixir of life	https://en.wikipedia.org/wiki/Samudra_Manthana
17	World Registry of Marine Species	https://marinespecies.org/index.php
18	Marine Metagenomics Portal	https://bio.tools/mmp
19	Ocean Gene Atlas	https://tara-oceans.mio.osupytheas.fr/ocean-gene-atlas/
20	Ocean Barcode Atlas	https://oba.mio.osupytheas.fr/ocean-atlas/
21	Tara Oceans Bacterial and Archaeal Genomes	https://www.genoscope.cns.fr/tara/
22	Ocean Microbiomics Database	https://microbiomics.io/ocean/
23	Extremophile scale up facility	https://marinebiotechnology.umbc.edu/extremophile-scale-up-facility/
24	Marine Biotechnology—International Journal	https://www.springer.com/journal/10126/
25	Tara Foundation	https://fondationtaraocean.org/en/home/
26	European Marine Biological Resource Centre	https://www.embrc.eu/

27	Institute of Marine and Environmental Technology	https://imet.usmd.edu/
28	Department of Marine Biotechnology. UMBC	https://marinebiotechnology.umbc.edu/
29	National Facility for Marine Cyanobacteria	https://www.bdu.ac.in/centers/NFMC/
30	National Institute of Ocean Technology- Marine Biotechnology Themes	https://www.niot.res.in/niot_mbtech_en.php
31	National Institute of Oceanography Publications-Marine Biotechnology	https://drs.nio.org/drs/handle/2264/1/discover?query=biotechnology&submit=&filtertype_0=dateIssued&filter_relational_operator_0=equals&filter_0=%5B1907+TO+1999%5D
32	CSMCRI Algal Research Station	https://www.csmcri.res.in/research-and-development/Marine-Algal-Research-Station-Mandapam?team=135
33	Blue Economy Working Group 3 Report (Economic Advisory Council to Prime Minister	https://eprints.cmfri.org.in/13370/1/Blue%20Economy_Working%20Group%203%20Final%20Report%20to%20EAC%20to%20the%20Prime%20Minister%20of%20India_June%202019.pdf
34	Deep Ocean Mission of Government of India	https://www.moes.gov.in/sites/default/files/PIB1797250.pdf
35	Drugs from the Sea	https://www.moes.gov.in/programmes/ocean-science-services/coastal-research/drugs-sea
36	Council of Scientific and Industrial Traditional Knowledge Digital Library	http://www.tkdl.res.in/tkdl/LangDefault/Common/Home.asp?GL=Eng
37	Fish Mitogenome Research, National Bureau of Fish Genetic Resources	https://mail.nbfgr.res.in/fmir/ https://journals.plos.org/plosone/article?id=10.1371/journal.pone.0136711
38	National Repository of Fish Cell Line, National Bureau of Fish Genetic Resources	https://mail.nbfgr.res.in/nrfc/
39	Aquatic Animal Health Laboratory, C Abdul Hakeem College	https://aahl.res.in/fish-cell-line/
40	Ministry of Petroleum and Natural Gas. Government of India	https://mopng.gov.in/en/page/11

Table 4.2 Key institutions in marine biotechnology research in India

No.	Institution	Web page link
1	National Institute of Ocean Technology	https://www.niot.res.in/
2	National Institute of Oceanography	https://www.nio.org/
3	Central Marine Fisheries Research Institute	https://www.cmfri.org.in/
4	Central Salt and Marine Chemical Research Institute	https://www.csmcri.res.in/
5	Central Institute of Brackishwater Aquaculture	https://ciba.icar.gov.in/
6	National Bureau of Fish Genetic Resources	https://www.nbfgr.res.in/
7	Bharathidasan University	https://www.bdu.ac.in/
8	Faculty of Marine Sciences, Annamalai University	https://annamalaiuniversity.ac.in/T00_info.php?fc=T00
9	National Centre for Polar and Oceanic Research	https://ncpor.res.in/
10	CSMCRI Algal Research Station	https://www.csmcri.res.in/research-and-development/Marine-Algal-Research-Station-Mandapam?team=135
11	Cochin University of Science and Technology	https://cusat.ac.in/
12	Central Drug Research Institute	https://www.cdri.res.in/
13	Bose Institute, Kolkata	http://jcbose.ac.in/
14	Pondicherry University	https://www.pondiuni.edu.in/department/department-of-ocean-studies-and-marine-biology/

Table 4.3 Indicative list of non-food Indian species with potential marine biotechnology applications

Resource	Classification	Uses	Opportunities/challenges	Application/commercialisation
Extremophiles	Bacteria, fungi	Exploiting novel pathways, production of industrial enzymes with high stability	Limitations in culture techniques and laboratory facilities	Initial stages
Seaweed-associated heterotrophs	*Bacillus, Pseudomonas*	Bioactive compounds, enzymes, cosmeceuticals, and novel antibiotics	Limitations in developing large-scale fermentation/ factory production and approval protocols	Under process to develop potential technologies from seaweed-associated heterotrophs
Microalgae	Chlorophyta, *Haematococcus, Dunaliella, Isochrysis, Nannochloropsis, Pavlova*	Pigments, bioactive compounds, polyunsaturated fatty acids, biofuel	Developing culture conditions (in open raceway/photobioreactor)	Under process to develop potential technologies
Fungi	Ascomycota, Basidiomycota, Mucoromycota, Chytridiomycota, etc.	Cellulose and chitin degradation, source of natural products	Developing culture conditions	New metabolites with antibacterial, antiviral, and anticancer activities
Thraustochytrids	Large-celled marine heterokonts	Production of docosahexaenoic (DHA) and eicosapentaenoic (EPA) ω-3-fatty acids	Limitations in culture techniques and development of large-scale fermentation	Initial stages
Eubacteria-Firmicutes	*Bacillus subtilis, Bacillus licheniformis, B. amyloliquefaciens*	Bioactivities against human ailments, drug delivery system, bioremediation	Yield and production in factory conditions, approval protocols	Under process to develop potential technologies
Molluscs	Bivalves (*Perna, Crassostrea, Saccostrea*), Cephalopods (*Octopus, Sepia*), and Gastropods	Bioactive compounds	Harvesting and culture under captivity	Bioactive compounds, such as anti-diabetic, anti-hyperlipidaemic, antidiarrheal, antimicrobial/ antiviral, antimalarial, and so on
Horseshoe crab	Crustacean *Tachypleus gigas*	Tachypleus amoebocyte lysate (TAL) used in endotoxin testing	Endangered species; viable farming limited	LAL kit for endotoxin testing not commercialised

(continued)

Table 4.3 (continued)

Resource	Classification	Uses	Opportunities/challenges	Application/commercialisation
Bioremediation	Microbial conglomerates, single-celled alga *Chlamydomonas reinhardtii*, cyanobacteria	Removal of oil spills, toxic chemicals/antibiotics/ pesticides, managing seafood processing, sewage disposal, and recuperation of minerals	Economic viability	Early stages
Seaweeds or marine macroalgae	*Chlorophyta*, *Rhodophyta*, *and Phaeophyta*	Bioactive compounds and nutraceuticals	Developing suitable engineering devices for large-scale culture in open sea and harvesting	Anti-diabetic, anti-dyslipidaemic, antiarthritic, steatohepatitis, immunoboost agent, etc.

4.3 Bioactive Compounds

4.3.1 Marine Bioprospecting

Marine bioprospecting remains largely unexplored and is perhaps the largest sector that remains untapped. Unprecedented opportunities for new applications are opening up at the genome, transcriptome, proteome, and metabolome levels. Marine organisms synthesise unique secondary metabolites, enzymes, and biopolymers in response to environmental stimuli. The Indian scenario presents unique opportunities.

Bhatia and Chugh (2015) analysed the role of marine bioprospecting contracts concerning access and benefit sharing while factoring in the interests of traditional knowledge holders. Pharmaceuticals are the primary focus of marine bioprospecting followed by nutraceuticals and cosmetics. The patent options are clear gateways for wealth generation in marine bioprospecting (Demunshi and Chugh 2009). Use of metabolites to counter multidrug-resistant *P. aeruginosa* (Sajayan et al. 2023), screening for antibiotic resistance genes (Hussain et al. 2023), and biofilm management approaches (Lahiri et al. 2023) could be used to develop a strategy against emerging drug resistance and related issues.

About 30,000 compounds of marine origin are known. Since 2008, more than 1000 new compounds have been discovered every year. The oldest marine product was the dye Tyrian purple extracted from marine molluscs by the Phoenicians around 1600 BC (Lindequist 2016). Marine bioactives that are anti-inflammatory and immunomodulatory have a bearing on the management of COVID-19 (Singh et al. 2021).

Cherry et al. (2022) have listed 12 natural products derived from marine organisms that have specific anti-inflammatory activity and enhance the immune system. There are 27 antibody-drug conjugates (ADCs) in the pipeline with eight undergoing clinical trials. With the emergence of ADCs, targeted therapy has emerged (Newman 2021).

There are 17 clinically approved drugs derived from marine organisms (Martins et al. 2014; Jimenez et al. 2020) and many are under clinical trials. More than 35,000 natural products have been identified from marine organisms (Rotter et al. 2020). Pawar et al. (2022) indicated the immense untapped deep-sea resources for developing novel marine drugs for targeting cancer and the emerging options for discovering new antineoplastic agents. Mayer et al. (2010) reported 13 marine-derived compounds to be in Phase 1, 2, and 3 trials, while hundreds continue to be in the pipeline. By 1974, cytarabine, Ara C, vidarabine, and Ara A were in the market. After 30 years, in 2004, ziconotide (Prialt) was approved.

Furthermore, Cherry et al. (2022) presented a very insightful analysis of the drug discovery journey and recognised that the emergence of natural product-derived drugs was more than those through the synthetic chemistry route with the leading role played by The International Union of Basic and Clinical Pharmacology (IUPHAR). Drug development is a challenging business that calls for huge investments over a long period. It takes about 15–20 years for an entity to successfully pass through all the clinical trial stages in the drug discovery pipeline (Taylor 2015).

4.3.2 Actinomycetes and Streptomycetes

Actinomycetes and Streptomycetes have been treasure troves for the discovery of new drugs. Several species of Streptomyces have been examined for *L. asparaginase* which is used in the treatment of leukaemia (Narayana et al. 2008). Marine actinobacteria from the Andaman coast of India are potential sources of pharmaceutically important bioactive molecules, and the possibility of discovering newer proteins makes a strong case for bioprospecting (Meena et al. 2013).

A total of 267 new natural products were derived from 96 rare marine actinomycetes during the decade 2007–2017, unequivocally showing that marine bioprospecting needs to be intensified (Subramani and Sipkema 2019). Marine actinomycetes are a huge untapped resource. It has been estimated that about 10% of all marine bacteria are actinomycetes, and only 1% of these have been identified (Subramani and Sipkema 2019; Subramani and Aalbersberg 2013).

While reviewing the array of bioactive compounds from marine actinomycetes, it is apparent that functional genomics will be crucial to drug discovery (Mahapatra et al. 2020). About 23,000 medicinally important compounds have been derived from marine microorganisms. A major portion of 70% was extracted from actinomycetes, 20% from fungi, and the remaining from bacilli and other microbes (Blunt et al. 2017; Falaise et al. 2016; Mehbub et al. 2016).

A sterol derivative from actinomycete *Nocardiopsis alba* has antibacterial activity against *Vibrio harveyi,* a pathogen in aquaculture and can be used as a probiotic in aquaculture (Sunish et al. 2021). Meena et al. (2015) focussed on process optimisation in the production of L-asparaginase from *Streptomyces griseus*. The engineered *Escherichia coli* harbouring the L-asparaginase gene for biosynthesis presents an interesting option for producing glutaminase-free anti-tumour enzyme. Padmanaban et al. (2017) showed that a deep-sea piezotolerant *Streptomyces* bacteria produced secondary metabolites effective against methicillin-resistant *Staphylococcus aureus*, underscoring the need for a greater focus on deep-sea organisms for bioprospecting.

4.3.3 Fungi

The discovery of pharmaceutical compounds from marine fungi has not matched the expectations. Metabolic engineering and genomics could throw up interesting candidates as drugs. The potential of marine fungi to produce novel metabolites gets amplified with metagenomics, rDNA techniques, and improved molecular tools (Devi and Thakur 2022). Devi et al. (2013) showed antifungal properties of bioactives isolated from *Trichoderma harzianum*. The production of broad-spectrum antifungal metabolite sesquiterpenes from deep-sea fungus *Ascotricha* sp makes a case for screening deep-sea organisms for drug prospecting (Kumar et al. 2019a).

Extracellular secondary metabolites produced by *Nigrospora* sp., a piezotolerant fungus isolated from the deep sea showed anticancer and antimicrobial activities. The production of secondary metabolites like griseofulvin, spirobenzofuran, and pyrone derivatives under optimised conditions shows prospects of scaling up (Arumugam et al. 2015).

A marine sediment-derived fungus, *Penicillium* sp. showed active fractions containing oleamide that displayed anticancer and antibacterial properties, suggesting the therapeutic value of these fractions (Farha and Hatha 2019). Kamat et al. (2020) screened 26 endophytic fungi from the Konkan Coast. Extracts from these were active against five pathogenic bacteria. The clear pointer is that algae-derived fungal endophytes are promising for pharmaceuticals.

4.3.4 Thraustochytrids

Thraustochytrids are a promising source of bioactive compounds and can be grouped with fungi as they are heterotrophic fungi like the clade of *Stramenopiles* (Leyland et al. 2017). Thraustochytrids are a group that secrete a range of useful enzymes. Production of extremely halophytic, stable alpha-amylases by mesohaline Thraustochytrids is possible under relatively easy conditions requiring salinity not more than that of seawater (Shirodkar et al. 2020).

The widely useful DHA needs a reliable source of production. The morphology and molecular profiles of a large assemblage of 127 thraustochytrids that were studied from the south Andaman Islands showed them to be promising (Kalidasan et al. 2021b).

Thraustochytrids from the mangrove system of Navi Mumbai are potential sources of palmitic acid (PA) and docosahexaenoic acid (DHA) as these were the dominant fatty acids that played a part in DHA enhancement in the marine food web (Bagul and Annapure 2021). Jaseera and Kaladharan (2020) have further profiled the thraustochytrids as a vast source of oil and nutraceuticals, especially PUFA-enriched products.

Further, studies on mangrove-derived thraustochytrid *Thraustochytrium kinnei* showed it to be a good source for DHA production and nanoparticle synthesis, nutraceuticals, antioxidants, and antimicrobials, calling for detailed studies (Kalidasan et al. 2021a).

Thraustochytrids are under focus due to their potential to generate enzymes with multiple hydrolytic actions and hence are of industrial value. They produce a range of enzymes like agarases, amylases, pectinases, and chitinases (Shirodkar et al. 2017). Production of extracellular sulfated polysaccharides reported in Thraustochytrids is noteworthy because of its application potential (Jain et al. 2005).

Kalidasan et al. (2022) showed the potential of developing an anti-colon cancer drug using unsaturated fatty acids from thraustochytrid *Schizochytrium* sp. This was confirmed through a cytotoxicity assay against colon cancer cell line HT29.

4.3.5 Algae/Seaweed

The promise of using the fatty acids carotenoid complex from *Chlorella vulgaris* against the multidrug-resistant strains of *Mycobacterium tuberculosis* has far-reaching implications that need to be explored fully (Kumar et al. 2020e).

Thambi and Chakraborty (2022) demonstrated the inhibitory properties of red and brown marine algae against antioxidants. The three species studied *(Sargassum plagiophyllum, Turbinaria decurrens, Hydropuntia edulis)* are good sources of pharmaceuticals and functional foods.

Bioactives against white spot shrimp virus (WSSV) in shrimp, *Fenneropenaeus indicus* were isolated from *Enteromorpha flexuosa,* pointing towards a potential remedy for a major scourge in shrimp farming (Velmurugan et al. 2015).

4.3.6 Extremophiles and Others

A rather enigmatic group of organisms largely bacteria and archaea, classified as extremophiles that include thermophilic, halophilic, alkalophilic, piezophilic, and psychrophilic have novel biochemical pathways and the potential to yield bioactive compounds and enzymes of industrial applications.

The unique characteristics and adaptations of extremophilic marine organisms make them very attractive for bioprospecting as they can throw up new molecules with surprising qualities (Shukla et al. 2020; Padmanaban et al. 2017). There is a need to intensify efforts in tapping marine resources for pharmaceuticals, with a focused approach to process development, especially for marine extremophilic organisms (Kanase and Singh 2018).

More than 120 genomes of hyperthermophiles have been fully sequenced. Taq DNA polymerase enzyme is the best example of an enzyme from a thermophile, though it was not from the marine environment (Krüger et al. 2018). It has been shown that thermostable archaeal enzymes have greater stability towards high-pressure, organic solvents, detergents, and proteolytic enzymes (Sana 2015). The extent to which enzymes of thermophilic or psychrophilic bacteria can result in applications is demonstrated by their role in modifying exopolysaccharides for tissue engineering (Poli et al. 2017).

Molecules extracted from jellyfish have demonstrated antimicrobial and wound-healing properties (Merquiol et al. 2019). The use of green fluorescent protein as a molecular marker from marine jellyfish *Aequorea victoria* led to the 2008 Nobel Prize in Chemistry for Shimomura, Chalfie, and Tsein.

Bacillus subtilis strains from deep-sea sediment produced antimicrobial peptides (AMPs) that had specific antifungal and antibacterial activities. This lead needs to be pursued as extremophilic organisms are most promising in terms of novel bioactive molecules (Kumar et al. 2020b).

4.3.7 Sponges

Sponge biomass is 40–60% microbial in nature (Yarden 2014). About 60% of the new marine natural products and bioactive molecules are derived from microorganisms (Carroll et al. 2019). This fast-growing field of marine biotechnology will help ocean-based industries outperform the growth of the global economy by 2030 (Rayner et al. 2019). An extensive review of marine sponges has been done by Pallela and Ehrlich (2016).

Sponges are a treasure trove as they have a diversity of secondary metabolites with biotechnological properties (Schröder et al. 2003; Müller et al. 2004; Perdicaris et al. 2013). They are sources of diverse chemical compounds that have therapeutic potential. Nucleosides derived from sponges formed the basis for the synthesis of antiviral drugs Ara-A and Ara-C for the treatment of leukaemia and lymphoma (Proksch et al. 2002). Allelochemicals from sponges are good candidates for developing cytotoxic compounds that have pharmaceutical applications (Singh and Thakur 2015).

Kumar and Pal (2016) reviewed the diverse class of bioactive compounds with a focus on those with therapeutic applications. Secondary metabolites from *Spongosorites halichondriodes* are known to show antibacterial, antifungal, and anti-inflammatory activities. To obtain substantial amounts of bioactives from sponges, culture has been suggested as an option that needs to be explored (Kiruba-Sankar et al. 2017).

Sponges present a unique opportunity for discovering bioactives because of their association with symbiotic bacteria (Kiran et al. 2018). The review on bioactives from marine sponge-associated fungi by Venkataraman et al. (2022) shows the promise of discovering pharmaceuticals. Lekshmi et al. (2020) isolated 19 endophytic fungi from 3 sponges that showed promising cytotoxic, anti-inflammatory, and antioxidant activities. There is a need to pursue the leads obtained.

4.3.8 Snails: Conotoxins

Gao et al. (2017) have made a case for conotoxins as a fertile source of drug discovery. With more than 700 species, cone snail venom has been promising as pharmacological probes and potential drugs because of their specificity and affinity to ion channels, receptors, and transporters of target prey and humans. Sudarslal et al. (2004) showed that acyclic peptides from the marine snail *Conus monile* targeted potassium channels and can be a candidate for inhibiting K+ channels.

Glutaminyl cyclase sequences extracted from nine cone snail species showed a high degree of homology between cone sequences and well-characterised animal sequences, Vijayasarathy et al. (2018). The diversity of conopeptides and coenzymes was reported by Pushpabai et al. (2021).

Conotoxins have been characterised for potential medical applications (Vijayasarathy and Balaram 2018, 2019; Vijayasarathy et al. 2018; Franklin et al.

2017). The occurrence of contryphan genes and mature peptides in nine species of cone snails suggests the bioactive nature and biological significance of their venom (Vijayasarathy et al. 2017).

Findings suggestive of the therapeutic potential of conopeptides of venom of cone snail *C. bayani* was reported by Pushpabai et al. (2021). Characterisation of venom of *C. figulinus* showing novel peptides indicates that these can be used as lead compounds for developing neuropharmacologic drugs (Rajesh 2015).

4.4 High-Value Products/Nutraceuticals

These range from applications in agriculture to stimulate plant growth (El Boukhari et al. 2020) and use as cosmeceuticals and cosmetics (Martins et al. 2014). Spirulina, *Arthrospira platensis,* a freshwater species adapted well to seawater and showed good growth and lutein production. This opens up the interesting possibility of deploying this strain in a larger system to scale up lutein production, Leema et al. (2010).

The influence of various nitrogen sources on the production of lutein by *C. vulgaris* was studied by Leema et al. (2011). Its ability to respond to different nitrogen sources opens the possibility of scaling up production through optimisation of culture conditions and also through the use of genetic manipulations. A novel microwave-assisted extraction of lutein from *Chlorella sorokiniana* showed a yield that was more than three times that of the conventional method. The rapidity of the method is a pointer to a cheaper method for industrial uses (Leema et al. 2022).

Leema et al. (2021) in yet another study demonstrated the importance of treatments to remove salts from the seawater culture medium to enhance production of high-value lutein and carotenoids, thereby lowering production costs.

The possibility of producing high-value products using marine bacteria is promising. It has been shown that the deployment of bacteria in pigment production units is attractive because of the speed and relative ease with which the processes can be scaled up (Velmurugan et al. 2020).

4.5 Biodiesel

The policy of the Government of India is to enhance the availability of biofuels in the market and increase the blending percentage. At present the ethanol blending percentage in petrol is around 2.0%, while the blending percentage in diesel is less than 0.1%. An ambitious target of 20% blending of ethanol in petrol and 5% blending of biodiesel in diesel is proposed by 2030. This is likely to be reached by a combination of increasing domestic production, setting up second-generation biorefineries, identifying new feedstock, development of new feedstock for biofuels and through technoeconomically viable new technologies https://mopng.gov.in/en/page/11.

Biodiesel production from marine sources is an area for an extreme focus on newer biomass platforms and enhancement of lipid production. A review of studies carried out so far will provide the contextual setting. The use of *C. vulgaris* for enhanced biomass production is due to the ease of cultivating it. Josephine et al. (2022a) have documented the potential opportunities based on enhanced biomass production in *C. vulgaris*. A significant increase in the expression of genes associated with lipid production and enhanced biomass production was shown by Josephine et al. (2022b). *Nodosilinea nodulosa* has been flagged as a promising microalgal species for biodiesel production with good potential for scaling up, Aravindh et al. (2023).

Enhanced biomass and lipid production linked to the carbon sources and growth stimulators used was evaluated by Josephine et al. (2015 and 2018). The use of chlorella growth factor (CGF) in combination with phytohormones was more effective than CGF and carbon sources. These pointers need further scaling up especially since *C. vulgaris* is one of the few microalgae suitable for biofuel production. Mahata et al. (2022) while examining the triple advantage of food, fuel, and feed from marine algae have advocated the use of genetic engineering to enhance the yields. Kumar et al. (2020a) have mapped the opportunities of microalgae bioengineering to produce biomass, lipids, and bioactive molecules.

Microalgae contain good levels of lipids and their doubling rates exceed most terrestrial plants, making them suitable for biofuel production (Veeramuthu and Ngamcharussrivichai 2020) and there is a need for intervention at the genetic level with the use of gene editing techniques (Maeda et al. 2018 and Mahata et al. 2022).

Random mutation of *C. vulgaris* was attempted for increased lipid levels and enhanced biodiesel production from algal biomass. It showed the role of the omega-6 FAD gene in altering the fatty acid saturation/desaturation ratio in the mutants of *C. vulgaris*. Alterations in lipid metabolising genes and the stability of these mutants would be promising for large-scale biodiesel production (Josephine et al. 2014).

As seen, chlorella growth factor was found to enhance the biomass and lipid production, thereby showing that the process can be scaled up for large-scale biomass production. Cytokinin-treated group appeared to be the ideal one for biodiesel production, and the biomass production was 5.2-fold higher in the cytokinin given group (Josephine et al. 2015).

Biodiesel blending with diesel is getting a push through the Government of India policy. *C. vulgaris* based biodiesel when tested for engine performance showed results comparable with diesel, accompanied by reduced carbon monoxide and hydrocarbon emissions (Mathimani et al. 2017a). Mathimani et al. (2015) demonstrated an efficient method for lipid transesterification in *Chlorella* sp. BDUG91771 with a biodiesel yield of 45%.

Techno-economic cultivation methods are key to biodiesel production. Cultivation of *C. vulgaris* in open pond showed the feasibility of a semi-continuous cultivation method. The fatty acid pool showed it to be comparable to high-quality biodiesel (Mathimani et al. 2017b).

The success of a microalgal biorefinery will depend on the mass production of biomass. Phytohormone supplementations showed a marked 2-to-3.5-fold increase in biomass. The lipid content showed an increase of 1.4-fold, while the PUFA content was enhanced two fold (Seemashree et al. 2022). Ravichandran et al. (2023) while assessing marine microalga *Dictyota bartayresiana* for biodiesel production, indicated the good techno-economic feasibility of the system.

Various methods have been adopted to enhance the production of high-value metabolites or lipids in microalgae for end use as biofuel or nutraceuticals. The use of methyl jasmonate (MeJA) promoted the growth of microalgae, while treatment with salicylic acid induced higher lipid production, Udayan et al. (2020). An efficient biorefinery system based on microalgae that would also produce high-value nutraceuticals would be a priority, Udayan et al. (2022).

Though microalgae are a candidates amenable to genetic engineering aimed at achieving the techno-economic viability of production systems, the limitations are mainly due to inadequate genetic information on commercially suitable strains. There is a need for using non-GM approaches and simultaneously bringing in better regulation for the use of GM algae (Kumar et al. 2020a).

Researchers have been focussing on interventions through CRISPR–Cas 9 systems. The establishment of an industrially suitable strain is possible with the optimisation of critical culture conditions and effective transformation tools (Tanwar and Kumar 2020). We can expect to have CRISPR toolkits that are specific to algae (Tanwar et al. 2018). New options emerging are those of CRISPR-based gene editing techniques for microbial biofuel production and the creation of microbial biorefineries (Shanmugam et al. 2020).

An IIT-Madras, alumni's startup for making crude oil out of seaweed is an important development. The Bengaluru-based startup Sea6 Energy envisages developing large-scale production technology of biofuel. According to them an estimated 4–5 kg of biomass would be needed to produce 1 litre of biocrude.

The company reckons that an area of 120,000–285,000 sq. km of open ocean is needed to cultivate the seaweed required for 200 million tonnes of crude oil. Sea-based systems for biomass production are more promising than the land based ones. https://yourstory.com/2022/10/iit-madras-alumni-startup-crude-oil-seaweed-sea6-energy; https://shaastramag.iitm.ac.in/special-feature/seaweed-fuel.

4.6 Biomaterial

Marine biomaterial holds immense promise for a wide range of applications. The technology generation potential of marine biomaterial calls for it to be designated as an area of extreme focus. Key findings of relevance are discussed.

A comprehensive review of marine biomaterials by Venkatesan and Kim (2015) maps the immense potential of emerging marine biomaterials for several biomedical applications. Interestingly, the structural organisation of marine organisms like sponges has triggered many technical solutions in the fabrication of biomaterial (Macha et al. 2019). Bone restoration following surgeries or injuries is a challenge. Repair of bone defects using growth factors (IGF 1 and BMP 2) loaded marine sponge skeleton as natural scaffolds show promise and are better than bare coral scaffolds (Nandi et al. 2015).

Marine-derived ceramics are used in bone-related treatments, an application that is bound to grow in view of the need for such techniques (Venkatesan and Kim 2015). Varghese et al. (2020) reviewed the various biomaterials that can be used in dental regeneration.

The limitations of conventional chemotherapy are low bioavailability, toxicity, and diffuse distribution of the drugs. A review by Jha et al. (2022) advocates the use of drug delivery systems based on marine biopolymers. This can also help in optimising or reducing the dose.

Alginates, chitosan, and carrageenan are three marine polymers of great interest for their utility as gel-forming biomaterial that is suitable for oral delivery of drugs. Marine biopolymer-based hydrogels will be the focus in developing controlled oral delivery systems (Manna et al. 2022). Also, the versatility of marine biopolymers is shown in their applications as dermal drug carriers and in the development of transdermal drug delivery systems (Kumar et al. 2022).

Sriram et al. (2018) showed that curcumin drug delivery in a controlled manner could be feasible using calcium ferrite nanoparticles embedded in chitosan as the carrier. The usefulness of the system is enhanced since the carrier was found to be biocompatible and non-toxic to normal cells while showing anticancer activity against MCF-7 cell lines.

Karthika et al. (2020) demonstrated that chitosan-polymer coated with reduced graphene oxide implanted with Fe_3O_4 nanoparticles is promising as a nanocomposite that has a high loading capacity with an anticancer drug, indicating that it can be deployed as a nanocarrier for drugs.

The advantages of using alginates in various applications are their biocompatibility, nontoxicity, and amenability to modifications. Its structure makes it very suitable to be used in a drug delivery system (Kirti and Khora 2022).

Eel skin is a promising source of collagen that can be incorporated into alginate hydrogel to create scaffolds using extrusion-based 3D printing technology. Collagen-embedded hydrogels showed enhanced cell proliferation and metabolic activity (Govindharaj et al. 2019).

The role of chitosan, a widely used marine biomaterial can be enhanced by the addition of polymers, metals, and other moieties to form functional chitosan nanocomposites. There is growing evidence of such products finding application in targeted drug delivery (Selvasudha et al. 2022). Based on the bio-adhesive nature of chitosan, synergistic transferrin receptor-targeted micelles using tocopheryl glycol succinate were generated. The ability of these to target over-expressed transferrin receptors of glioma cells makes them a potent weapon in the treatment of brain cancer (Agrawal et al. 2017).

Chitosan-based nanocomposites are gaining increased interest for a wide range of applications such as in regenerative medicine, drug delivery, and biosensors (Murugesan and Scheibel 2021). There is growing interest in metal oxide nanocomposites in nanomedicine. Bharathi et al. (2019) showed that chitosan-copper oxide nanocomposites checked the proliferation of A549 cancer cells and also induced apoptosis in them, indicating their potential use as an anticancer agent.

Chitosan-pectin-titanium dioxide-based nano dressing showed good promise in wound healing due to its antibacterial activity, biocompatibility, and wound closure (Archana et al. 2013). An excellent review (Asmathunisha and Kathiresan 2013) anchored in Indian work on the biosynthesis of nanoparticles by marine organisms documents studies on 27 species producing metallic nanoparticles. These reveal the versatility of marine organisms in producing nanoparticles.

Nanoparticles from gold and silver have found promise in biomedical and pharmaceutical applications. Yang et al. (2021) have examined the various safety issues and outlined a framework to minimise harm to humans while harnessing the full potential of a technology that is increasingly defining the contours of many applications.

Seaweeds have been candidate species of choice in the preparation of nanoparticles. Suriya et al. (2012) reported biosynthesis of silver nanoparticles using seaweed *Urospora* sp. and showed their antibacterial activity against *S. aureus, E. coli,* and *P. aeruginosa,* thus suggesting pharmacological applications of the nanoparticles.

Kumar et al. (2013) showed the synthesis of silver nanoparticles from seaweed *Gracilaria corticata* that had antifungal activity against *Candida* spp. The antifouling properties of silver and gold nanoparticles synthesised using another seaweed *Turbinaria conoides* showed promise in being used as coatings on aquaculture installations to prevent biofouling (Vijayan et al. 2014).

Synthesis of silver nanoparticles was shown using an extract of seaweed *Sargassum polycystum.* The particles showed antimicrobial activity against human pathogens. The cytotoxicity against breast cancer cell line MCF 7 indicated considerable promise in developing anticancer protocols (Thangaraju et al. 2012).

Enteromorpha compressa, a seaweed has joined the list of platforms for producing nanoparticles. The silver nanoparticles produced showed antimicrobial properties against human pathogens and fungal strains. In vitro cytotoxicity against Ehrlich Ascites Carcinoma cells has potential application in developing anticancer treatments, Ramkumar et al. (2017). This must be viewed in the context of the range of nanoparticles produced in marine species and the diverse applications as elaborated (Asmathunisha and Kathiresan 2013).

The use of marine alga *Padina pavonica* for synthesis of silver nanoparticles was demonstrated by Sahayaraj et al. (2012). These showed inhibitory action against two cotton pests. A related study by Bhuyar et al. (2020) showed antibacterial activity of silver nanoparticles against *S. aureus* and *P. aeruginosa.*

Kalidasan et al. (2021a) identified *T. kinnei* as a promising platform for synthesising silver and gold nanoparticles. A review by Ponnuchamy and Jacob (2016) showed seaweeds to be a promising platform for producing metal nanoparticles and their scaling up for pharmaceutical applications. A range of organisms from bacteria to fungi used in nanoparticle production has been documented by Singh et al. (2015), reinforcing the view that marine organisms are reliable platforms for nanoparticle production.

An interesting approach adopted by Inbakandan et al. (2012) was the use of extracts of marine sponge *Acanthella elongata* for silver nanoparticle synthesis. Sharma et al. (2012) showed the possibility of using *Marinobacter pelagius* to achieve a fast rate of gold nanoparticle synthesis.

Shunmugam et al. (2021) showed that *Vibrio alginolyticus* can be a good platform for the biosynthesis of gold nanoparticles. The microbe-mediated biosynthesis was relatively non-toxic. The antioxidant and anticancer activity of these nanoparticles is of added interest.

The marine biomaterial manufacturing process is an intrinsic part of regenerative medicine, and there are many developments in the healthcare sector because of its advantages like creating composite-based scaffolds, matching the mechanical properties of the in vivo tissue, or being able to precisely create architectures favourable for developing regenerative medicine.

Marine biomanufacturing can contribute to the development of new drugs, therapeutics and diagnostic tools, for addressing global health challenges and expanding treatment options. The sector offers an opportunity to sustainably harness marine resources, thereby reducing reliance on non-renewable resources and mitigating environmental impacts associated with traditional manufacturing processes.

4.7 Bioremediation

Bioremediation applications stretch across a wide landscape of technologies. These are brought out through the findings that are briefly reviewed here.

Chitosan nanofilms with antimicrobial properties can be used in setting up filter purification systems for eliminating pathogens (Dananjaya et al. 2017).

An experiment designed to assess the efficiency of marine cyanobacterium *Synechococcus elongatus* in siderophore mediated uranium sequestration showed siderophore complexation with uranium (Rashmi et al. 2013). It was shown that culture filtrate derived from cyanobacterium *S. elongatus* could absorb 75% of uranium and 51% of hydrocarbons, both of which are major contributors to energy production (Rashmi et al. 2021). With the fine-tuning of the protocols towards techno-economic viability, we may have an effective protocol for bioremediation.

Heavy metal toxicity is a challenge to the marine systems due to their long-term effects. *Streptomyces* was shown to be a good candidate for bioabsorption of mercury and lead in the marine environment (Sanjenbam et al. 2012).

Polystyrenes are hazardous to the marine environment through bioaccumulation and the possibility of unknown long-term effects. A study conducted on a deep-sea strain of *B. paralicheniformis* showed that 34% of polystyrene was degraded within 60 days of treatment. Further studies on optimising the process could lead to a reliable technology for bioremediation of widespread polymers (Kumar et al. 2021c).

The deep-sea marine organism repositories find applications in diverse ways. Marine pollution with spent engine oil (SE) is a concern. Studies by Kumar et al. (2021a) showed that a high degradation of polycyclic aromatic hydrocarbons and spent engine oil was possible through *Aspergillus sydowii*, a deep-sea hydrocarbonoclastic and piezotolerant fungus. The findings hold promise in developing a technology for handling spent engine oil spills.

About 80% of plastics found in the oceans are estimated to be from land-based sources and the remaining are from vessels, discards from fishing gear, and anthropogenic activities, and their long-term effect though formation of nanoparticles is of concern (Cozar et al. 2014; Mattsson et al. 2015). Kumar et al. (2020c) while reviewing the scenario have suggested the use of newer strains and technologies including the use of genetically engineered organisms that target specific synthetic polymers. Unique marine bacterial strains have been identified for biodegradation of polypropylene microplastics, Jeyavani et al. (2024). *B. cereus* was found to be promising in the microbial degradation of low-density polyethylene and polystyrene, Jebashalomi et al. (2024).

Eco-friendly bioremediation is the need of the hour. A bacterial consortium of nine isolates from deep-sea sediments was tested on crude oil. The immobilised consortia were able to bring about the complete breakdown and degradation of crude oil leading to an eco-friendly technology for tackling oil spills (Kumar et al. 2019c).

The use of a deep-sea piezotolerant isolate *Nesiotobacter exalbescens* in degrading toluene suggests the possibility of incorporating the findings for designing a protocol for bioremediation of hydrocarbon-polluted deep-sea environments caused by oil spills (Kumar et al. 2019b). Kumar et al. (2023) demonstrated that hydrocarbonoclastic piezotolerant *B. subtilis* EB1 could degrade phenanthrene rapidly. This can be potentially used as a platform for mitigation strategies. A bacterial consortium isolated from deep-sea sediment enriched with spent engine (SE) oil appeared very promising in the bioremediation of marine environments contaminated with spent engine oil. The protocol can be scaled up for application (Kumar et al. 2014).

Eco-restoration involves a series of complex changes that are essential. Biological methods are eco-friendly and best suited for the same. Recycling of leaf litter through decomposition by thraustochytrids in a detritus-based mangrove ecosystem is a critical step that enhances soil fertility and plays a key role in the mangrove ecosystem (Raghukumar et al. 1995; Kalidasan et al. 2019).

4.8 Mariculture

For productivity enhancement, a multipronged approach is essential with strong underpinnings of selective breeding based on genetic and genomic tools, together with interventions through nutrigenomics (Zenger et al. 2019; Athanasios 2014; Narasimapallavan et al. 2022). Cataloguing and characterisation of marine genetic resources must be accelerated. The next-generation sequencing methods and genomics need rapid deployment. Gene modifying technologies involving transgenesis and the emerging CRISPR Cas9 techniques must be dovetailed into production enhancement strategies (Diwan et al. 2017).

Since mariculture is not a fully commercial activity in India, broodstock banks of important species must be set up as in the case of cobia and pompano. Costs and expertise in new-generation genomic technologies are the limiting factors in the development of marine biotechnologies. This again inescapably reiterates the need for major investment and a large national programme with multi-stakeholder participation. *Penaeus vannamei* farming has been quite successful, and the approach can be replicated in other species also https://krishi.icar.gov.in/jspui/bitstream/123456789/23385/1/NFDB%20training%20final%20manual.pdf. Such technologies together with broodstock banks for commercially important species as in cobia and pompano will lead to the much-needed transformation in mariculture. Studies have shown that there is much potential for sea cage farming in India, Divu et al. (2023).

4.8.1 Marker-Assisted Breeding, Genomic Selection, and Genome Editing

There is a need for a long-term plan of genetic characterisation of marine species with a focus on their farming using genetically improved strains. Studies so far have been promising, and there is a need to intensify things further. Studies show that *P. monodon* holds much promise for mariculture. Their populations on the west coast are genetically more structured and differentiated than those on the east coast. The high level of genetic diversity and the population structure show good possibilities for genetic improvement for economic traits. The best founding stocks from the wild should be used for mariculture (Kumar et al. 2007; Mandal et al. 2012).

Divya et al. (2015) deployed ATPase 6/8 gene sequences to assess the genetic diversity and population structure of *P. argenteus* populations in the Arabian Sea and Bay of Bengal. These baseline values are useful in the conservation and management of these stocks. Studies show that the 16S ribosomal RNA region is the best mitochondrial marker for biogeographical phylogenetic studies in *P. monodon*. Such studies are significant for the establishment of selective breeding strategies to develop genetically improved broodstock and to maintain wild stock diversity, Mondal et al. (2020).

Genetic variation in wild stocks of an important commercial shrimp, *F. indicus,* from the Indian Ocean was analysed using polymorphic microsatellite loci and mitochondrial COI gene. The information will be useful in developing selective breeding programmes for the species, Sajeela et al. (2019).

An important development in the breeding techniques for the Indian pompano, *Trachinotus mookalee* led to the establishment of a broodstock bank for the species. Such developments are clear pointers to the feasibility of taking up mariculture with genetically improved strains (Ranjan et al. 2018; Anil et al. 2021).

Another critical area needing attention is the enhancement of biodiesel production through gene editing of strains (Mahata et al. 2022). Microalgae, with their good lipid levels and doubling rates make them suitable for biofuel production (Veeramuthu and Ngamcharussrivichai 2020). With alterations in genes involved in lipid metabolism, genes stabilising such mutants would be the new tools for large-scale biodiesel production (Josephine et al. 2022b).

CRISPR toolkits that are specific to algae can be developed (Tanwar et al. 2018). These are the new options that are emerging for microbial biofuel production and the creation of microbial biorefineries (Shanmugam et al. 2020). Optimisation of culture conditions together with effective transformation tools for industrially suitable strains will take us closer to setting up biorefineries (Tanwar and Kumar 2020).

4.8.2 Cellular Aquaculture/Lab-Grown Fish Meat

It is believed that cultured meat can meet the increasing nutritional needs of the growing human population. This is projected as a responsible method of meat consumption as animals are not sacrificed in its production. Goswami et al. (2022) made a fairly comprehensive review of cellular aquaculture, the emerging approaches to producing fish meat without fish. The mass production of muscle cells in bioreactors will lead towards viable technologies. The cell lines developed at NBFGR https://mail.nbfgr.res.in/nrfc/ and the cell lines at CAH College https://aahl.res.in/fish-cell-line/ are excellent starting points for developing cellular aquaculture. Co-culture of different cell types as found in fish muscle needs development with structured scaffolding, Goswami et al. (2022).

The techno-economic viability of cellular aquaculture will remain a key question for quite some time (Waltz 2021). While it may be possible to overcome technical bottlenecks in fish muscle growth, it seems unlikely that the system will make economic sense. Treich (2021) stressed the need for more research on the economics of the process while in parallel trying to understand the demand for cultured meat, its environmental impact, regulatory issues and food safety concerns that are likely to be thrown up (Ong et al. 2021). Since marine cell and tissue culture is a neglected area (Rubio et al. 2019), the prospects are somewhat speculative but there is a clear need to actively work on the production of seafood without marine animals.

4.9 A Strategy Reworked

4.9.1 Focus on the Non-food Sector

The above review makes a strong case for greater focus on the non-food sector where immense promise lies. The benefits can be fully harnessed only if multi-stakeholder efforts converge. There is a need to capture all initiatives under an umbrella that will provide the necessary driving force to reach endpoints. The need for setting up a specialised institute of marine biotechnology with an extreme focus in the non-food sector was recognised by the Department of Biotechnology, Government of India, in the 1990s with a proposal for setting up a Marine Biotechnology Application Centre (MBAC) and later on with a proposal for setting up a National Institute of Marine Biotechnology (NIMB). However, constraints in funds and policy did not see them through. The Deep Ocean Mission of the Ministry of Earth Sciences approved in 2021 has envisaged an Advanced Marine Station for

Ocean Biology https://pib.gov.in/PressReleasePage.aspx?PRID=1727525. This is a welcome step for enhancing the development of new technologies. This station, however, should be redesigned and upgraded as an independent institute to focus exclusively on the non-food sector of marine biotechnology including development of drugs from the sea. India with its long coastline and rich marine diversity should have an institution devoted exclusively to the non-food sector of marine biotechnology. Considering the expanse and duration, the results of the 'Drugs from the Sea' programme have been somewhat limited, but the valuable experience gained through its implementation which started in the 1980s will be very helpful in developing a robust blueprint for the nation.

The National Institute of Ocean Technology (NIOT) has some facilities for studying deep-sea marine organisms. The National Institute of Oceanography (NIO) has also been a pioneer in this direction. To understand the effect of pressure on microbial gene regulation and to culture novel pressure-loving piezophiles from deep sea, NIOT has developed a system to collect and bring deep-sea water samples to the laboratory, and segregate and cultivate the microbes in the laboratory under in situ conditions. This consists of three major systems, viz. 600 bar pressure-rated high-pressure retainable water sampler system, 350 bar pressure-rated high-pressure low-temperature serial dilution system, and high-pressure low-temperature fermenter system. Using this facility, bacteria, fungi, and deep-sea actinomycetes were characterised from the water and sediment samples collected from the Andaman Sea, Arabian Sea, Bay of Bengal, and the Indian Ocean. This facility is expected to be a part of the Advanced Marine Station for Ocean Biology. Such programmes should be integrated with an independent institution that needs to be set up for extreme focus on the non-food sector of marine biotechnology and development of drugs from the sea.

4.9.2 Enhancing Food Production Through Mariculture

An additional seven million metric tonnes of fish are required by 2030. A major portion of this is expected from mariculture. Therefore, there is a need for a comprehensive strategy that will address multiple issues towards sustainable development in active collaboration with coastal states and union territories. This can be achieved only by an independent body such as a 'Mariculture Authority' or 'Sea Farming Authority'. The Working Group 3 Report on Blue Economy of the Economic Advisory Council to Prime Minister which was submitted to the Government of India also reflects this http://eprints.cmfri.org.in/13370/.

The remarkable success in semi-intensive farming of *Penaeus monodon* and *Penaeus indicus* demonstrated in the early 1990s through a project of the Department of Biotechnology, Government of India, implemented by the Marine Products Export Development Authority shows that large-scale farming activities are technically feasible. This frontline demonstration led to the growth of shrimp farming in the country. Lessons can also be drawn from the large-scale farming of *Penaeus vannamei*. A national plan is needed to develop broodstock banks for commercially important marine species as done in the case of cobia and pompano.

References

Abreu A, Bourgois E, Gristwood A, Trouble R, Acinas SG, Bork P, Boss E, Bowler C, Budinich M, Chaffron S, de Vargas C, Delmont TO, Eveillard D, Guidi L, Iudicone D, Kandels S, Morlon H, Lombard F, Pepperkok R, Karlusich JJP, Piganeau G, Régimbeau A, Sommeria-Klein G, Stemmann L, Sullivan MB, Sunagawa S, Wincker P, Zablocki O, Arendt D, Bilic J, Bork P, Finn R, Heard E, Kandels S, Pepperkok R, Rouse B, Vamathevan J, Casotti R, Cancio I, Cunliffe M, Kervella AE, Kooistra WHCF, Obst M, Pade N, Power DM, Santi I, Tsagaraki TM, Vanaverbeke J (2022) Priorities for ocean microbiome research. Nat Microbiol 7:937–947

Agrawal P, Singh RP, Sharma G, Mehata AK, Singh S, Rajesh CV, Pandey BL, Koch B, Muthu MS (2017) Bioadhesive micelles of d-α-tocopherol polyethylene glycol succinate 1000: synergism of chitosan and transferrin in targeted drug delivery. Colloids Surf B: Biointerfaces 152:277–288

Anil MK, Ambarish GP, Surya S, Gomathi P, Swathi Lekshmi PS, Joseph S, Raju B, Praveen P, Krishna Priya PM, Shalin O, Anand V, Gopalakrishnan A (2021) National broodbank of pompano serving the marine finfish farming sector of India. Aquac Spectr 4(9):29–42

Aravindh SP et al (2023) Exploring the biodiesel production potential of marine microalgae *Nodosilinea nodulosa* SNMVBTAR001 from Kerala coastal region. India. Biocatal Agric Biotechnol 53. https://doi.org/10.1016/j.bcab.2023.102882

Archana D, Dutta J, Dutta PK (2013) Evaluation of chitosan nano dressing for wound healing: characterization, *in vitro* and *in vivo* studies. Int J Biol Macromol 57:193–203

Arumugam GK, Srinivasan SK, Joshi G, Gopal D, Ramalingam K (2015) Production and characterization of bioactive metabolites from piezotolerant deep sea fungus *Nigrospora* sp. in submerged fermentation. J Appl Microbiol 118(1):99–111

Asmathunisha N, Kathiresan K (2013) A review on biosynthesis of nanoparticles by marine organisms. Colloids Surf B: Biointerfaces 103:283–287

Athanasios E (2014) Nutrigenomics in aquaculture research. Fish Aquac J 5:2. https://doi.org/10.4172/2150-3508.1000e107

Bagul VP, Annapure US (2021) Isolation of fast-growing thraustochytrids and seasonal variation on the fatty acid composition of thraustochytrids from mangrove regions of Navi Mumbai, India. J Environ Manag 290:112597

Bharathi D, Ranjithkumar R, Chandarshekar B, Bhuvaneshwari V (2019) Bio-inspired synthesis of chitosan/copper oxide nanocomposite using rutin and their anti-proliferative activity in human lung cancer cells. Int J Biol Macromol 141:476–483

Bhatia P, Chugh A (2015) Role of marine bioprospecting contracts in developing access and benefit sharing mechanism for marine traditional knowledge holders in the pharmaceutical industry. Global Ecol Conserv 3:176–187

Bhuyar P, Rahim MH, Sundararaju S, Ramaraj R, Maniam GP, Govindan N (2020) Synthesis of silver nanoparticles using marine macroalgae *Padina* sp. and its antibacterial activity towards pathogenic bacteria. Beni-Suef Univ J Basic Appl Sci 9:1–5

Blunt JW, Copp BR, Keyzers RA, Munro M, Prinsep MR (2017) Marine natural products. Nat Prod Rep 34:235–294

Broggiato A, Lallier LE, Sirakaya A, Vanagt T (2018) Utilisation of marine genetic resources (GRS): the access and benefit-sharing legal framework. In: Rampelotto PH, Trincone A (eds) Grand challenges in marine biotechnology. Springer, New York, pp 579–599. https://doi.org/10.1007/978-3-319-69075-9_15

Carroll AR, Copp BR, Davis RA, Keyzers RA, Prinsep MR (2019) Marine natural products. Nat Prod Rep 36(1):122–173

Cherry L et al (2022) Future directions for the discovery of natural product-derived immunomodulating drugs: an IUPHAR positional review. https://doi.org/10.1016/j.phrs.2022.106076

Cozar A et al (2014) Plastic debris in the open sea. PNAS, 111 (28) 10239–10244. https://www.pnas.org/cgi/doi/10.1073/pnas.1314705111

Dananjaya SH, Kulatunga DC, Godahewa GI, Nikapitiya C, Oh C, Edussuriya M, Lee J, De Zoysa M (2017) Preparation, characterization, and antimicrobial properties of chitosan–silver nanocomposites films against fish pathogenic bacteria and fungi. Indian J Microbiol 57:427–437

Demunshi Y, Chugh A (2009) Patenting trends in marine bioprospecting based pharmaceutical sector. J Intellec Prop Rights 14:131–141

Devi P, Thakur N (2022) Bioprospecting of marine fungi. In: Progress in mycology: biology and biotechnological applications. Springer Nature Singapore, Singapore, pp 545–571

Devi P, Ciavatta ML, Wahidullah S, Vuppala S, D'Souza L (2013) Fungicidal compounds from a marine ascidian-associated fungus *Trichoderma harzianum*. Discov Biotechnol 4(10):12–17

Divya PR, Gopalakrishnan A, Basheer VS, Swaminathan R, Mohitha C, Joy L, Kumar R, Manoj P, Jena JK (2015) Mitochondrial ATPase 6/8 genes to infer the population genetic structure of silver pomfret fish *Pampus argenteus* along the Indian waters. Mitochondrial DNA 26(2):189–194. https://doi.org/10.3109/19401736.2013.879655

Divu D et al (2023) Mapping of potential sea-cage farming sites through spatial modelling: Preliminary operative suggestions to aid sustainable mariculture expansion in India. 2023. Indian Journal of Fisheries 70(4):10.21077/ijf.2023.70.4.132866-01

Diwan AD et al. (2017) Gene editing (CRISPR-Cas) technology and fisheries sector. Can J Biotech 1(2): 65–72. https://doi.org/10.24870/cjb.2017-000108

El Boukhari ME, Barakate M, Bouhia Y, Lyamlouli K (2020) Trends in seaweed extract based biostimulants: manufacturing process and beneficial effect on soil-plant systems. Plants 9(3):359, 1–23

Falaise C, François C, Travers MA, Morga B, Haure J, Tremblay R, Turcotte F, Pasetto P, Gastineau R, Hardivillier Y, Leignel V (2016) Antimicrobial compounds from eukaryotic microalgae against human pathogens and diseases in aquaculture. Mar Drugs 14(9):159, 1–28

Farha AK, Hatha AM (2019) Bioprospecting potential and secondary metabolite profile of a novel sediment-derived fungus *Penicillium sp.* ArCSPf from continental slope of eastern Arabian Sea. Mycology 10(2):109–117

Franklin JB, Rajesh RP, Vinithkumar NV, Kirubagaran R (2017) Identification of short single disulfide-containing contryphans from the venom of cone snails using *de novo* mass spectrometry-based sequencing methods. Toxicon 132:50–54

Gao B, Peng C, Yang J, Yi Y, Zhang J, Shi Q (2017) Cone snails: a big store of conotoxins for novel drug discovery. Toxins 9(12):397, 1–23

Goswami M, Belathur Shambhugowda Y, Sathiyanarayanan A, Pinto N, Duscher A, Ovissipour R, Lakra WS, Chandragiri Nagarajarao R (2022) Cellular aquaculture: prospects and challenges. Micromachines 13(6):828, 1–20

Govindharaj M, Roopavath UK, Rath SN (2019) Valorization of discarded marine eel fish skin for collagen extraction as a 3D printable blue biomaterial for tissue engineering. J Clean Prod 230:412–419, 1–8

Hussain A, Kumar SHK, Prathiviraj R et al (2023) The genome of *Symbiodiniaceae*-associated *Stutzerimonas frequens* CAM01 reveals a broad spectrum of antibiotic resistance genes indicating anthropogenic drift in the Palk Bay coral reef of south-eastern India. Arch Microbiol 205:319. https://doi.org/10.1007/s00203-023-03656-z

Inbakandan D, Sivaleela G, Peter DM, Kiurbagaran R, Venkatesan R, Khan SA (2012) Marine sponge extract assisted biosynthesis of silver nanoparticles. Mater Lett 87:66–68

Indap MM (2009) Drugs from the sea. Indian J Mar Sci 38:7–9

Jain R, Raghukumar S, Tharanathan R, Bhosle NB (2005) Extracellular polysaccharide production by thraustochytrid protists. Mar Biotechnol 7:184–192

Jaseera KV, Kaladharan P (2020) Overview of systematics, morphology, biodiversity and potential utilisation of Thraustochytrids. J Mar Biol Assoc India 62(2):13–21

Jebashalomi V et al (2024) Microbial degradation of low-density polyethylene (LDPE) and polystyrene using *Bacillus cereus*. (OR268710) isolated from plastic-polluted tropical coastal environment. Sci Total Environ. https://doi.org/10.1016/j.scitotenv.2024.171580

Jeyavani J et al (2024) Bacterial screening in Indian coastal regions for efficient polypropylene microplastics biodegradation. Sci Total Environ. https://www.sciencedirect.com/science/article/pii/S0048969724006363

Jha A, Kumar M, Mishra B (2022) Marine biopolymer-based anticancer drug delivery systems. In: Marine biomaterials: drug delivery and therapeutic applications. Springer, Berlin, pp 351–401

Jimenez PC, Wilke DV, Branco PC, Bauermeister A, Rezende-Teixeira P, Gaudêncio SP, Costa-Lotufo LV (2020) Enriching cancer pharmacology with drugs of marine origin. Br J Pharmacol 177(1):3–27

Josephine A, Rangamaran VR, Gopal D, Shivasankarasubbiah KT, Thilagam MLJ, Dhassiah MP, Padinjattayil DSM, Valsalan VKN, Manambrakat V, Dakshinamurthy S, Thirunavukkarasu S, Ramalingam K (2014) Ultraviolet and 5′ fluorodeoxyuridine induced random mutagenesis in *Chlorella vulgaris* and its impact on fatty acid profile: a new insight on lipid-metabolizing genes and structural characterization of related proteins. Mar Biotechnol 17:66–80. https://doi.org/10.1007/s10126-014-9597-5

Josephine A, Niveditha C, Radhika A, Shali AB, Kumar TS, Dharani G, Kirubagaran R (2015) Analytical evaluation of different carbon sources and growth stimulators on the biomass and lipid production of *Chlorella vulgaris*—implications for biofuels. Biomass Bioenergy 75:170–179

Josephine A, Sivashankarasubbiah KT, Thonthula S, Rangamaran VR, Gopal D, Ramalingam K (2018) An efficient method for the sequential production of lipid and carotenoids from the chlorella growth factor-extracted biomass of *Chlorella vulgaris*. J Appl Phycol 30:2325. https://doi.org/10.1007/s10811-018-1430-5

Josephine A, Kumar TS, Surendran B, Rajakumar S, Kirubagaran R, Dharani G (2022a) Evaluating the effect of various environmental factors on the growth of the marine microalgae, *Chlorella vulgaris*. Front Mar Sci 9:954622. https://doi.org/10.3389/fmars.2022.954622

Josephine A, Rangamaran VR, Sivasankarasubbiah KT, Nagaraj N, Gopal D, Ramalingam K (2022b) An insight into the influence of random mutagenesis on growth and lipid accumulation in *Chlorella vulgaris*—a transcriptome study. Biomass Convers Biorefinery. https://doi.org/10.1007/s13399-022-03249-8

Kalidasan K, Phusit H, Kathiresan K (2019) Enumeration of thraustochytrids in decomposing leaves of mangroves as influenced by physicochemical and microbial factors. Curr Res Environ Appl Mycol 9:288–300

Kalidasan K, Asmathunisha N, Gomathi V, Dufossé L, Kathiresan K (2021a) Isolation and optimization of culture conditions of *Thraustochytrium kinnei* for biomass production, nanoparticle synthesis, antioxidant and antimicrobial activities. J Mar Sci Eng 9(6):678, 1–19

Kalidasan K, Vinithkumar NV, Peter DM, Dharani G, Dufossé L (2021b) Thraustochytrids of mangrove habitats from Andaman Islands: species diversity, PUFA profiles and biotechnological potential. Mar Drugs 19(10):571, 1–25

Kalidasan K, Dufossé L, Manivel G, Senthilraja P, Kathiresan K (2022) Antioxidant and anti-colorectal cancer properties in methanolic extract of mangrove-derived *Schizochytrium* sp. J Mar Sci Eng 10(3):431, 1–20

Kamat S, Kumari M, Taritla S, Jayabaskaran C (2020) Endophytic fungi of marine alga from Konkan coast, India, a rich source of bioactive material. Front Mar Sci 7(31):1–11

Kanase HR, Singh KN (2018) Marine pharmacology: potential, challenges, and future in India. J Med Sci 38(2):49–53

Karthika V, AlSalhi MS, Devanesan S, Gopinath K, Arumugam A, Govindarajan M (2020) Chitosan overlaid Fe3O4/rGO nanocomposite for targeted drug delivery, imaging, and biomedical applications. Sci Rep 10(1):18912

Kiran S, Sekar S, Ramasamy P, Thinesh T, Hassan S, Lipton AN, Ninawe AS, Selvin J (2018) Marine sponge microbial association: towards disclosing unique symbiotic interactions. Mar Environ Res 140:169–179. https://doi.org/10.1016/j.marenvres.2018.04.017

Kirti, Khora SS (2022) Alginate: a promising biopolymer in drug delivery system. In: Jana S, Jana S (eds) Marine biomaterials. Springer, Singapore. https://doi.org/10.1007/978-981-16-4787-1_3

Kiruba-Sankar R, Vinod K, Dam Roy S, Krishnan P, Chadha NK, Sawant PB, Saharan N (2017) Cultivation of marine sponges with pharmaceutical value: status and future prospects in India. Int J Curr Microbiol Appl Sci 6(12):4334–4351

Krüger A, Schäfers C, Schröder C, Antranikian G (2018) Towards a sustainable biobased industry—highlighting the impact of extremophiles. Nat Biotechnol 40:144–153. https://doi.org/10.1016/J.NBT.2017.05.002

Kumar MS, Pal AK (2016) A review of bioactive compounds from marine organisms with special mention on the potential of marine sponges in pharmacological applications. www.mbai.org.in. https://doi.org/10.6024/jmbai.2016.58.1.1862-11

Kumar N, Lakra WS, Majumdar KC, Goswami M, Ravinder K (2007) Genetic diversity in the Indian population of *Penaeus monodon* (Fabricius, 1798) as revealed by mtDNA sequence analysis. Aquac Res 38(8):862–869

Kumar P, Senthamil Selvi S, Govindaraju M (2013) Seaweed-mediated biosynthesis of silver nanoparticles using *Gracilaria corticata* for its antifungal activity against *Candida* spp. Appl Nanosci 3:495–500

Kumar AG, Vijayakumar L, Joshi G, Peter DM, Dharani G, Kirubagaran R (2014) Biodegradation of complex hydrocarbons in spent engine oil by novel bacterial consortium isolated from deep sea sediment. Bioresour Technol 170:556–564

Kumar AG, Balamurugan K, Vijaya Raghavan R, Dharani G, Kirubagaran R (2019a) Studies on the antifungal and serotonin receptor agonist activities of the secondary metabolites from piezotolerant deep-sea fungus *Ascotricha* sp. Mycology 10(2):92–108

Kumar AG, Mathew NC, Sujitha K, Kirubagaran R, Dharani G (2019b) Genome analysis of deep sea piezotolerant *Nesiotobacter exalbescens* COD22 and toluene degradation studies under high pressure condition. Sci Rep 9(1):18724, 1–14

Kumar AG, Rajan NN, Kirubagaran R, Dharani G (2019c) Biodegradation of crude oil using self-immobilized hydrocarbonoclastic deep sea bacterial consortium. Mar Pollut Bull 146:741–750

Kumar G, Shekh A, Jakhu S, Sharma Y, Kapoor R, Sharma TR (2020a) Bioengineering of microalgae: recent advances, perspectives, and regulatory challenges for industrial application. Front Bioeng Biotechnol 8(914):1–9. https://doi.org/10.3389/fbioe.2020.00914

Kumar AG, Dharani G, Kirubagaran R, Atmanand MA (2020b) Production and characterization of antimicrobial peptides from *Bacillus subtilis* isolated from deep-sea core samples. Curr Sci 118(11):1725–1730

Kumar AG, Anjana K, Hinduja M, Sujitha K, Dharani G (2020c) Review on plastic wastes in marine environment—biodegradation and biotechnological solutions. Mar Pollut Bull 150:110733

Kumar MS, Kushwaha B, Mishra AK, Kumar R, Lal KK (2020d) National repository of fish cell lines: accelerating fish cell line research in fisheries and aquaculture. Curr Sci 119(4):589–590

Kumar TS, Josephine A, Sreelatha T, Dusthackeer VA, Mahizhaveni B, Dharani G, Kirubagaran R, Kumar SR (2020e) Fatty acids-carotenoid complex: an effective anti-TB agent from the chlorella growth factor-extracted spent biomass of *Chlorella vulgaris*. J Ethnopharmacol 249:112392

Kumar AG, Manisha D, Sujitha K, Magesh Peter D, Kirubagaran R, Dharani G (2021a) Genome sequence analysis of deep sea *Aspergillus sydowii* BOBA1 and effect of high pressure on biodegradation of spent engine oil. Sci Rep 11(1):9347, 1–9

Kumar AG, Anjana K, Hinduja M, Sujitha K, Dharani G (2021b) Review on plastic wastes in marine environment—biodegradation and biotechnological solutions. Nat Sci Rep 11:9347

Kumar AG, Hinduja M, Sujitha K, Rajan NN, Dharani G (2021c) Biodegradation of polystyrene by deep-sea *Bacillus paralicheniformis* G1 and genome analysis. Sci Total Environ 774:145002

Kumar M, Jha A, Mishra B (2022) Marine biopolymers for transdermal drug delivery. In: Marine biomaterials: drug delivery and therapeutic applications. Springer Nature Singapore, Singapore, pp 157–207

Kumar AG, Manisha D, Nivedha Rajan N, Sujitha K, Magesh Peter D, Kirubagaran R, Dharani G (2023) Biodegradation of phenanthrene by piezotolerant *Bacillus subtilis* EB1 and genomic insights for bioremediation. Mar Pollut Bull 194:115151

Lahiri D, Nag M, Dey A et al (2023) Marine bioactive compounds as antibiofilm agent: a metabolomic approach. Arch Microbiol 205:54. https://doi.org/10.1007/s00203-022-03391-x

Lallier LE, McMeel O, Greiber T, Vanagt T, Dobson AD, Jaspars M (2014) Access to and use of marine genetic resources: understanding the legal framework. Nat Prod Rep 31(5):612–616

Leema JM, Kirubagaran R, Vinithkumar NV, Dheenan PS, Karthikayulu S (2010) High value pigment production from *Arthrospira* (*Spirulina*) *platensis* cultured in seawater. Bioresour Technol 101(23):9221–9227

Leema JM, Kirunagaran R, Abdul Nazar AK, Kathiroli S (2011) Heterotrophic production of Lutein and biomass by *Chlorella vulgaris* with different nitrogen sources. Algae Biofuel:91–101

Leema JTM, Kumar TS, Dharani G (2021) A critical look into different salt removal treatments for the production of high value pigments and fatty acids from marine microalgae *Chlorella vulgaris* (NIOT-74). Biotechnol Rep 30:e00627, 1–9

Leema JTM, Jothy TP, Dharani G (2022) Rapid green microwave assisted extraction of lutein from *Chlorella sorokiniana* (NIOT-2)–process optimization. Food Chem 372:131151, 1–10

Lekshmi N, Umar M, Dhaneesha M, Joseph R, Ravinesh R, Sajeevan TP (2020) Endophytic fungi isolated from the marine sponges as a source of potential bioactive compounds. J Aquat Biol Fish 8:58–66

Leyland B, Leu S, Boussiba S (2017) Are Thraustochytrids algae? Fungal Biol 121:835–840

Lindequist U (2016) Marine-derived pharmaceuticals—challenges and opportunities. Biomol Ther 24(6):561–571

Macha IJ, Ben-Nissan B, Müller W (2019) Marine-based biomaterials for tissue engineering applications. In: Choi AH, Ben-Nissan B (eds) Marine-derived biomaterials for tissue engineering applications. Springer, Singapore, pp 99–111. https://doi.org/10.1007/978-981-13-8855-2_5

Maeda et al (2018) Marine microalgae for production of biofuels and chemicals. Curr Opin Biotechnol 50:111–120. https://doi.org/10.1016/j.copbio.2017.11.018

Mahapatra GP, Raman S, Nayak S, Gouda S, Das G, Patra JK (2020) Metagenomics approaches in discovery and development of new bioactive compounds from marine actinomycetes. Curr Microbiol 77:645–656

Mahata C, Das P, Khan S, Thaher MI, Abdul Quadir M, Annamalai SN, Al Jabri H (2022) The potential of marine microalgae for the production of food, feed, and fuel (3F). Fermentation 8(7):316. https://doi.org/10.3390/fermentation8070316

Mandal A et al (2012) Population genetic structure of *Penaeus monodon,* in relation to monsoon current patterns in Southwest, East and Andaman coastal waters of India. Gene 491(2):149–157. https://doi.org/10.1016/j.gene.2011.10.002

Manna S, Jana S, Mandal D, Bhowmik M (2022) Marine biopolymer-based oral in situ gel: drug delivery application. In: Marine biomaterials: drug delivery and therapeutic applications. Springer Nature Singapore, Singapore, pp 139–155

Martins A, Vieira H, Gaspar H, Santos S (2014) Marketed marine natural products in the pharmaceutical and cosmeceutical industries: tips for success. Mar Drugs 12(2):1066–1101

Mathimani T, Uma L, Prabaharan D (2015) Homogeneous acid catalysed transesterification of marine microalga *Chlorella* sp. BDUG 91771 lipid—an efficient biodiesel yield and its characterization. Renew Energy 81:523–533

Mathimani T, Kumar TS, Chandrasekar M, Uma L, Prabaharan D (2017a) Assessment of fuel properties, engine performance and emission characteristics of outdoor grown marine *Chlorella vulgaris* BDUG 91771 biodiesel. Renew Energy 105:637–646

Mathimani T, Bhumathi D, Ahamed TS, Dineshbabu G, Deviram G, Uma L, Prabaharan D (2017b) Semicontinuous outdoor cultivation and efficient harvesting of marine *Chlorella vulgaris* BDUG 91771 with minimum solid co-precipitation and high floc recovery for biodiesel. Energy Convers Manag 149:13–25

Mattsson K, Hansson L-A, Cedervall T (2015) Nanoplastics in the aquatic environment. Environ Sci Process Impacts 17:1712. https://doi.org/10.1039/c5em00227c

Mayer AMS, Glaser KB, Cuevas C, Jacobs RS, William Kem R, Daniel Little J, McIntosh M, Newman DJ, Potts BC, Shuster DE (2010) The odyssey of marine pharmaceuticals: a current pipeline perspective. Trends Pharmacol Sci 31:255–265

Meena et al (2013) Novel marine actinobacteria from emerald Andaman & Nicobar Islands: a prospective source for industrial and pharmaceutical byproducts. BMC Microbiol 13:145. https://doi.org/10.1186/1471-2180-13-145

Meena B, Anburajan L, Sathish T, Vijaya Raghavan R, Dharani G, Valsalan Vinithkumar N, Kirubagaran R (2015) L-asparaginase from *Streptomyces griseus* NIOT-VKMA29: optimization of process variables using factorial designs and molecular characterization of L-asparaginase gene. Sci Rep 5(1):1–12

Mehbub MF, Perkins MV, Zhang W, Franco CM (2016) New marine natural products from sponges (Porifera) of the order *Dictyoceratida* (2001 to 2012); a promising source for drug discovery, exploration and future prospects. Biotechnol Adv 34(5):473–491

Merquiol L, Romano G, Ianora A, D'Ambra I (2019) Biotechnological applications of Scyphomedusae. Mar Drugs 17(11):604, 1–26

Mondal D, Dutta S, Mallik A, Mandal N (2020) Mitochondrial DNA diversity: insight into population diversity, structure and demographic history of *Penaeus monodon* along the entire coastal region of India. Aquac Res 51:4649. https://doi.org/10.1111/are.14812

Müller WE, Grebenjuk VA, Le Pennec G, Schröder HC, Brümmer F, Hentschel U, Müller IM, Breter HJ (2004) Sustainable production of bioactive compounds by sponges—cell culture and gene cluster approach: a review. Mar Biotechnol 6:105–117

Murugesan S, Scheibel T (2021) Chitosan-based nanocomposites for medical applications. J Polym Sci 59(15):1610–1642

Nagpure NS, Rashid I, Pathak AK, Singh M, Pati R, Singh SP, Sarkar UK (2015) FMiR: a curated resource of mitochondrial DNA information for fish. PLoS One 10(8):e0136711

Nandi SK, Kundu B, Mahato A, Thakur NL, Joardar SN, Mandal BB (2015) *In vitro* and *in vivo* evaluation of the marine sponge skeleton as a bone mimicking biomaterial. Integr Biol 7(2):250–262

Narasimapallavan IG (2022) Nutrigenomics tools to address the emerging issues in marine fish larviculture. Advances in agricultural, animal and fisheries sciences. In: Dayal D, Shamsudheen M (eds) Society for technology, environment, science & people. ZNAN Publishers, Kozhikode, India, 20–29. ISBN : 978-81-956227-2-6

Narayana KJP, Kumar KJ, Vijalakshimi M (2008) L-asparaginase production by *Streptomyces albidoflavus.* Indian J Microbiol 48:331–336

Newman DJ (2021) Natural product based antibody drug conjugates: clinical status as of November 9, 2020. J Nat Prod 84(3):917–931

Ong KJ, Johnston J, Datar I, Sewalt V, Holmes D, Shatkin JA (2021) Food safety considerations and research priorities for the cultured meat and seafood industry. Compr Rev Food Sci Food Saf 20(6):5421–5448. https://doi.org/10.1111/1541-4337.12853. Epub 2021 Oct 10

Padmanaban VP, Verma P, Venkatabaskaran S, Keppayan T, Gopal D, Sekar AK, Ramalingam K (2017) Antimicrobial potential and taxonomic investigation of piezotolerant *Streptomyces* sp. NIOT-Ch-40 isolated from deep-sea sediment. World J Microbiol Biotechnol 33:27–32

Pallela R, Ehrlich H (eds) (2016) Marine sponges: chemicobiological and biomedical applications. Springer, Berlin. https://doi.org/10.1007/978-81-322-2794-6

Parveen A, Parveen B, Parveen R, Ahmad S (2015) Challenges and guidelines for clinical trial of herbal drugs. J Pharm Bioallied Sci 7(4):329–333

Patwardhan B, Mashelkar RA (2009) Traditional medicine-inspired approaches to drug discovery: can Ayurveda show the way forward ? Drug Discov Today 14(15–16):804–811

Pawar VM, Raut GS, Gite PG, Paricharak SP (2022) Review on anticancer agents from the deep of ocean. Eur J Pharm Med Res 9(4):128–138

Perdicaris S, Vlachogianni T, Valavanidis A (2013) Bioactive natural substances from marine sponges: new developments and prospects for future pharmaceuticals. Nat Prod Chem Res 1(3):2329–6836. https://doi.org/10.4172/2329-6836.1000115

Poli A, Finore I, Romano I, Gioiello A, Lama L, Nicolaus B (2017) Microbial diversity in extreme marine habitats and their biomolecules. Microorganisms 5(2):25, 1–30

Ponnuchamy K, Jacob JA (2016) Metal nanoparticles from marine seaweeds—a review. Nanotechnol Rev 5(6):589–600

Proksch P, Edrada R, Ebel R (2002) Drugs from the seas—current status and microbiological implications. Appl Microbiol Biotechnol 59:125–134
Pushpabai RR et al (2021) Diversity of conopeptides and conoenzymes from the venom duct of the marine cone snail *Conus bayani* as determined from transcriptomic and proteomic analyses. Drugs 19(4):202. https://doi.org/10.3390/md19040202
Raghukumar S, Sathe-Pathak V, Sharma S, Raghukumar C (1995) Thraustochytrid and fungal component of marine detritus. III. Field studies on decomposition of leaves of the mangrove *Rhizophora apiculata*. Aquat Microb Ecol 9(2):117–125
Rajesh RP (2015) Novel M-superfamily and T-superfamily conotoxins and contryphans from the vermivorous snail *Conus figulinus*. J Pept Sci 21(1):29–39
Ramkumar VS, Pugazhendhi A, Gopalakrishnan K, Sivagurunathan P, Saratale GD, Dung TN, Kannapiran E (2017) Biofabrication and characterization of silver nanoparticles using aqueous extract of seaweed *Enteromorpha compressa* and its biomedical properties. Biotechnol Rep 14:1–7
Ranjan R, Megarajan S, Xavier B, Ghosh S, Santhosh B, Gopalakrishnan A (2018) Broodstock development, induced breeding and larval rearing of Indian pompano, *Trachinotus mookalee*, (Cuvier, 1832)—a new candidate species for aquaculture. Aquaculture 495:550–557
Rashmi V, Shylaja Naciyar M, Rajalakshmi R, D'Souza SF, Prabaharan D, Uma L (2013) Siderophore mediated uranium sequestration by marine cyanobacterium *Synechococcus elongatus* BDU 130911. Bioresour Technol 130:204–210
Rashmi V, Darshana A, Bhuvaneshwari T, Saha SK, Uma L, Prabaharan D (2021) Uranium adsorption and oil emulsification by extracellular polysaccharide (EPS) of a halophilic unicellular marine cyanobacterium *Synechococcus elongatu*s BDU130911. Curr Res Green Sustain Chem 4:100051, 1–9
Ravichandran P et al (2023) Investigations on evaluation of marine macroalgae *Dictyota bartayresiana* oil for industrial scale production of biodiesel through technoeconomic analysis. Bioresour Technol 374. https://doi.org/10.1016/j.biortech.2023.128769
Rayner R, Jolly C, Gouldman C (2019) Ocean observing and the blue economy. Front Mar Sci 6:330. https://doi.org/10.3389/fmars.2019.00330
Rotter A, Bacu A, Barbier M, Bertoni F, Bones AM, Cancela ML, Carlsson J, Carvalho MF, Cegłowska M, Dalay MC, Dailianis T, Deniz I, Drakulovic D, Dubnika A, Einarsson H, Erdoğan A, Eroldoğan OT, Ezra D, Fazi S, FitzGerald RJ, Gargan LM, Gaudêncio SP, Ivošević DeNardis N, Joksimovic D, Kataržyte M, Kotta J, Mandalakis M, Matijošyte I, Mazur-Marzec H, Massa-Gallucci A, Mehiri M, Nielsen SL, Novoveská L, Overlinge D, Portman ME, Pyrc K, Rebours C, Reinsch T, Reyes F, Rinkevich B, Robbens J, Rudovica V, Sabotič J, Safarik I, Talve S, Tasdemir D, Schneider XT, Thomas OP, Toruńska-Sitarz A, Varese GC, Vasquez MI (2020) A new network for the advancement of marine biotechnology in Europe and beyond. Front Mar Sci 7:278. https://doi.org/10.3389/fmars.2020.00278
Rubio N, Datar I, Stachura D, Kapla D, Krueger K (2019) Cell-based fish: a novel approach to seafood production and an opportunity for cellular agriculture. Front Sustain Food Syst 3:43. www.frontiersin.org
Sahayaraj K, Rajesh S, Rathi JM (2012) Silver nanoparticles biosynthesis using marine alga *Padina pavonica* (Linn.) and its microbicidal activity. Digest J Nanomater Biostruct 7(4):1557–1567
Sana B (2015) Marine microbial enzymes: current status and future prospects. In: Kim SK (ed) Springer handbook of marine biotechnology. Springer handbooks. Springer, Berlin, Heidelberg. https://doi.org/10.1007/978-3-642-53971-8_38
Sajayan A, Ravindran A, Selvin J, Ragothaman P, Kiran GS (2023) An antimicrobial metabolite n-hexadecenoic acid from marine sponge-associated bacteria *Bacillus subtilis* effectively inhibited biofilm forming multidrug-resistant *P. aeruginosa*. Biofouling 39(5):502–515. https://doi.org/10.1080/08927014.2023.2232722. The Journal of Bioadhesion and Biofilm Research. Volume 39, 2023, Issue 5
Sajeela KA, Gopalakrishnan A, Basheer VS, Mandal A, Bineesh KK, Grinson G, Gopakumar SD (2019) New insights from nuclear and mitochondrial markers on the genetic diversity and

structure of the Indian white shrimp *Fenneropenaeus indicus* among the marginal seas in the Indian Ocean. Mol Phylogenet Evol 136:53–64

Sanjenbam P, Saurav K, Kannabiran K (2012) Biosorption of mercury and lead by aqueous *Streptomyces* VITSVK9 sp. isolated from marine sediments from the Bay of Bengal, India. Front Chem Sci Eng 6:198–202

Schröder HC, Brümmer F, Fattorusso E, Aiello A, Menna M, De Rosa S, Batel R, Müller WE (2003) Sustainable production of bioactive compounds from sponges: primmorphs as bioreactors. Sponges (Porifera):163–197

Seemashree MH, Chauhan VS, Sarada R (2022) Phytohormone supplementation mediated enhanced biomass production, lipid accumulation, and modulation of fatty acid profile in *Porphyridium purpureum* and *Dunaliella salina* cultures. Biocatal Agric Biotechnol 39:102253

Selvasudha N, Sweety JP, Dhanalekshmi UM, Ruckmani K (2022) Chitosan-based nanocomposite as targeted drug delivery carrier. In: Marine biomaterials: drug delivery and therapeutic applications. Springer Nature Singapore, Singapore, pp 97–137

Shanmugam S, Ngo HH, Wu YR (2020) Advanced CRISPR/Cas-based genome editing tools for microbial biofuels production: a review. Renew Energy 149:1107–1119

Sharma N, Pinnaka AK, Raje M, Fnu A, Bhattacharyya MS, Choudhury AR (2012) Exploitation of marine bacteria for production of gold nanoparticles. Microb Cell Factories 11:1–6

Shirodkar PV, Muraleedharan UD, Raghukumar SE (2017) Production of extracellular polysaccharases by the marine protists, Thraustochytrids, with special reference to alpha-amylase activity. Int J Pharm Bio Sci 8(2): (B) 453–462

Shirodkar PV, Muraleedharan UD, Damare S, Raghukumar S (2020) A mesohaline thraustochytrid produces extremely halophilic alpha-amylases. Mar Biotechnol 22:403–410

Shukla PJ, Bhatt VD, Suriya J, Mootapally C (2020) Marine extremophiles: adaptations and biotechnological applications. In: Encyclopedia of marine biotechnology, vol 3. Wiley, New York, pp 1753–1771

Shunmugam R, Balusamy SR, Kumar V, Menon S, Lakshmi T, Perumalsamy H (2021) Biosynthesis of gold nanoparticles using marine microbe (*Vibrio alginolyticus*) and its anticancer and antioxidant analysis. J King Saud Univ Sci 33(1):101260, 1–7

Singh A, Thakur NL (2015) Significance of investigating allelopathic interactions of marine organisms in the discovery and development of cytotoxic compounds. Chem Biol Interact 243:135–147

Singh CR, Kathiresan K, Anandhan S (2015) A review on marine based nanoparticles and their potential applications. Afr J Biotechnol 14(18):1525–1532

Singh R, Chauhan N, Kuddus M (2021) Exploring the therapeutic potential of marine-derived bioactive compounds against COVID-19. Environ Sci Pollut Res 28(38):52798–52809

Sriram K, Uma Maheswari P, Meera Sheriffa Begum KM, Arthanareeswaran G (2018) Functionalized chitosan with super paramagnetic hybrid nanocarrier for targeted drug delivery of curcumin. Iran Polym J 27:469–482

Subramani R, Aalbersberg W (2013) Culturable rare Actinomycetes: diversity, isolation and marine natural product discovery. Appl Microbiol Biotechnol 97:9291–9321

Subramani R, Sipkema D (2019) Marine rare actinomycetes: a promising source of structurally diverse and unique novel natural products. Mar Drugs 17(5):249, 1–35

Sudarslal S, Singaravadivelan G, Ramasamy P, Ananda K, Sarma SP, Sikdar SK, Krishnan KS, Balaram P (2004) A novel 13 residue acyclic peptide from the marine snail, *Conus monile*, targets potassium channels. Biochem Biophys Res Commun 317(3):682–688

Sunish KS, Sreedharan P, Daniel S, Biji M, Rosamma P, Sukumaran V, Mohandas A, Singh IB (2021) A novel substituted derivative of sterol from marine actinomycetes *Nocardiopsis alba* MCCB 110 antagonistic to the aquaculture pathogen *Vibrio harveyi*. Microb Pathog 157:104967

Suriya J, Raja SB, Sekar V, Rajasekaran R (2012) Biosynthesis of silver nanoparticles and its antibacterial activity using seaweed *Urospora* sp. Afr J Biotechnol 11(58):12192–12198

Tanwar A, Kumar S (2020) Genome editing of algal species by CRISPR Cas9 for biofuels. In: Genome engineering via CRISPR-Cas9 system. Academic Press, New York, pp 163–176. https://doi.org/10.1016/B978-0-12-818140-9.00015-5

Tanwar A, Sharma S, Kumar S (2018) Targeted genome editing in algae using CRISPR/Cas9. Indian J Plant Physiol 23:653–669

Taylor D (2015) The pharmaceutical industry and the future of drug development. In: Hester RE, Harrison RM (eds) Pharmaceuticals in the environment. Royal Society of Chemistry, London, pp 1–33. https://doi.org/10.1039/9781782622345-00001

Thambi A, Chakraborty K (2022) Brown and red marine macroalgae as novel bioresources of promising medicinal properties. J Aquat Food Prod Technol 31(3):227–241

Thangaraju N, Venkatalakshmi RP, Chinnasamy A, Kannaiyan P (2012) Synthesis of silver nanoparticles and the antibacterial and anticancer activities of the crude extract of *Sargassum polycystum C. Agardh*. Nano Biomed Eng 4(2):89–94

Townsed M, Davies K, Hanley N, Hewitt JE, Lundquist CJ, Lohrer AM (2018) The challenge of implementing the marine ecosystem service concept. Front Mar Sci 5:359. https://doi.org/10.3389/fmars.2018.00359

Treich N (2021) Cultured meat: promises and challenges. Environ Resour Econ 2021(79):33–61. https://doi.org/10.1007/s10640-021-00551-3

Udayan A, Sabapathy H, Arumugam M (2020) Stress hormones mediated lipid accumulation and modulation of specific fatty acids in *Nannochloropsis oceanica* CASA CC201. Bioresour Technol 310:123437

Udayan A, Pandey AK, Sirohi R, Sreekumar N, Sang BI, Sim SJ, Kim SH, Pandey A (2022) Production of microalgae with high lipid content and their potential as sources of nutraceuticals. Phytochem Rev:1–28. https://doi.org/10.1007/s11101-021-09784-y

Varghese MG, Thomas George V, Thomas NG, Pyas AM, Francis AM (2020) Marine based biomaterials in dental regeneration. Int J Innov Res Med Sci (IJIRMS) 5(10):1–6. https://doi.org/10.23958/ijirms/vol05-i10/965

Veeramuthu A, Ngamcharussrivichai C (2020) Potential of microalgal biodiesel: challenges and applications. Renew Energy Technol Appl 8(316):1–15

Velmurugan S, Jerin N, Michael Babu M, Bindhu F, Albin Dhas S, Citarasu T (2015) Screening and characterization of antiviral compounds from *Enteromorpha flexuosa* against white spot syndrome virus (WSSV) and its in vivo influence on Indian white shrimp *Fenneropenaeus indicus*. Aquac Int 23:65–80

Velmurugan P, Venil CK, Veera Ravi A, Dufossé L (2020) Marine bacteria is the cell factory to produce bioactive pigments: a prospective pigment source in the ocean. Front Sustain Food Syst 4:589655, 1–7

Venkataraman V, Vaithi K, Singaram J (2022) Bioactive novel natural products from marine sponge: associated fungi. IntechOpen, London. https://doi.org/10.5772/intechopen.101403

Venkatesan J, Kim SK (2015) Marine biomaterials. In: Kim SK (ed) Springer handbook of marine biotechnology. Springer handbooks. Springer, Berlin. https://doi.org/10.1007/978-3-642-53971-8_53

Vijayan SR, Santhiyagu P, Singamuthu M, Kumari Ahila N, Jayaraman R, Ethiraj K (2014) Synthesis and characterization of silver and gold nanoparticles using aqueous extract of seaweed, *Turbinaria conoides*, and their antimicrofouling activity. Sci World J 2014:1–11

Vijayasarathy M, Balaram P (2018) Mass spectrometric identification of bromotryptophan containing conotoxin sequences from the venom of *Conus amadis*. Toxicon 144:68–74

Vijayasarathy M, Balaram P (2019) Cone snail prolyl-4-hydroxylase α-subunit sequences derived from transcriptomic data and mass spectrometric analysis of variable proline hydroxylation in *Conus amadis* venom. J Proteome 194:37–48

Vijayasarathy M, Basheer SM, Franklin JB, Balaram P (2017) Contryphan genes and mature peptides in the venom of nine cone snail species by transcriptomic and mass spectrometric analysis. J Proteome Res 16(2):763–772

Vijayasarathy M, Basheer SM, Balaram P (2018) Cone snail glutaminyl cyclase sequences from transcriptomic analysis and mass spectrometric characterization of two pyroglutamyl conotoxins. J Proteome Res 17(8):2695–2703

Waltz E (2021) No bones, no scales, no eyeballs: appetite grows for cell-based seafood. Nat Biotechnol 39(12):1483–1485

Yang W, Wang L, Mettenbrink EM, DeAngelis PL, Wilhelm S (2021) Nanoparticle toxicology. Annu Rev Pharmacol Toxicol 61:269–289

Yarden O (2014) Fungal association with sessile marine invertebrates. Front Microbiol 5:228. https://doi.org/10.3389/fmicb.2014.00228

Zenger et al (2019) Genomic selection in aquaculture: application, limitations and opportunities with special reference to marine shrimp and pearl oysters. Front Genetics. https://doi.org/10.3389/fgene.2018.00693

Part III

Medical Biotechnology

Unveiling Fundamentals and Research Opportunities of mRNA Technology

5

Durgesh Kumar, Ajay Singh, Swarnendu Kaviraj, and Sanjay Singh

Abstract

This chapter presents a comprehensive overview of messenger RNA (mRNA) technology, highlighting its historical evolution, underlying mechanisms, diverse platforms, manufacturing intricacies, and inherent advantages. Initially conceptualized within the framework of the central dogma of molecular biology in the 1950s–1960s, mRNA serves as an intermediary for genetic information translation into functional proteins. This chapter delves into the various mRNA platforms, encompassing conventional and self-amplifying mRNA types, elucidating their distinct attributes and applications. Manufacturing intricacies and the scalability of mRNA synthesis are discussed, outlining the crucial steps involved in production. Despite promising attributes, the technology faces challenges related to stability, immunogenicity, and delivery systems, which are analyzed in detail. Platform optimization strategies focusing on structural modifications, novel delivery systems, and adjuvant use are highlighted to enhance efficacy and stability, paving the way for future improvements. The narrative specifically addresses emerging opportunities in mRNA technology, emphasizing its pivotal role in combating infectious diseases, cancer, and metabolic disorders. This chapter further underscores the potential impact of mRNA technology in India, acknowledging the nation's vast population and diverse health challenges. The discussion revolves around the need for infrastructure enhancement, cost-effective production, and skill development to harness mRNA technology's potential effectively. Conclusively, this chapter provides insights into the transformative role of mRNA technology, outlining its promising future as a robust therapeutic and preventive tool. It underscores the imperative for continued research, collabora-

D. Kumar · A. Singh · S. Kaviraj · S. Singh (✉)
Gennova Biopharmaceutical Ltd., Pune, India
e-mail: Sanjay.Singh@gennova.co.in

G. John, A. K. Panda (eds.), *Biotechnology in India - Reworking A Strategy*,
https://doi.org/10.1007/978-981-97-0896-3_5

tions, and technology refinement to address global health challenges and accelerate the integration of mRNA technology into clinical applications.

5.1 Introduction

Messenger RNA (mRNA) therapeutics have evolved from the central dogma concept by harnessing the role of mRNA in protein synthesis. mRNA therapeutics leverage the ability to modify or introduce specific mRNA molecules to treat diseases. This includes the development of mRNA vaccines, as well as novel therapies for genetic disorders, cancer, and other conditions. By designing and delivering synthetic mRNA molecules into cells, we can instruct the production of therapeutic proteins or trigger immune responses, revolutionizing the field of medicine and providing a versatile platform for targeted and customizable treatments. Moreover, in vitro transcribed mRNA offers a programmable production method that is not only more efficient in design and manufacture but also provides greater flexibility and cost-effectiveness compared to traditional approaches. Capitalizing on these exceptional advantages, mRNA vaccines elicit the swiftest immune response to large-scale infectious disease outbreaks, as exemplified by their rapid deployment in countering the ongoing and devastating COVID-19 pandemic. Scientists are currently working to enhance the stability, immunogenicity, translation efficiency, and delivery mechanisms for the effective and secure conveyance of mRNA. Encouragingly, these scientific aspirations are progressively materializing remarkable advancements in molecular biology, RNA technology, the field of vaccinology, and the application of nanotechnology. In this chapter, we comprehensively describe mRNA-based therapeutics, including their history of development, mechanism of action, type of platforms with respect to pros and cons, scope of optimization, and research opportunities.

5.2 History of mRNA

The history of mRNA technology is marked by significant milestones in the understanding and application of this genetic material (Fig. 5.1). In the late 1980s, Robert Malone conducted a pioneering experiment involving the amalgamation of mRNA with lipid-based nanoparticles, effectively creating a molecular complex (Malone et al. 1989; Wolff et al. 1990). This experiment drew inspiration from previous endeavors dating back to 1978, where researchers employed lipid-based structures known as liposomes to transport mRNA into cells, thereby facilitating protein expression (Dimitriadis 1978). In 1984, Tom Maniatis and Michael Green at Harvard University succeeded in generating biologically active mRNA in the laboratory using an RNA-synthesis enzyme derived from a virus (Krieg and Melton 1984). In 1997, Eli Gilboa proposed using synthetic mRNA encoding tumor proteins to stimulate immune cells, which, when reintroduced into the body, would mount an immune response against cancer (Boczkowski et al. 1996). However, this

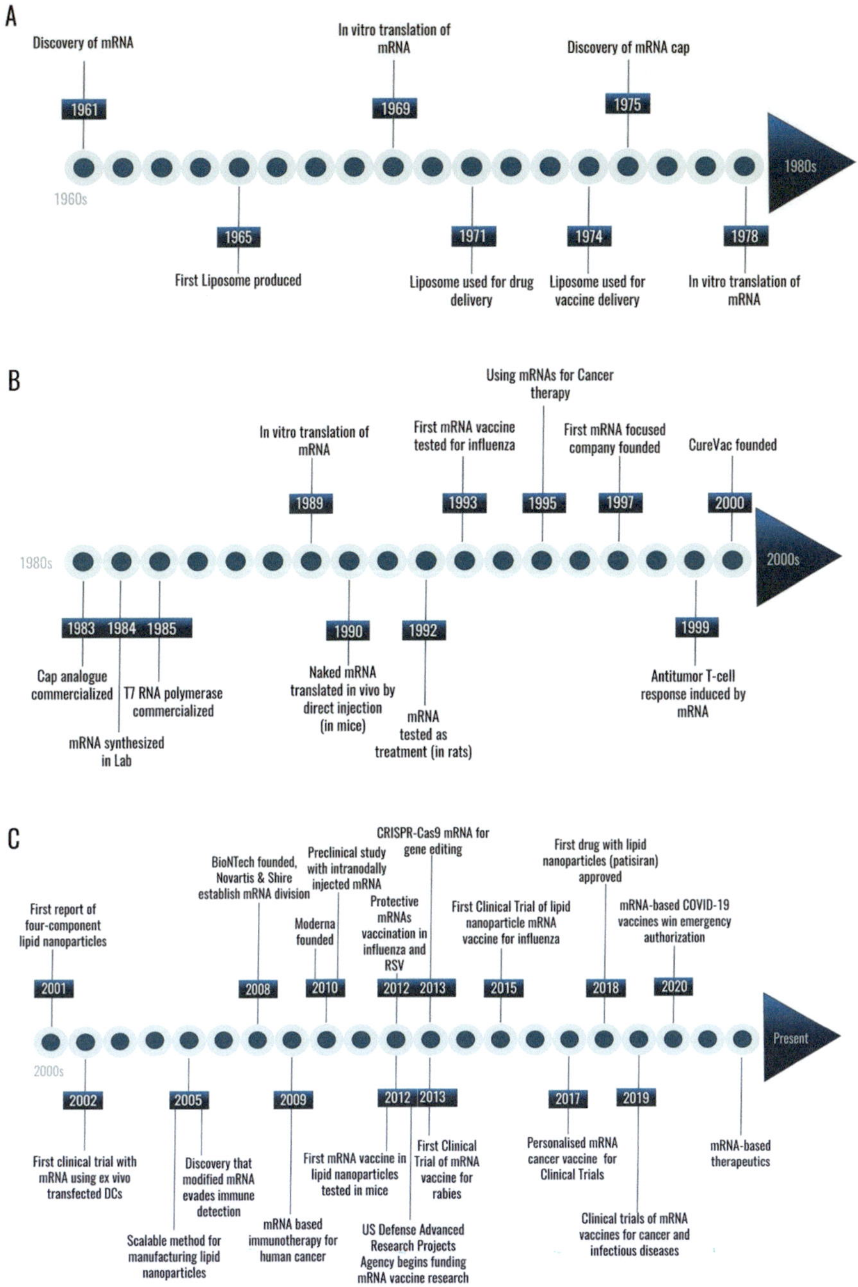

Fig. 5.1 History of mRNA including major milestones from (**a**) 1960s to 1980s, (**b**) 1980s to 2000s, and (**c**) 2001 to till date

approach faced challenges, and late-stage clinical trials yielded disappointing results. Despite this setback, Gilboa's work served as an inspiration for the founders of prominent mRNA companies, CureVac and BioNTech. Ingmar Hoerr and Uğur Şahin, the founders of these companies, were motivated by Gilboa's concept but chose to administer mRNA directly into the body. Katalin Karikó and Drew Weissman made a crucial discovery that modified segments of the mRNA code enabled synthetic mRNA to bypass the cell's innate immune defenses (Kariko et al. 2005). Recently, Katalin Karikó and Drew Weissman have been honored with the 2023 Nobel Prize in Physiology or Medicine. Moderna, a company built upon this foundational work, gained prominence by providing mRNA vaccine for mass immunization during COVID-19 pandemic.

The successful deployment of mRNA vaccines during the COVID-19 pandemic has brought mRNA technology into the spotlight and holds promise for its broader application in medicine and vaccine development, marking a significant milestone in the history of this innovative technology. In 2015, Gennova initiated exploration of the mRNA platform as part of its neo-antigen therapeutic strategy. Gennova has developed two indigenous thermostable mRNA-based vaccines against COVID-19 named as GEMCOVAC®-19 (targeting B.1 variant of SARS-CoV-2) and GEMCOVAC®-OM (as a booster, targeting BA.1 Omicron variant of SARS-CoV-2). Apart from Gennova, Biological E Ltd. has received mRNA technology from the WHO technology transfer hub. Hyderabad-based Bharat Biotech and Akshaya Biotech are also exploring the mRNA vaccine platform.

5.3 Mechanism of Action of mRNA Vaccine

The mechanism of action of mRNA vaccines involves a two-step process: first, the delivery of synthetic mRNA into cells and the subsequent production of antigens to elicit an immune response (Bettini and Locci 2021; Pardi et al. 2018). Briefly, mRNA vaccines contain synthetic messenger RNA molecules that encode a portion of the target pathogen, such as a viral spike protein in the case of COVID-19 vaccine. The mRNA is complexed with lipid nanoparticles. These lipid nanoparticles are crucial for the stability and delivery of the mRNA into the cell. After immunization, the synthetic mRNA is translated by the cellular machinery into the protein, which, in the case of COVID-19 mRNA vaccines, is the spike protein of the SARS-CoV-2 virus. During this translation, the poly-A tail engages with the poly-A-binding protein (PABP), and translation initiation and elongation factors (eIFs) connect with the 5′UTR cap to initiate the translation (Fabbri et al. 2021). The cells then express this viral protein on their surfaces (Fig. 5.2).

The immune system recognizes the foreign antigen displayed on the cell's surface. Immune cells, such as dendritic cells and macrophages, engulf and process the antigen. They present parts of the antigen (spike protein), called epitopes, to helper T cells. Helper T cells become activated upon recognizing the epitopes, and they release chemical signals (cytokines) that instruct other immune cells. Cytokines stimulate the maturation and activation of B cells and cytotoxic T cells. B cells are

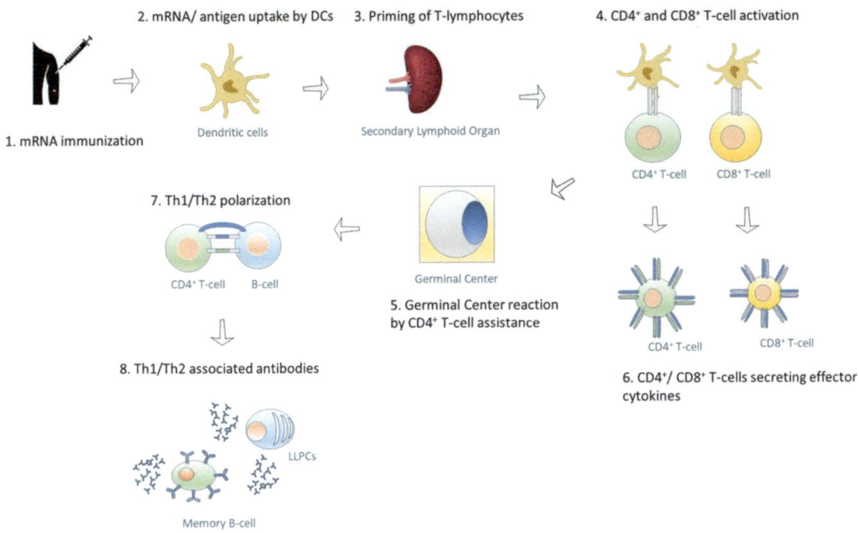

Fig. 5.2 Mechanism of action of mRNA vaccines

activated and differentiate into plasma cells. Plasma cells produce antibodies specific to the encoded antigen. Some activated B and T cells become memory cells. These cells "remember" the antigen and remain in the body for a long time.

5.4 Type of mRNA Platforms

Certainly, mRNA platforms have evolved, and different types are used for various applications, including vaccines and therapies. Structural representation of different mRNA platforms is depicted in Fig. 5.3. Here is a basic overview on the types of mRNA platforms.

5.4.1 Unmodified/Conventional mRNA Platform

This platform consists of single-stranded mRNA encoding specific proteins. This platform has certain limitations, including potentially lower protein expression and reduced stability compared to modified mRNA platforms (Pardi et al. 2018).

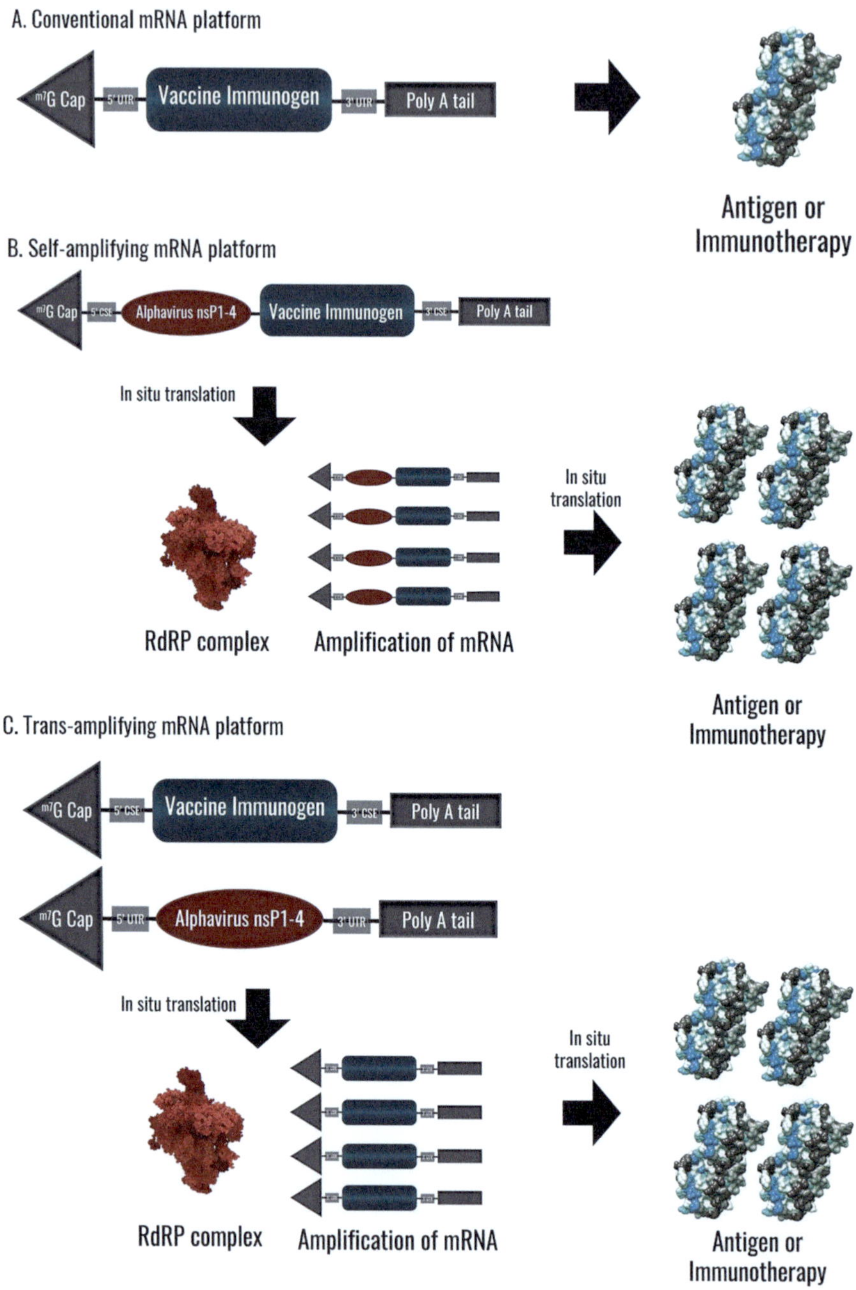

Fig. 5.3 Types of mRNA platforms. Structural components of (**a**) conventional, (**b**) self-amplifying, and (**c**) trans-amplifying mRNA platform. UTR untranslated region, CSE conserved sequence elements, nsP1–4 nonstructural proteins 1–4, RdRP RNA-dependent RNA polymerase

5.4.2 Nucleoside-Modified mRNA Platform

The nucleoside-modified mRNA platform represents a significant advancement in mRNA technology. In this platform, synthetic mRNA is chemically engineered to include modified nucleosides, such as pseudouridine and 5-methylcytidine. These modifications are pivotal in enhancing the stability of the mRNA molecule, reducing its immunogenicity, and improving its translational efficiency through innate escape (Pardi et al. 2018). The fundamental principle is to make the mRNA more resilient, ensuring that it remains intact within the host cell for longer. However, the platform may require a slightly more complex manufacturing process compared to unmodified mRNA.

5.4.3 Self-Amplifying mRNA (saRNA) Platform

This platform introduces an RNA replicase gene alongside the desired genetic sequence, thus enabling the mRNA to replicate within the host cell. Fundamentally, this saRNA can generate multiple copies of itself, leading to higher and sustained protein expression levels (McKay et al. 2020; Bloom et al. 2021; Avogadri et al. 2014). The platform may also yield long-lasting immune responses. However, it is worth noting that this platform is more complex to engineer and manufacture than conventional mRNA due to the need to incorporate the replicase gene (Bloom et al. 2021).

5.4.4 Trans-Amplifying RNA (taRNA) Platform

taRNA represents an evolved form of saRNA with significant potential as a vaccine platform (Perkovic et al. 2023). The primary distinction between saRNA and taRNA lies in their genetic content. In saRNA, there is an inherent sequence comprising nonstructural proteins 1–4 (nsP 1–4) in cis form, while taRNA necessitates the simultaneous delivery of two RNA components (Papukashvili et al. 2022), because nsPs are present in trans positions.

5.4.5 Circular mRNA Platform

The basic principle of this platform involves circularizing the mRNA to enhance its stability and efficiency (Chen et al. 2023). The circular structure reduces the susceptibility to degradation, allowing for prolonged mRNA persistence within the host cell. This can lead to sustained and extended protein expression (Chen et al. 2023). However, the circular mRNA platform may have some challenges. The process of circularization adds complexity to the mRNA synthesis process (Niu et al. 2023). This complexity can potentially lead to increased manufacturing costs. Furthermore, the delivery of circular mRNA into host cells may require specialized techniques (Niu et al. 2023).

5.5 Manufacturing of mRNA

mRNA platform offers several advantages, including the ability to design and manufacture them quickly, the absence of live or inactivated pathogens, and the potential for customized vaccine development. The manufacturing overview of Gennova's mRNA vaccine typically involves various key steps as shown in Fig. 5.4. This whole process can be subdivided into parts, namely plasmid design, mRNA synthesis, purification, formulation, and fill-finish.

Briefly, in the manufacturing of in vitro transcription (IVT) mRNA, linearized plasmid DNA templates or PCR templates are employed, necessitating the presence of a promoter and the corresponding mRNA construct sequence (Sahin et al. 2014; Baiersdorfer et al. 2019). IVT mRNA synthesis is achieved by introducing polymerases like T7, T3, or SP6, but it requires an additional step for capping (Baiersdorfer et al. 2019). There are two primary methods for capping IVT mRNA: co-transcriptional capping and posttranscriptional capping (Muttach et al. 2017; Ouranidis et al. 2021). Capping enzymes derived from the vaccinia virus are widely used due to their higher capping efficiency. The addition of poly(A) tails to IVT mRNAs is typically accomplished by encoding them in the DNA template or by enzymatic polyadenylation, with the former providing more precise control over the length of poly(A) tails (Sahin et al. 2014). The downstream processes in mRNA manufacturing commence with the purification of plasmid DNA. Subsequently, after IVT, mRNAs are mixed with RNA polymerase and DNA templates

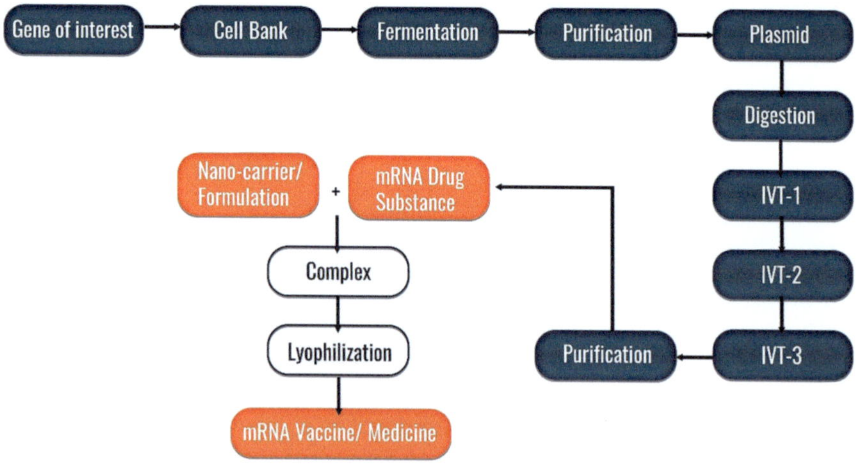

Fig. 5.4 Flowchart illustrating the essential stages in the production of mRNA vaccines. IVT-1 comprises in vitro transcription, IVT-2 comprises DNA degradation, and IVT-3 comprises capping steps

(post-synthesis). It is imperative to purify IVT mRNA during this stage, involving the removal of immune-stimulatory contaminants, free ribonucleotides, short mRNA, and DNA templates (Granados-Riveron and Aquino-Jarquin 2021). Typically, DNase is employed to degrade excess DNA templates, followed by mRNA precipitation with high concentrations of LiCl or alcohol-based precipitation, or the utilization of chromatographic methods like molecular exclusion chromatography, ion-exchange chromatography, or affinity chromatography with immobilized oligo-dT. Alternatively, elution from a silica membrane column is employed, which effectively eliminates proteins, free nucleotides, or other components while preserving RNA integrity but may not remove double-stranded RNA impurities (Martins et al. 2014). Following the purification process, the mRNA drug substance can be combined with a lipid nano-carrier system to create the mRNA drug product. This drug product is subsequently processed through filling and lyophilization procedures. Regardless of the chosen aforesaid methods, strict quality control measures at each step of manufacturing, viz. plasmid (Table 5.1), drug substance (Table 5.2), and drug products (Table 5.3) are integral to maximize the advantages of mRNA therapeutics.

5.6 Advantages of mRNA Platform

The advantages of mRNA platform over conventional vaccine platforms are discussed below.

5.6.1 Manufacturing

mRNA vaccines offer a significant advantage in terms of manufacturing. The production process is more straightforward and adaptable compared to traditional vaccine manufacturing methods. It allows for faster development and production, making it ideal for rapid response to emerging diseases, as exemplified by the COVID-19 pandemic (Whitley et al. 2022; Krammer 2020).

5.6.2 Scalability

The mRNA platform provides exceptional scalability. This adaptability is particularly valuable during pandemics when large quantities of vaccines are required quickly. mRNA vaccine production can be rapidly scaled up, addressing global vaccine needs efficiently (Lurie et al. 2020). India plays a vital role in global healthcare, including vaccine manufacturing and distribution. Scalable mRNA production can contribute to the global efforts in equitable vaccine distribution, helping address not only national but also international health challenges.

Table 5.1 Proposed testing for plasmid DNA prior to release

Quality	Attribute	Method
Identity	Sequence	Sequencing
	Restriction map	Restriction enzyme analysis with agarose gel electrophoresis
Concentration	Plasmid concentration (A260)	Ultraviolet spectroscopy (UV)
Purity	Plasmid purity (A260/280)	Ultraviolet spectroscopy (UV)
	% supercoiled	Capillary electrophoresis (CE) or high-performance liquid chromatography (HPLC)
	Residual host RNA	High-performance liquid chromatography (HPLC) or agarose gel electrophoresis
	Residual host DNA	Quantitative PCR (qPCR)
	Residual protein	SDS-PAGE or bicinchoninic acid assay (BCA)
	Host cell protein	Enzyme-linked immunosorbent assay (ELISA)
	Residual kanamycin	Enzyme-linked immunosorbent assay (ELISA)
Safety	Endotoxin	USP <85>
	Bioburden	USP <61>
	Sterility	USP <71>
Other	Appearance	USP <1>, <790>
	pH	USP <791>

*Focuses on the quality control protocols for plasmid production, ensuring the integrity, purity, and identity of the plasmid DNA used as the template for mRNA synthesis

This table is extracted from United States Pharmacopeia (USP) guidelines entitled 'Analytical Procedures for mRNA Vaccine Quality'—2nd Edition

Table 5.2 Characterization and release testing for mRNA drug substance

Quality	Attribute	Method
Identity	mRNA sequence identity confirmation	High-throughput sequencing (HTS)
		Sanger sequencing
		Reverse transcriptase-PCR (RT-PCR)
Content	RNA concentration	Quantitative PCR (qPCR)
		Digital PCR (dPCR)
		Ultraviolet spectroscopy (UV)
Integrity	mRNA intactness	Capillary electrophoresis
		Capillary gel electrophoresis
		Agarose gel electrophoresis
Purity	5′ capping efficiency	Reversed-phase liquid chromatography mass spectroscopy (RP-LC-MS/MS)
		Ion pair reversed-phase high-performance liquid chromatography (IP-RP-HPLC)
	3′ poly(A) tail length	Ion pair reversed-phase high-performance liquid chromatography (IP-RP-HPLC)
	Product-related impurities—dsRNA	Immunoblot
		Enzyme-linked immunosorbent assay (ELISA)
	Product-related impurities—aggregate quantitation	Size exclusion-high-performance liquid chromatography (SEC-HPLC)
	Product-related impurities—percentage of fragment mRNA	Reversed-phase HPLC (RP-HPLC)
	Process-related impurities—residual DNA template	Quantitative PCR (qPCR)
	Process-related impurities—quantitation of free/non-incorporated nucleosides	Reversed-phase liquid chromatography mass spectroscopy (RP-LC-MS/MS)
	Process-related impurities—residual T7 RNA polymerase content	Enzyme-linked immunosorbent assay (ELISA)

(continued)

Table 5.2 (continued)

Quality	Attribute	Method
Potency	Expression of target protein	Cell-based assay
Safety	Endotoxin	USP <85>
	Bioburden	USP <61>, <62>, <1115>
Other	Appearance	USP <790>
	Residual solvents	USP <467>
	pH	USP <791>

*Addresses the quality control steps for the mRNA drug substance, highlighting crucial parameters such as concentration, purity, integrity, potency, and safety of the synthesized mRNA

This table is extracted from United States Pharmacopeia (USP) guidelines entitled 'Analytical Procedures for mRNA Vaccine Quality'—2nd Edition

Table 5.3 Characterization and release testing for mRNA drug product

Quality	Attribute	Method
Identity	mRNA sequence identity confirmation	Sanger sequencing
		Reverse transcriptase-PCR (RT-PCR)
	Identity of lipids	Reversed-phase high-performance liquid chromatography with charged aerosol detector (RP-HPLC-CAD)
Content	RNA concentration/RNA encapsulation efficiency	Fluorescence-based assay
	Lipid content	Reversed-phase high-performance liquid chromatography with charged aerosol detector (RP-HPLC-CAD)
Integrity	LNP size and polydispersity	Dynamic light scattering (DLS)
	RNA size and integrity	Capillary gel electrophoresis (CGE)D
Purity	Product-related impurities—aggregate quantitation	Size exclusion-high-performance liquid chromatography (SEC-HPLC)
	Product-related impurities—percentage of fragment mRNA	Ion pair reversed-phase high-performance liquid chromatography (IP-RP-HPLC)
Potency	Expression of target protein	Cell-based assay
Safety	Endotoxin	USP <85>
	Sterility	USP <71>
Other	Appearance	USP <790>
	Residual solvents	USP <467>
	Osmolality	USP <785>
	Subvisible particles	USP <787>
	Residual solvents	USP <467>
	Extractable volume	USP <1>, <698>
	Container closure integrity	USP <1207>
	pH	USP <791>

*Details the quality control measures for the final drug products, which include the mRNA formulated with lipid nanoparticles (LNPs) as carriers. This stage ensures the final mRNA-vaccine product meets all required standards for identity, content, integrity, purity, potency, safety, and formulation evaluation tests before it is approved for distribution and use

This table is extracted from United States Pharmacopeia (USP) guidelines entitled 'Analytical Procedures for mRNA Vaccine Quality'—2nd Edition

5.6.3 Safety

mRNA vaccines have a robust safety profile. They do not contain live viruses or require the use of adjuvants, reducing the risk of adverse reactions associated with traditional vaccines. This safety feature is appealing for widespread vaccination campaigns (Pardi et al. 2018).

5.6.4 Variability Over Mutating Strains

mRNA vaccines can be swiftly adapted to new variants of viruses. This platform's flexibility allows for the modification of the mRNA sequence to match emerging virus strains, providing an advantage in addressing the challenge of rapidly mutating viruses (Freyn et al. 2021). India, like many other countries, faces the constant threat of infectious disease outbreaks. The ability to rapidly scale up mRNA vaccine production allows for a swift response to emerging infectious diseases, enhancing the country's pandemic preparedness and response capabilities.

5.6.5 Effectiveness

mRNA vaccines have demonstrated high effectiveness, as observed with the COVID-19 mRNA vaccines (Polack et al. 2020; Baden et al. 2021). They can induce potent immune responses, including strong antibody and T cell responses, leading to robust protection against the target pathogen (Polack et al. 2020; Baden et al. 2021). India has a substantial burden of infectious diseases, including vector-borne diseases, respiratory infections, and emerging diseases. In summary, the need for efficacious vaccines in India is multifaceted, ranging from safeguarding public health to addressing economic and global health implications.

5.7 Challenges with mRNA Platform

Addressing the challenges is crucial for harnessing the full potential of mRNA vaccines in preventing and controlling infectious diseases while ensuring accessibility and acceptance among diverse populations. The key challenges associated with mRNA platform are discussed below.

5.7.1 Thermostability

The thermostability challenges of mRNA vaccines stem from the inherent fragility of the mRNA molecule. Unlike traditional vaccines that can often be stored at less stringent temperatures, mRNA vaccines, particularly lipid nanoparticle (LNP)-encapsulated mRNA, typically require ultra-low temperatures for storage and transportation. Thus, maintaining temperatures around −70 °C to −80 °C is often necessary to preserve the integrity and efficacy of mRNA vaccines (Uddin and Roni 2021). Developing more thermostable formulations or storage solutions for mRNA vaccines is an ongoing area of research to address this challenge. The increased susceptibility of mRNA to degradation in the presence of water, attributed to mRNA oxidation and hydrolysis (Uddin and Roni 2021), led us to utilize a desiccation technique known as freeze-drying or lyophilization for both of our thermostable mRNA vaccines, namely GEMCOVAC®-19 and GEMCOVAC®-OM.

5.7.2 Delivery

Delivery challenges associated with mRNA vaccines primarily revolve around ensuring the efficient and targeted delivery of the mRNA to the cells for translation. mRNA is a large, negatively charged molecule that cannot easily cross cell membranes. Consequently, the broader utilization of mRNA-based therapeutics is restricted by the demand for enhanced vectors or drug delivery systems. Advanced delivery systems have the potential to overcome issues related to mRNA's stability, cellular targeting, and translational efficacy when used without protective mechanisms (Deering et al. 2014).

5.7.3 Acceptability

Acceptability challenges concerning mRNA vaccines pertain to public perception, regulatory approval, and education. mRNA technology, while groundbreaking, is relatively new, and there might be hesitancy or skepticism among the public (Yarlagadda et al. 2022). Effective communication and education campaigns are vital to address these concerns and ensure acceptance. Additionally, obtaining regulatory approvals and trust from authorities is essential to enable widespread usage of mRNA vaccines. These challenges emphasize the need for comprehensive public outreach and scientific communication efforts to inform and reassure the public about the safety and effectiveness of mRNA vaccines (Costantini et al. 2022).

5.7.4 Accessibility

Accessibility challenges related to mRNA vaccines encompass ensuring equitable distribution, especially in resource-constrained regions, and addressing infrastructure limitations (Yarlagadda et al. 2022). The ultra-low temperature storage requirements, typically around −70 °C to −80 °C, pose logistical and financial challenges, making it difficult to reach underserved areas with limited cold chain infrastructure (Uddin and Roni 2021). Moreover, ensuring equitable access to mRNA vaccines in low-income regions is crucial to address global health disparities. Strategies for equitable vaccine distribution and investment in cold chain infrastructure are key aspects of addressing accessibility challenges (Costantini et al. 2022).

5.7.5 Innate Immune Responses

mRNA vaccines are designed to harness the host's cellular machinery to express a specific antigen and trigger an adaptive immune response. However, the inherent features of mRNA molecules, such as their structural characteristics and sequences, can also stimulate the innate immune system (Pardi et al. 2018). Consequently, it leads to decreased mRNA translation and hence reduced immunogenicity. The primary mechanisms through which mRNA vaccines can activate the innate immune system include the recognition of foreign RNA by cellular sensors, particularly Toll-like receptors (TLRs) and retinoic acid-inducible gene I (RIG-I)-like receptors (RLRs). TLRs, such as TLR3, TLR7, and TLR8, recognize double-stranded RNA or single-stranded RNA with certain motifs, while RLRs, like RIG-I and MDA5, sense viral RNA (Kariko et al. 2005; Nance and Meier 2021; Minnaert et al. 2021). Upon recognition of these foreign RNA molecules, these sensors trigger signaling cascades that lead to the production of pro-inflammatory cytokines and interferons (Diamond and Kanneganti 2022). This challenge requires fine-tuning the mRNA sequence, incorporating modified nucleosides to reduce immune recognition, and optimizing the formulation of lipid nanoparticles used for mRNA delivery.

5.8 Optimization of mRNA Platform

Despite the numerous benefits offered by mRNA therapies, there is still untapped potential for optimizing various aspects of mRNA drug development. This includes the design of plasmid, refining the composition of the mRNA drug substance, enhancing the carrier system used in drug product formulation, and optimizing immunization techniques.

5.8.1 Plasmid Design

Designing a plasmid for optimized mRNA expression involves several steps to ensure efficient transcription, translation, stability, and immunogenicity of the mRNA molecule (Schlake et al. 2012). Figure 5.5 provides an overview of the typical workflow for this process. Below, we will delve into the essential components in the following subsections.

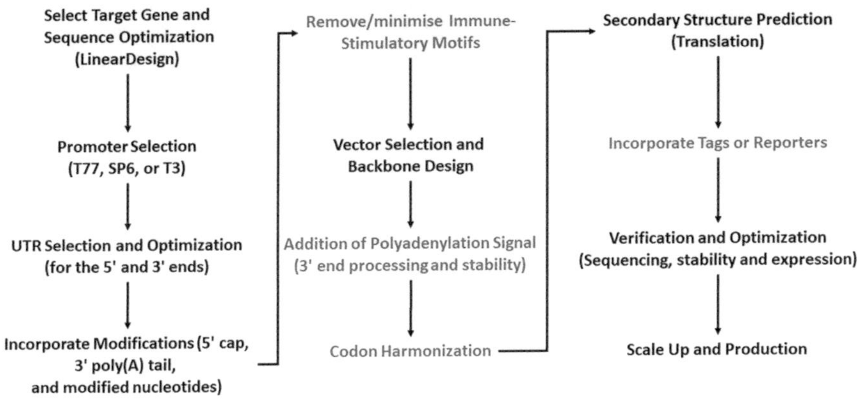

Fig. 5.5 Flowchart illustrating the essential steps in plasmid design

5.8.2 Optimization of mRNA Itself

It is crucial to understand successful mRNA translation and immunogenicity outcomes. Basically, these depend on the presence of specific elements in the synthesized mRNA vaccine, including the 5′ cap, 5′UTR, translated region main coding sequence (Open reading frame/ORF), 3′UTR, and poly-A tail, all of which are recognized by the ribosome for the effective translation of mRNA into the required antigen (Qin et al. 2022). Figure 5.6 illustrates the structural elements of mRNA drug substance, their primary functions (depicted in blue/gray boxes), and areas where further improvements can be made (highlighted in white boxes) (Ouranidis et al. 2021). While there is a wealth of experimental knowledge regarding components like the 5′ cap, 5′UTR, 3′UTR, and poly-A tail (Ouranidis et al. 2021), our emphasis here is on the main coding sequence (ORF), which holds a crucial role in mRNA drug development.

Two frequently employed strategies for modifying the open reading frame (ORF) regions are codon optimization and the incorporation of base analogs (Chabanovska et al. 2021; Kim et al. 2022). Codon optimization involves the substitution of codons with synonymous counterparts, aligning with the codon preferences of the host organism into which the mRNA will be introduced. This optimization is grounded in the principle that codons corresponding to more abundant tRNA species can expedite the translation process, resulting in increased protein production before mRNA degradation occurs (Hia and Takeuchi 2021; Cannarozzi et al. 2010). Furthermore, the selection of optimal codons may also impact the stability of the mRNA molecule, as noted in (Presnyak et al. 2015). A comprehensive optimization effort encompasses considerations beyond mere codon selection, including the GC content of the final transcript and the deliberate introduction of suboptimal sequences. Additionally, base analogs render the mRNA less susceptible to the host cell's innate immune response against foreign mRNAs (Minnaert et al. 2021; Mauger et al. 2019). Commonly employed base analogs to evade TLR recognition

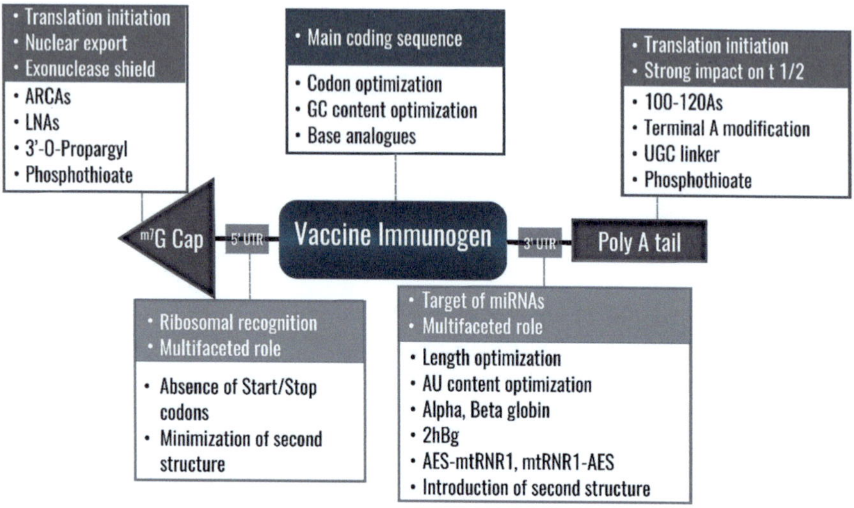

Fig. 5.6 Structural elements of mRNA, their primary functions (depicted in blue/gray boxes), and areas of improvements (highlighted in white boxes) (Ouranidis et al. 2021)

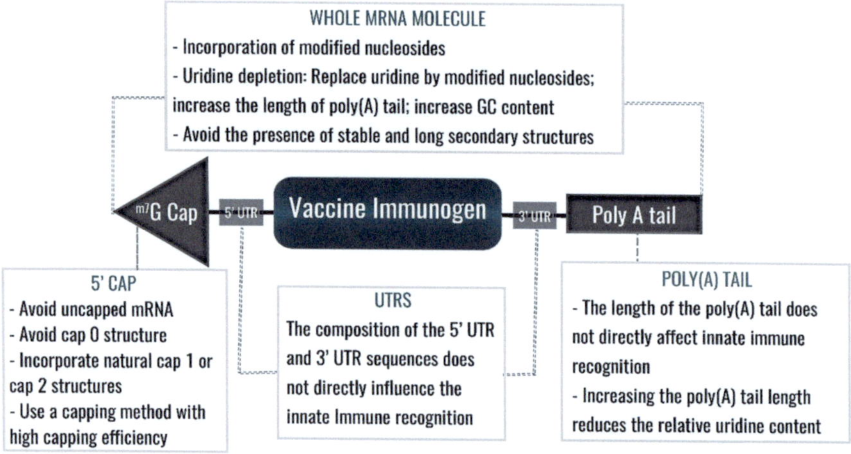

Fig. 5.7 Approaches to manage innate immune responses following mRNA delivery (Minnaert et al. 2021)

include pseudouridine (Ψ), N1-methylpseudouridine (m1 Ψ), 2-thiouridine (S2U), 5-methyluridine (m5U), 5-methylcytidine (m5C), and N6-methyladenosine (m6A) (Chabanovska et al. 2021). Various approaches for managing innate immune responses following mRNA delivery are presented in Fig. 5.7.

Importantly, a recent algorithm named LinearDesign has been proposed for both structural stability and codon usage in mRNA design (Zhang et al. 2023). Briefly, this model jointly optimizes mRNA stability and codon optimality using lowest minimum-free-energy change (MFE) and codon adaptation index (CAI), respectively. To enhance stability, the algorithm selects the mRNA sequence with the lowest MFE from all possible sequences. Codon optimality within this model is assessed using CAI, which represents the geometric mean of the relative adaptiveness of each codon within the mRNA.

5.8.3 Optimization of mRNA Drug Product

Effective mRNA delivery systems are imperative to achieve optimal mRNA potency, provide protection, facilitate cellular uptake, and ensure endosome escape. The mechanisms of mRNA loading are believed to involve interactions such as electrostatic forces, hydrogen bonding, or coordination interactions achieved through techniques like thin-film hydration, nanoprecipitation, or microfluidic mixing. Various vectors have been engineered and synthesized to enhance mRNA delivery, including lipid nanoparticles (LNPs), polymeric nanoparticles, cationic lipid nanoemulsions (CLNEs), and other delivery systems (Ho et al. 2021; Wadhwa et al. 2020). Ongoing progress in formulating and delivering mRNA using diverse nanomaterials has the potential to expand the applications of mRNA in the treatment and prevention of infectious diseases and cancers. In addition, adjuvant research for mRNA therapeutics could also be an important research area to harness the correlates of protection (CoP) based advantages. Although mRNA itself act as adjuvant; additionally, other immune activating agents like CD70, CD40 L, TLR agonists, inflammasome agonists, pathogen molecular patterns, lipids, GM-CSF and Fms-like tyrosine kinase 3 ligands were also explored as adjuvant in mRNA products (Qin et al. 2022; Christensen et al. 2011).

5.8.4 Selection of Delivery Route

The choice of administration routes is a critical factor in mRNA delivery systems, as certain diseases necessitate specific routes for effective treatment, although intravenous administration can suffice for most conditions. For example, pulmonary diseases benefit from inhaled or intratracheal administration (Qiu et al. 2019). Cerebral conditions can be addressed with intra-cerebroventricular or intrathecal injections (Kim et al. 2019), while liver diseases may be managed through intravenous, intraperitoneal, subcutaneous, or intramuscular administration (Hajj et al. 2020). In the context of infectious diseases and cancer, the intradermal (ID) route emerges as a promising approach, offering enhanced effectiveness and efficiency (Schnyder et al. 2022; Intapiboon et al. 2021). The dermis hosts a higher concentration of antigen-producing cells in comparison to muscle tissue, resulting in a more robust immune response, which is advantageous for vaccine recipients. In a similar vein, GEMCOVAC™-OM received approval as an intradermal booster for combating the Omicron variant.

5.9 Emerging Opportunities of mRNA Medicines

mRNA-based therapeutics are anticipated to emerge as a potent treatment option for an array of challenging medical conditions, encompassing infectious diseases, cancer, metabolic and genetic disorders (Fig. 5.8). In the realm of mRNA vaccines for infectious diseases and cancer, one approach involves encoding the antigens associated with pathogens or cancer cells. Additionally, mRNA therapeutics are employed to address diseases directly by delivering functional proteins derived from mRNA. These proteins can include native proteins that are either deficient or diminished, foreign functional proteins, antibodies, and proteins used for gene editing purposes (Qin et al. 2022). Furthermore, the approach of directly expressing proteins within specialized "cell factories" can be harnessed for cell engineering, such as the modification of T cells (Billingsley et al. 2020). Table 5.4 provides an overview of registered clinical trials related to mRNA-based vaccines for infectious diseases. Exploring key mRNA research opportunities within the context of India's healthcare landscape is outlined as follows.

5.9.1 mRNA Vaccine for Tuberculosis (TB)

In 2022, TB reclaimed its position as the leading infectious cause of death worldwide, surpassing COVID-19 and HIV/AIDS (Larsen et al. 2023a). Among TB vaccine candidates, M72/AS01E stands out as the only vaccine meeting the WHO's stringent criteria and demonstrating clinical efficacy >50% (Tait et al. 2019; Ghebreyesus and Lima 2023). Although the correlates of protection (CoP) for TB are not yet well-established, a comparative study involving individuals with or without M.tb infection who received various TB vaccines, including M72/AS01E, indicated that M72/AS01E not only stimulated diverse and functional T cell responses

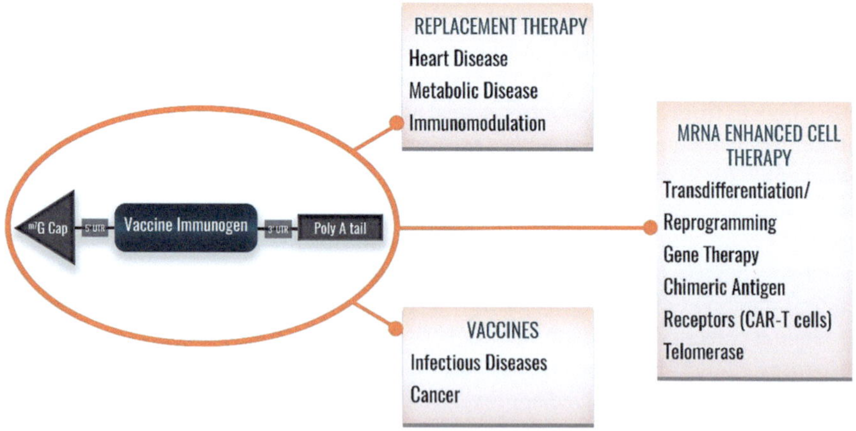

Fig. 5.8 Emerging areas of mRNA medicines (Ouranidis et al. 2021)

Table 5.4 List of clinical trials for mRNA vaccines targeting infectious diseases

Study product	Antigen	Administration	Phase	NCT identifier	Status	Target
–	–	Intradermal	I/II	NCT00833781	Completed	HIV
iHIVARNA-01	HTI	Inguinal-intranodal	I	NCT02413645	Completed	HIV
iHIVARNA-01	HTI	Intranasal	II	NCT02888756	Terminated	HIV
mRNA-1644	eOD-GT8 60mer	Intramuscular	I	NCT05414786	Active, not recruiting	HIV
Trimer mRNA	BG505 MD39.3 gp151	Intramuscular	I	NCT05217641	Active, not recruiting	HIV
mRNA-1647/mRNA-1443	CMV-associated antigens	–	I	NCT03382405	Active, not recruiting	CMV
mRNA-1647	gB, pentamer complex	–	II	NCT04232280	Recruiting	CMV
mRNA-1273	Spike protein	Intramuscular	I	NCT04283461	Active, not recruiting	SARS-CoV-2
mRNA-1273	Spike protein	–	II	NCT04405076	Active, not recruiting	SARS-CoV-2
BNT162a1/BNT162b1/BNT162b2/BNT162c2	Spike protein	Intramuscular	I/II	NCT04380701	Recruiting	SARS-CoV-2
BNT162a1/BNT162b1/BNT162b2/BNT162c2	Spike protein	Intramuscular	I/II	NCT04368728	Recruiting	SARS-CoV-2
CVnCoV vaccine	Spike protein	Intramuscular	I	NCT04449276	Recruiting	SARS-CoV-2
Sanofi vaccine (quadrivalent)	–	Intramuscular	I	NCT05426174	Active, not recruiting	Influenza
GSK4382276A	–	Intramuscular	I	NCT05446740	Recruiting	Influenza
mRNA-1020 and 1030	HA and NA	Intramuscular	I/II	NCT05333289	Completed	Influenza
Quadrivalent influenza modRNA vaccine (qIRV)	HA	Intramuscular	III	NCT05540522	Recruiting	Influenza

(continued)

Table 5.4 (continued)

Study product	Antigen	Administration	Phase	NCT identifier	Status	Target
VAL-506440	H10N8 HA	Intramuscular/ intradermal	I	NCT03076385	Completed	Influenza
VAL-339851	H7N9 HA	Intramuscular	I	NCT03345043	Active, not recruiting	Influenza
mRNA-1653	hMPV, PIV3	–	I	NCT03392389	Completed	hMPV
mRNA-1653	hMPV, PIV3	–	I	NCT04144348	Recruiting	hMPV
CV7201	Rabies virus glycoprotein	–	I	NCT02241135	Completed	Rabies
CV7202	RABV-G protein antigens	Intramuscular	I	NCT03713086	Active, not recruiting	Rabies
mRNA-1325	–	–	I	NCT03014089	Completed	Zika
mRNA-1893	Zika virus-associated antigen	–	I	NCT04064905	Recruiting	Zika
BNT164a1 and BNT164	–	Intramuscular	I	NCT05547464	Not yet recruiting	TB

but also induced the highest memory CD4 T cell response (Rodo et al. 2019). An essential aspect of an RNA-based vaccine approach is its potential to elicit T cell responses, a challenging feat with traditional subunit vaccines (Larsen et al. 2023b). Notably, the initial proof of concept for an mRNA vaccine targeting tuberculosis (TB) emerged in 2004. In this pioneering work, researchers demonstrated that mice receiving repeated injections of naked in vitro transcribed (IVT) mRNA encoding the immune-dominant antigen MPT83 at three-week intervals mounted specific humoral and cellular responses. This approach resulted in modest yet significant protection against TB upon challenge (Xue et al. 2004). More recently, a replicating mRNA-based vaccine was developed, encoding a fusion protein comprising four M.tb antigens (Rv3619, Rv2389, Rv3478, Rv1886) and formulated with a lipid nano-carrier. This vaccine induced cellular immune responses and demonstrated protective effects against M.tb and *M. avium* challenges in mice, particularly when using a heterologous RNA-prime and protein-boost vaccination approach (Larsen et al. 2023b; Rais et al. 2023). Recently, BioNTech, Germany, announced a phase I clinical trial that aims to evaluate the safety and immunogenicity of a potential mRNA TB vaccine candidate (ClinicalTrials.gov Identifier: NCT05547464). Given that approximately 40% of the Indian population is infected with TB bacteria, with an annual risk of infection at 1.5% (Pai et al. 2005; Chauhan et al. 2023), developing an mRNA vaccine and selecting suitable antigens presents a significant research opportunity for identifying CoP and creating therapeutic and prophylactic TB vaccines.

5.9.2 mRNA Vaccine for Human Immuno-Deficiency Virus (HIV)

The Government of India estimates that approximately 2.4 million individuals in the country are living with HIV (Kumar et al. 2022). The proposed correlates of protection, which have been identified in animal models and clinical studies, primarily involve neutralizing antibodies that target specific epitopes of the HIV envelope (Env) glycoprotein (Mascola and Haynes 2013; Sliepen and Sanders 2016; Suryawanshi et al. 2021). Designing a vaccine that effectively targets the highly glycosylated Env glycoprotein is challenged by the virus's rapid mutation rate, resulting in a wide array of viral variants (pseudoviruses). Moreover, eliciting the development of broad neutralizing B cell precursors faces difficulties due to immune tolerance mechanisms, such as auto- or poly-reactivity (Sliepen and Sanders 2016; Finney and Kelsoe 2018). Consequently, optimizing vaccine design becomes imperative to address multiple conserved epitopes from the Env glycoprotein. In pursuit of this goal, the HIV mRNA vaccine eOD-GT8 60mer (mRNA-1644) is currently in a phase I clinical trial, developed by the International AIDS Vaccine Initiative (IAVI) in collaboration with Moderna, to assess safety and immunogenicity in healthy adult volunteers (Matarazzo and Bettencourt 2023). The National Institute of Allergy and Infectious Diseases (NIAID) is also investigating the safety and immunogenicity of three HIV trimer mRNA vaccines (BG505 MD39.3, BG505 MD39.3 gp151, and BG505 MD39.3 gp151 CD4KO) in healthy individuals (Matarazzo and

Bettencourt 2023). Notably, inducing high-affinity antibody production involves generating germinal center (GC) responses in secondary lymphoid organs, a critical component of the immune response. mRNA vaccines, as demonstrated in their approval for COVID-19, have shown potential in driving robust GC responses (Lederer et al. 2022). When coupled with efficient delivery systems, mRNA technology holds promise for overcoming the challenges faced in HIV vaccine development.

5.9.3 mRNA Vaccine for Flu

The development of mRNA vaccines for influenza is of significant importance due to the virus's high mutation rate and its ability to undergo frequent antigenic changes. Influenza viruses are notorious for their antigenic variability, which stems from the mutation and reassortment of viral genes. These changes result in new strains of the virus emerging regularly, leading to the need for annual vaccine updates to match the predominant circulating strains (Neuzil 2023). The influenza virus can alter its surface proteins, hemagglutinin (HA) and neuraminidase (NA), through a process known as antigenic drift (Neuzil 2023). Nachbagauer et al. selected the conserved components of the influenza virus, including the HA stalk domain, matrix-2 ion channel, nucleoprotein, and the broadly reactive neuraminidase, with the goal of offering broad protection against the virus. mRNA vaccines offer several advantages in addressing the challenges posed by the influenza virus.

5.9.4 mRNA Vaccine for Dengue

Dengue vaccine development and accessibility are vital for India and other LMICs to address this persistent public health challenges. Dengue is caused by four distinct serotypes, namely DENV 1–4. The advantage of mRNA platforms is the ability to encode multiple antigens simultaneously. This means an mRNA dengue vaccine can be designed to target all four serotypes, providing broad protection against the disease. Dengue vaccines should ideally induce neutralizing antibodies against all serotypes to avoid antibody-dependent enhancement (ADE), a phenomenon in which prior infection with one serotype increases the risk of severe disease upon infection with another serotype. mRNA vaccines can be engineered to express conserved epitopes of each serotype, reducing the risk of ADE. In addition, the fine-tuning of the immune response is possible with mRNA vaccines. mRNA vaccines can be designed to promote a balanced response while reducing the risk of over-reactivity. Richner et al. successfully established a proof of concept for a dengue mRNA vaccine by encoding the structural proteins of DENV-1 (Wollner et al. 2021).

5.9.5 mRNA Vaccine for Cancer

For the year 2022, the estimated count of newly diagnosed cancer cases in India stood at around 1.5 million, with a crude rate of 100.4 per 100,000 individuals (Sathishkumar et al. 2022). It is worth noting that in India, the probability of one in nine individuals developing cancer during their lifetime is a concerning statistic (Sathishkumar et al. 2022). mRNA cancer vaccines leverage the inherent characteristics of messenger RNA to stimulate the patient's immune system to recognize and target cancer cells. The advantages of mRNA cancer vaccines include their ability to be personalized, elicit a strong immune response, and adapt to the dynamic nature of cancer (Diken et al. 2017). However, several challenges persist in the development and implementation of mRNA cancer vaccines. These challenges include optimizing the vaccine's delivery, ensuring safety, addressing potential immune-related adverse events, and establishing standardized protocols for clinical trials. Ongoing research and clinical trials are continually refining the mRNA cancer vaccine approach (Table 5.5). Early results from pancreatic cancer clinical trials are promising (Rojas et al. 2023; Lorentzen et al. 2022). Clinical evaluation of mRNA vaccines has been conducted for castration-resistant prostate cancer and non-small-cell lung cancer (Sebastian et al. 2019). Additionally, mRNA vaccines targeting melanoma, glioblastoma, acute myeloid leukemia (AML), and renal cell carcinoma (RCC) have shown encouraging responses to immunotherapy. This suggests the need for in-depth exploration within the mRNA vaccine domain (Liu 2019). These advances, combined with personalized medicine approaches, hold the potential to revolutionize cancer treatment and provide new hope for patients.

5.9.6 Other Emerging Opportunities of mRNA

Application of mRNA technology holds the potential to address various hematologic disorders by delivering the corresponding factors through mRNA templates, as detailed in (Russick et al. 2020). Furthermore, mRNA-based therapies offer a promising avenue for treating metabolic diseases, such as hepatorenal tyrosinemia, acute intermittent porphyria, Fabry disease, glycogen storage disease type 1 A, Crigler-Najjar syndrome type 1, and ornithine transcarboxylase deficiency, for which effective treatments are currently lacking as indicated by (Trepotec et al. 2019; Berraondo et al. 2019). In parallel, mRNA-based gene editing therapeutics are gaining momentum across different domains, driven by the rapid advancement of programmable nucleases and the CRISPR/Cas nuclease system (Qin et al. 2022). These technologies find applications in addressing conditions like cancer, infectious diseases, primary immune system defects, muscular dystrophy, and hematological disorders (Nabhan et al. 2016; Schumann et al. 2018; Gaj et al. 2013). Additionally, emerging opportunities include mRNA-based autologous T cell therapies (such as CAR-T and CD4+ T cells), stem cell therapeutics, and combination-based therapies, capitalizing on the advantages offered by the mRNA platform technology (Qin

Table 5.5 List of clinical trials for mRNA vaccines targeting cancer

Antigen	Administration	Phase	NCT Identifier	Status	Target
WT1	Intradermal	I	NCT00834002	Completed	AML
WT1	Intradermal	II	NCT01686334	Recruiting	AML
Leukemia-associated antigens, CMV antigen	Intradermal	I/II	NCT01734304	Completed	AML
WT1	–	I/II	NCT03083054	Active, not recruiting	AML
hTERT, LAMP-1	–	II	NCT00510133	Completed	AML
Leukemia-associated antigens	–	I	NCT00514189	Terminated	AML
–	–	I	NCT02808416	Active, not recruiting	Brain metastases
CEA	Intravenous/intradermal	I/II	NCT00228189	Completed	Colorectal cancer, liver metastases
MUC1, survivin	–	I/II	NCT02693236	Not updated	Esophagus cancer
–	Intradermal	I/II	NCT00846456	Completed	GBM
Human CMV pp65-LAMP, HIV-gag	Intradermal	II	NCT03688178	Suspended	GBM
Human CMV pp65-LAMP	Intradermal	II	NCT02366728	Active, not recruiting	GBM
WT1	Intradermal	I/II	NCT02649582	Recruiting	GBM
–	Intravenous/intradermal	I	NCT02709616	Active, not recruiting	GBM
Glioma-associated antigens	–	I	NCT02808364	Active, not recruiting	GBM
hTERT, survivin, autologous tumor antigens	Intradermal	II/III	NCT03548571	Recruiting	GBM
pp65	Subcutaneous	II	NCT02465268	Recruiting	GBM
Human CMV pp65-LAMP	Intradermal	II	NCT03927222	Recruiting	GBM
CMV pp65-LAMP	Intradermal	I	NCT00639639	Active, not recruiting	GBM
WT1	Intradermal	I/II	NCT01291420	Not updated	GBM
Brain tumor stem cell specific antigens	Intradermal	I	NCT00890032	Completed	GBM
MiHA	Intradermal	I/II	NCT02528682	Recruiting	Hematological malignancies

(continued)

Table 5.5 (continued)

Antigen	Administration	Phase	NCT Identifier	Status	Target
CMV pp65-LAMP	Intradermal	I	NCT00626483	Completed	Malignant neoplasms brain
WT1	Intradermal	I/II	NCT02649829	Recruiting	Malignant pleural mesothelioma
–	Intradermal/ intranasal	I/II	NCT01278940	Completed	Melanoma
gp100, tyrosinase	–	I/II	NCT00243529	Completed	Melanoma
Melan-A, mage-A1, mage-A3, survivin, gp100, tyrosinase	Intradermal	I/II	NCT00204607	Completed	Melanoma
Melan-A, mage-A1, mage-A3, survivin, gp100, tyrosinase	Intradermal	I/II	NCT00204516	Completed	Melanoma
gp100, tyrosinase	Intradermal/ intravenous	I/II	NCT00940004	Completed	Melanoma
hTERT, survivin	–	I/II	NCT00961844	Terminated	Melanoma
hTERT, survivin, p53	Intradermal	I	NCT00978913	Completed	Melanoma
–	Intravenous/ intranasal	I	NCT01066390	Completed	Melanoma
Gp 100, tyrosinase	Intranasal	I/II	NCT01530698	Completed	Melanoma
Gp 100, tyrosinase	Intradermal/ intravenous	II	NCT02285413	Completed	Melanoma
TRP2	Subcutaneous	I	NCT01456104	Active, not recruiting	Melanoma
Multiple neoantigens	–	II	NCT03897881	Recruiting	Melanoma
–	Intramuscular	I/II	NCT03480152	Terminated	Melanoma
CT7, mage-A3, WT1	Subcutaneous	I	NCT01995708	Active, not recruiting	Multiple myeloma
–	–	I/II	NCT00923312	Completed	NSCLC
OCS 1, MUC1, survivin	–	I/II	NCT02688686	Not updated	NSCLC
–	–	I/II	NCT03164772	Recruiting	NSCLC
hTERT, survivin	Intradermal	I/II	NCT01334047	Terminated	Ovarian cancer
–	Intravenous	I	NCT04163094	Recruiting	Ovarian cancer
TERT	–	I	NCT01456065	Not updated	Ovarian cancer
hTERT, survivin	–	I/II	NCT01197625	Active, not recruiting	Prostate cancer

(continued)

Table 5.5 (continued)

Antigen	Administration	Phase	NCT Identifier	Status	Target
NY-ESO-1, MUC1 PepTivator®	Intranasal	II	NCT02692976	Completed	Prostate cancer
hTERT, survivin, PSA, PAP	Intradermal	II	NCT01446731	Completed	Prostate cancer
Prostate-associated antigens	Intradermal	II	NCT02140138	Terminated	Prostate cancer
Multiple neoantigens	Intramuscular	I	NCT03313778	Recruiting	Solid tumors

Table 5.6 List of clinical trials for mRNA vaccines targeting genetic, metabolic, and cardiovascular disorders

Antigen	Administration	Phase	NCT identifier	Status	Target
Methylmalonyl-CoA mutase (mRNA- 3704)	Unknown/i.v.	I/II	NCT03810690	Terminated	Methylmalonic acidemia
Propionyl-CoA carboxylase (mRNA-3927)	Unknown/i.v.	I/II	NCT04159103	Recruiting	Propionic acidemia
Ornithine transcarbamylase (MRT5201)	LNP/i.v.	I/II	NCT03767270	Terminated	OTD
CFTCR (MRT5005)	Unknown/ inhalation	I/II	NCT03375047	Recruiting	Cystic fibrosis
CRISPR–Cas9 (NTLA-2001)	LNP/i.v	I	NCT04601051	Active, not recruiting	TAP
VEGFA (AZD8601)	Naked mRNA/i.d.	I	NCT02935712	Completed	Type II diabetes
VEGFA (AZD8601)	Naked mRNA/ epicardial injection	II	NCT03370887	Completed	Heart failure

et al. 2022). For an overview of clinical trials related to these conditions, refer to Table 5.6.

5.10 The Way Forward for India

The landscape of mRNA-based therapeutics presents a dynamic interplay between opportunities and challenges, marked by numerous inquiries that seek elucidation. These encompass pivotal questions like (1) optimization of mRNA construct, (2) enhancing the efficient delivery of mRNA macromolecules, and (3) harnessing the desired immune responses. Buoyed by robust enthusiasm and the immense potential in this domain, there is a confident anticipation of an accelerated pace in mRNA therapy research and development over the coming years. The way forward for mRNA technology in India can be categorized in the following three subsections.

5.10.1 Indian Challenges

India has a vast and diverse population, and delivering mRNA vaccines to remote areas can be challenging mainly because of ultra-low storage temperatures requirements. India's existing cold chain infrastructure is not well-equipped to handle such extreme temperature requirements, making storage, transportation, and distribution of mRNA vaccines challenging. In addition, reducing production costs and making these vaccines affordable to the Indian population is a significant challenge. Importantly, developing, manufacturing, and administering mRNA vaccines require a skilled workforce with expertise in biotechnology, molecular biology, and vaccine production. India needs to invest in training and developing such expertise.

5.10.2 Gennova's mRNA Vaccine Technology

Gennova has developed saRNA based two indigenous mRNA-based vaccines against COVID-19 named as GEMCOVAC®-19 and GEMCOVAC®-OM, which overcome the limitations associated with the currently approved mRNA vaccines. Both the Gennova's vaccines were stable at 2–8 °C and can be considered superior in terms of deployability in India and other LMICs compared to other approved mRNA vaccines. Gennova's vaccines utilize a unique approach by adsorbing mRNA onto the surface of a CLNE, in contrast to other mRNA vaccines where mRNA is encapsulated within LNPs. This innovation streamlines the manufacturing process, minimizes losses, and ensures scalability, making it conducive for technology transfer. Such developments are essential for the global democratization of mRNA-based technology, aiming to address vaccine distribution disparities on a global scale.

5.10.3 Considerations to Establish mRNA Technology in India

Here are the essential recommendations to promote the adoption and extensive utilization of mRNA vaccine technology in India:

1. India must invest in research and development to adapt and optimize mRNA vaccine technology for diseases prevalent in the country, such as tuberculosis, dengue, cancer, and others.
2. Developing the manufacturing capacity to produce mRNA vaccines at scale is a significant challenge. India needs to invest in expanding its vaccine manufacturing capabilities to meet the demand.

3. mRNA vaccines must go through stringent regulatory approval processes to ensure their safety and efficacy. India's regulatory authorities must adapt to this new technology and ensure a streamlined approval process without compromising safety.
4. Ensuring equitable access to mRNA vaccines is a challenge, as there are disparities in healthcare access in India. The government and healthcare organizations need to address these disparities to reach marginalized populations.
5. Building public trust in mRNA vaccines is essential. Public awareness campaigns and addressing vaccine hesitancy are significant challenges, especially in a diverse country like India.
6. The intellectual property landscape for mRNA vaccine technology is complex. India will need to navigate patent and licensing issues to access and produce these vaccines.

In the Indian perspective, mRNA technology offers a range of significant advantages. Firstly, it enables rapid response to emerging infectious diseases, allowing India to swiftly develop and deploy vaccines and therapies in the face of health crises. Its versatility and adaptability make it suitable for addressing the diverse healthcare challenges prevalent in India and other low- and middle-income countries (LMICs), including a wide range of infectious diseases. The scalability of mRNA technology facilitates cost-effective large-scale production, a critical factor given India's vast population. Customization of mRNA vaccines to regional health needs can be particularly impactful. Additionally, the reduced cold chain requirements simplify distribution, especially in remote and underserved areas. Embracing mRNA technology also supports capacity building, strengthening healthcare infrastructure and research capabilities.

References

Avogadri F et al (2014) Combination of alphavirus replicon particle-based vaccination with immunomodulatory antibodies: therapeutic activity in the B16 melanoma mouse model and immune correlates. Cancer Immunol Res 2:448–458. https://doi.org/10.1158/2326-6066.CIR-13-0220

Baden LR et al (2021) Efficacy and safety of the mRNA-1273 SARS-CoV-2 vaccine. N Engl J Med 384:403–416. https://doi.org/10.1056/NEJMoa2035389

Baiersdorfer M et al (2019) A facile method for the removal of dsRNA contaminant from in vitro-transcribed mRNA. Mol Ther Nucleic Acids 15:26–35. https://doi.org/10.1016/j.omtn.2019.02.018

Berraondo P, Martini PGV, Avila MA, Fontanellas A (2019) Messenger RNA therapy for rare genetic metabolic diseases. Gut 68:1323–1330. https://doi.org/10.1136/gutjnl-2019-318269

Bettini E, Locci M (2021) SARS-CoV-2 mRNA vaccines: immunological mechanism and beyond. Vaccines (Basel) 9:147. https://doi.org/10.3390/vaccines9020147

Billingsley MM et al (2020) Ionizable lipid nanoparticle-mediated mRNA delivery for human CAR T cell engineering. Nano Lett 20:1578–1589. https://doi.org/10.1021/acs.nanolett.9b04246

Bloom K, van den Berg F, Arbuthnot P (2021) Self-amplifying RNA vaccines for infectious diseases. Gene Ther 28:117–129. https://doi.org/10.1038/s41434-020-00204-y

Boczkowski D, Nair SK, Snyder D, Gilboa E (1996) Dendritic cells pulsed with RNA are potent antigen-presenting cells *in vitro* and *in vivo*. J Exp Med 184:465–472. https://doi.org/10.1084/jem.184.2.465

Cannarozzi G et al (2010) A role for codon order in translation dynamics. Cell 141:355–367. https://doi.org/10.1016/j.cell.2010.02.036

Chabanovska O, Galow AM, David R, Lemcke H (2021) mRNA—a game changer in regenerative medicine, cell-based therapy and reprogramming strategies. Adv Drug Deliv Rev 179:114002. https://doi.org/10.1016/j.addr.2021.114002

Chauhan A et al (2023) The prevalence of tuberculosis infection in India: a systematic review and meta-analysis. Indian J Med Res 157:135–151. https://doi.org/10.4103/ijmr.ijmr_382_23

Chen R et al (2023) Engineering circular RNA for enhanced protein production. Nat Biotechnol 41:262–272. https://doi.org/10.1038/s41587-022-01393-0

Christensen D, Korsholm KS, Andersen P, Agger EM (2011) Cationic liposomes as vaccine adjuvants. Expert Rev Vaccines 10:513–521. https://doi.org/10.1586/erv.11.17

Costantini A et al (2022) COVID-19 pandemic distress among a sample of Italian psycho-oncologists: risk of isolation and loneliness. Tumori Journal 108:77–85. https://doi.org/10.1177/0300891621992129

Deering RP, Kommareddy S, Ulmer JB, Brito LA, Geall AJ (2014) Nucleic acid vaccines: prospects for non-viral delivery of mRNA vaccines. Expert Opin Drug Deliv 11:885–899. https://doi.org/10.1517/17425247.2014.901308

Diamond MS, Kanneganti TD (2022) Innate immunity: the first line of defense against SARS-CoV-2. Nat Immunol 23:165–176. https://doi.org/10.1038/s41590-021-01091-0

Diken M, Kranz LM, Kreiter S, Sahin U (2017) mRNA: a versatile molecule for cancer vaccines. Curr Issues Mol Biol 22:113–128. https://doi.org/10.21775/cimb.022.113

Dimitriadis GJ (1978) Translation of rabbit globin mRNA introduced by liposomes into mouse lymphocytes. Nature 274:923–924. https://doi.org/10.1038/274923a0

Fabbri L, Chakraborty A, Robert C, Vagner S (2021) The plasticity of mRNA translation during cancer progression and therapy resistance. Nat Rev Cancer 21:558–577. https://doi.org/10.1038/s41568-021-00380-y

Finney J, Kelsoe G (2018) Poly- and autoreactivity of HIV-1 bNAbs: implications for vaccine design. Retrovirology 15:53. https://doi.org/10.1186/s12977-018-0435-0

Freyn AW et al (2021) Antigen modifications improve nucleoside-modified mRNA-based influenza virus vaccines in mice. Mol Ther Methods Clin Dev 22:84–95. https://doi.org/10.1016/j.omtm.2021.06.003

Gaj T, Gersbach CA, Barbas CF 3rd. (2013) ZFN, TALEN, and CRISPR/Cas-based methods for genome engineering. Trends Biotechnol 31:397–405. https://doi.org/10.1016/j.tibtech.2013.04.004

Ghebreyesus TA, Lima NT (2023) The TB vaccine accelerator council: harnessing the power of vaccines to end the tuberculosis epidemic. Lancet Infect Dis 23:1222. https://doi.org/10.1016/S1473-3099(23)00589-3

Granados-Riveron JT, Aquino-Jarquin G (2021) Engineering of the current nucleoside-modified mRNA-LNP vaccines against SARS-CoV-2. Biomed Pharmacother 142:111953. https://doi.org/10.1016/j.biopha.2021.111953

Hajj KA et al (2020) A potent branched-tail lipid nanoparticle enables multiplexed mRNA delivery and gene editing in vivo. Nano Lett 20:5167–5175. https://doi.org/10.1021/acs.nanolett.0c00596

Hia F, Takeuchi O (2021) The effects of codon bias and optimality on mRNA and protein regulation. Cell Mol Life Sci 78:1909–1928. https://doi.org/10.1007/s00018-020-03685-7

Ho W et al (2021) Next-generation vaccines: nanoparticle-mediated DNA and mRNA delivery. Adv Healthc Mater 10:e2001812. https://doi.org/10.1002/adhm.202001812

Intapiboon P et al (2021) Immunogenicity and safety of an intradermal BNT162b2 mRNA vaccine booster after two doses of inactivated SARS-CoV-2 vaccine in healthy population. Vaccines (Basel) 9:1375. https://doi.org/10.3390/vaccines9121375

Kariko K, Buckstein M, Ni H, Weissman D (2005) Suppression of RNA recognition by toll-like receptors: the impact of nucleoside modification and the evolutionary origin of RNA. Immunity 23:165–175. https://doi.org/10.1016/j.immuni.2005.06.008

Kim HJ et al (2019) Fine-tuning of hydrophobicity in amphiphilic Polyaspartamide derivatives for rapid and transient expression of messenger RNA directed toward genome engineering in brain. ACS Cent Sci 5:1866–1875. https://doi.org/10.1021/acscentsci.9b00843

Kim SC et al (2022) Modifications of mRNA vaccine structural elements for improving mRNA stability and translation efficiency. Mol Cell Toxicol 18:1–8. https://doi.org/10.1007/s13273-021-00171-4

Krammer F (2020) SARS-CoV-2 vaccines in development. Nature 586:516–527. https://doi.org/10.1038/s41586-020-2798-3

Krieg PA, Melton DA (1984) Functional messenger RNAs are produced by SP6 in vitro transcription of cloned cDNAs. Nucleic Acids Res 12:7057–7070. https://doi.org/10.1093/nar/12.18.7057

Kumar P et al (2022) Diversity in HIV epidemic transitions in India: an application of HIV epidemiological metrices and benchmarks. PLoS One 17:e0270886. https://doi.org/10.1371/journal.pone.0270886

Larsen SE, Baldwin SL, Coler RN (2023a) Tuberculosis vaccines update: is an RNA-based vaccine feasible for tuberculosis? Int J Infect Dis 130(Suppl 1):S47–S51. https://doi.org/10.1016/j.ijid.2023.03.035

Larsen SE et al (2023b) An RNA-based vaccine platform for use against *Mycobacterium tuberculosis*. Vaccines (Basel) 11:130. https://doi.org/10.3390/vaccines11010130

Lederer K et al (2022) Germinal center responses to SARS-CoV-2 mRNA vaccines in healthy and immunocompromised individuals. Cell 185:1008–1024 e1015. https://doi.org/10.1016/j.cell.2022.01.027

Liu MA (2019) A comparison of plasmid DNA and mRNA as vaccine technologies. Vaccines (Basel) 7:37. https://doi.org/10.3390/vaccines7020037

Lorentzen CL, Haanen JB, Met O, Svane IM (2022) Clinical advances and ongoing trials on mRNA vaccines for cancer treatment. Lancet Oncol 23:e450–e458. https://doi.org/10.1016/S1470-2045(22)00372-2

Lurie N, Sharfstein JM, Goodman JL (2020) The development of COVID-19 vaccines: safeguards needed. JAMA 324:439–440. https://doi.org/10.1001/jama.2020.12461

Malone RW, Felgner PL, Verma IM (1989) Cationic liposome-mediated RNA transfection. Proc Natl Acad Sci USA 86:6077–6081. https://doi.org/10.1073/pnas.86.16.6077

Martins R, Queiroz JA, Sousa F (2014) Ribonucleic acid purification. J Chromatogr A 1355:1–14. https://doi.org/10.1016/j.chroma.2014.05.075

Mascola JR, Haynes BF (2013) HIV-1 neutralizing antibodies: understanding nature's pathways. Immunol Rev 254:225–244. https://doi.org/10.1111/imr.12075

Matarazzo L, Bettencourt PJG (2023) mRNA vaccines: a new opportunity for malaria, tuberculosis and HIV. Front Immunol 14:1172691. https://doi.org/10.3389/fimmu.2023.1172691

Mauger DM et al (2019) mRNA structure regulates protein expression through changes in functional half-life. Proc Natl Acad Sci USA 116:24075–24083. https://doi.org/10.1073/pnas.1908052116

McKay PF et al (2020) Self-amplifying RNA SARS-CoV-2 lipid nanoparticle vaccine candidate induces high neutralizing antibody titers in mice. Nat Commun 11:3523. https://doi.org/10.1038/s41467-020-17409-9

Minnaert AK et al (2021) Strategies for controlling the innate immune activity of conventional and self-amplifying mRNA therapeutics: getting the message across. Adv Drug Deliv Rev 176:113900. https://doi.org/10.1016/j.addr.2021.113900

Muttach F, Muthmann N, Rentmeister A (2017) Synthetic mRNA capping. Beilstein J Org Chem 13:2819–2832. https://doi.org/10.3762/bjoc.13.274

Nabhan JF et al (2016) Intrathecal delivery of frataxin mRNA encapsulated in lipid nanoparticles to dorsal root ganglia as a potential therapeutic for Friedreich's ataxia. Sci Rep 6:20019. https://doi.org/10.1038/srep20019

Nance KD, Meier JL (2021) Modifications in an emergency: the role of N1-Methylpseudouridine in COVID-19 vaccines. ACS Cent Sci 7:748–756. https://doi.org/10.1021/acscentsci.1c00197
Neuzil KM (2023) An mRNA influenza vaccine—could it deliver? N Engl J Med 388:1139–1141. https://doi.org/10.1056/NEJMcibr2215281
Niu D, Wu Y, Lian J (2023) Circular RNA vaccine in disease prevention and treatment. Signal Transduct Target Ther 8:341. https://doi.org/10.1038/s41392-023-01561-x
Ouranidis A et al (2021) mRNA therapeutic modalities design, formulation and manufacturing under pharma 4.0 principles. Biomedicines 10:50. https://doi.org/10.3390/biomedicines10010050
Pai M et al (2005) Mycobacterium tuberculosis infection in health care workers in rural India: comparison of a whole-blood interferon gamma assay with tuberculin skin testing. JAMA 293:2746–2755. https://doi.org/10.1001/jama.293.22.2746
Papukashvili D et al (2022) Self-amplifying RNA approach for protein replacement therapy. Int J Mol Sci 23:12884. https://doi.org/10.3390/ijms232112884
Pardi N, Hogan MJ, Porter FW, Weissman D (2018) mRNA vaccines—a new era in vaccinology. Nat Rev Drug Discov 17:261–279. https://doi.org/10.1038/nrd.2017.243
Perkovic M et al (2023) A trans-amplifying RNA simplified to essential elements is highly replicative and robustly immunogenic in mice. Mol Ther 31:1636–1646. https://doi.org/10.1016/j.ymthe.2023.01.019
Polack FP et al (2020) Safety and efficacy of the BNT162b2 mRNA Covid-19 vaccine. N Engl J Med 383:2603–2615. https://doi.org/10.1056/NEJMoa2034577
Presnyak V et al (2015) Codon optimality is a major determinant of mRNA stability. Cell 160:1111–1124. https://doi.org/10.1016/j.cell.2015.02.029
Qin S et al (2022) mRNA-based therapeutics: powerful and versatile tools to combat diseases. Signal Transduct Target Ther 7:166. https://doi.org/10.1038/s41392-022-01007-w
Qiu Y et al (2019) Effective mRNA pulmonary delivery by dry powder formulation of PEGylated synthetic KL4 peptide. J Control Release 314:102–115. https://doi.org/10.1016/j.jconrel.2019.10.026
Rais M et al (2023) Immunogenicity and protection against *Mycobacterium avium* with a heterologous RNA prime and protein boost vaccine regimen. Tuberculosis (Edinb) 138:102302. https://doi.org/10.1016/j.tube.2022.102302
Rodo MJ et al (2019) A comparison of antigen-specific T cell responses induced by six novel tuberculosis vaccine candidates. PLoS Pathog 15:e1007643. https://doi.org/10.1371/journal.ppat.1007643
Rojas LA et al (2023) Personalized RNA neoantigen vaccines stimulate T cells in pancreatic cancer. Nature 618:144–150. https://doi.org/10.1038/s41586-023-06063-y
Russick J et al (2020) Correction of bleeding in experimental severe hemophilia a by systemic delivery of factor VIII-encoding mRNA. Haematologica 105:1129–1137. https://doi.org/10.3324/haematol.2018.210583
Sahin U, Kariko K, Tureci O (2014) mRNA-based therapeutics—developing a new class of drugs. Nat Rev Drug Discov 13:759–780. https://doi.org/10.1038/nrd4278
Sathishkumar K, Chaturvedi M, Das P, Stephen S, Mathur P (2022) Cancer incidence estimates for 2022 & projection for 2025: result from National Cancer Registry Programme, India. Indian J Med Res 156:598–607. https://doi.org/10.4103/ijmr.ijmr_1821_22
Schlake T, Thess A, Fotin-Mleczek M, Kallen KJ (2012) Developing mRNA-vaccine technologies. RNA Biol 9:1319–1330. https://doi.org/10.4161/rna.22269
Schnyder JL, de Jong HK, Grobusch MP (2022) Intradermal immunization-a dose-sparing strategy to combat global shortages of severe acute respiratory syndrome coronavirus 2 vaccines? Clin Microbiol Infect 28:6–8. https://doi.org/10.1016/j.cmi.2021.08.020
Schumann C et al (2018) Increasing lean muscle mass in mice via nanoparticle-mediated hepatic delivery of follistatin mRNA. Theranostics 8:5276–5288. https://doi.org/10.7150/thno.27847
Sebastian M et al (2019) A phase I/IIa study of the mRNA-based cancer immunotherapy CV9201 in patients with stage IIIB/IV non-small cell lung cancer. Cancer Immunol Immunother 68:799–812. https://doi.org/10.1007/s00262-019-02315-x

Sliepen K, Sanders RW (2016) HIV-1 envelope glycoprotein immunogens to induce broadly neutralizing antibodies. Expert Rev Vaccines 15:349–365. https://doi.org/10.1586/14760584.2016.1129905

Suryawanshi P, Bagul R, Shete A, Thakar M (2021) Anti-HIV-1 ADCC and HIV-1 Env can be partners in reducing Latent HIV reservoir. Front Immunol 12:663919. https://doi.org/10.3389/fimmu.2021.663919

Tait DR et al (2019) Final analysis of a trial of M72/AS01(E) vaccine to prevent tuberculosis. N Engl J Med 381:2429–2439. https://doi.org/10.1056/NEJMoa1909953

Trepotec Z, Lichtenegger E, Plank C, Aneja MK, Rudolph C (2019) Delivery of mRNA therapeutics for the treatment of hepatic diseases. Mol Ther 27:794–802. https://doi.org/10.1016/j.ymthe.2018.12.012

Uddin MN, Roni MA (2021) Challenges of storage and stability of mRNA-based COVID-19 vaccines. Vaccines (Basel) 9:1033. https://doi.org/10.3390/vaccines9091033

Wadhwa A, Aljabbari A, Lokras A, Foged C, Thakur A (2020) Opportunities and challenges in the delivery of mRNA-based vaccines. Pharmaceutics 12:102. https://doi.org/10.3390/pharmaceutics12020102

Whitley J et al (2022) Development of mRNA manufacturing for vaccines and therapeutics: mRNA platform requirements and development of a scalable production process to support early phase clinical trials. Transl Res 242:38–55. https://doi.org/10.1016/j.trsl.2021.11.009

Wolff JA et al (1990) Direct gene transfer into mouse muscle in vivo. Science 247:1465–1468. https://doi.org/10.1126/science.1690918

Wollner CJ et al (2021) A dengue virus serotype 1 mRNA-LNP vaccine elicits protective immune responses. J Virol 95:e02482. https://doi.org/10.1128/JVI.02482-20

Xue T et al (2004) RNA encoding the MPT83 antigen induces protective immune responses against *Mycobacterium tuberculosis* infection. Infect Immun 72:6324–6329. https://doi.org/10.1128/IAI.72.11.6324-6329.2004

Yarlagadda H et al (2022) COVID-19 vaccine challenges in developing and developed countries. Cureus 14:e23951. https://doi.org/10.7759/cureus.23951

Zhang H et al (2023) Algorithm for optimized mRNA design improves stability and immunogenicity. Nature 621:396–403. https://doi.org/10.1038/s41586-023-06127-z

6 Vaccine Research and Development in India for COVID-19 and Influenza

Kawkab Kanjo, Ghadiyaram Chakshusmathi, and Raghavan Varadarajan

Abstract

The COVID-19 pandemic has clearly demonstrated that respiratory viruses can cause immense loss of lives and livelihoods. India is the home to some of the world's largest vaccine manufacturers and has successfully vaccinated a large fraction of its population to protect against COVID-19 disease caused by the sarbecovirus, SARS-CoV-2. However, much of this was reliant on formulations developed elsewhere, and before vaccines could be widely deployed, a large fraction of the population was infected. There is a clear need to develop and nurture indigenous expertise to design, develop, and rapidly produce efficacious vaccines against respiratory viruses, including SARS-CoV-2 and influenza virus.

6.1 Introduction

The COVID-19 pandemic serves as a clear demonstration of the enormous threat that respiratory viruses, capable of infecting multiple species, pose to human health and livelihood. Vaccines are one of the most effective modalities to combat infectious diseases. Pandemics typically arise from zoonotic transmission of pathogens from other species to humans (Piret and Boivin 2021). Increases in human population, environmental destruction, and climate change all enhance the probability of pandemic occurrence. In this chapter, we discuss vaccine development with a focus on the Indian context for two pandemic respiratory diseases, influenza and COVID-19. We summarize current efforts and make suggestions for strategies to

K. Kanjo · R. Varadarajan (✉)
Molecular Biophysics Unit (MBU), Indian Institute of Science, Bengaluru, India
e-mail: Varadar@iisc.ac.in

G. Chakshusmathi
Mynvax Private Limited, Bengaluru, India

G. John, A. K. Panda (eds.), *Biotechnology in India - Reworking A Strategy*,
https://doi.org/10.1007/978-981-97-0896-3_6

encourage and hasten vaccine research and development to enhance our ability to combat future pandemics.

6.2 Vaccines Against SARS-CoV-2 from India

The ongoing COVID-19 pandemic is caused by the sarbecovirus SARS-CoV-2, with the first human cases reported in Wuhan, China, in late 2019 (Roberts et al. 2021; Huang et al. 2020). The virus spread rapidly globally and containment efforts were unsuccessful. Below we describe vaccine development and evaluation of COVID-19 vaccines in India. Commonly used animal models for testing COVID-19 vaccines include Balb/c mice, hACE2 transgenic mice, and Syrian golden hamsters (Moreau et al. 2020; Huang et al. 2021; Wang et al. 2020; Gu et al. 2020; Dinnon et al. 2020; Imai et al. 2020; Li et al. 2022; Sia et al. 2020). In India, Translational Health Science and Technology Institute (THSTI) has carried out extensive work with the hamster model (Rizvi et al. 2022), and NCBS-InStem has developed an indigenous hACE2-transgenic mouse strain (Mittal et al. 2023).

Covaxin (BBV152, whole-virion, BPL inactivated vaccine) and Covishield (ChAdOx1/AZD1222—replication-deficient adenovirus vectored vaccine) were the first vaccines to obtain emergency use approval in India, following which several Indian manufacturing companies geared up and established partnerships to manufacture vector based, protein-subunit, DNA as well as mRNA vaccines for COVID-19 (Central Drugs Standard Control Organization 2022) and received emergency use approvals for different age groups. So far, a total of about 2.18 billion vaccine doses have been distributed in India, the division of which is shown in Table 6.1.

Covaxin/BBV152, developed by Bharat Biotech International in collaboration with ICMR, is a Vero cell produced, β-propiolactone-inactivated whole-virus (strain NIV-2020-770 (spike variant D614G)) vaccine formulated with a toll-like receptor 7/8 agonist molecule adsorbed to alum (Algel-IMDG) or Algel alone. Three whole virion inactivated vaccine candidates BBV152A [3 mg + Aluminum hydroxide (Algel)-Imidazoquinoline (IMDG)], BBV152B (6 mg + Algel-IMDG), and BBV152C (6 mg + Algel) were tested. All three were immunogenic (mean Spike antibody titers greater than 10^4 and mean MNT_{50} (micro neutralization test) neutralization titers greater than 10^4 in Balb/C mice upon IM route of administration) and found to be safe in rodents (mice, rats, and rabbits) (Ganneru et al. 2021). In Syrian

Table 6.1 Number of vaccinations in India as a function of brand as on December 2022. Source: https://vaccinate-india.in/dashboard

Brand	Vaccination doses (millions)
Covishield	1744.3
Covaxin	361.1
Sputnik V	1.2
Corbevax	73.7
Covovax	0.03

hamsters, these vaccine candidates induced effective humoral immune responses after three immunizations (mean $PRNT_{50}$ neutralization titers greater than 10^3) and protected animals from wild-type SARS-CoV-2-induced pneumonia with minimal or no pathological or immune-histochemical findings (Mohandas et al. 2021). Similar observations were reported in immunogenicity and challenge studies conducted in Rhesus macaques (Yadav et al. 2021). All three candidates showed tolerable safety outcomes and enhanced immune responses in a double-blind, multicenter, randomized, controlled Phase 1 trial (Ella et al. 2021a). The two Algel-IMDG formulations were selected for Phase 2 immunogenicity trials involving about 380 participants. The BBV152B (6 µg + Algel-IMDG) that was forwarded to Phase 3 trials elicited $PRNT_{50}$ geometric mean titers of 197 [95% CI 155·6–249·4] by Day 56 following two doses of immunization, which was comparable to the titers in human convalescent sera (Ella et al. 2021b). In the Phase 3 trial, consisting of 24,419 participants, BBV152 showed an overall estimated efficacy of 77·8% (95% CI 65·2–86·4) with no clinically significant differences in the distributions of solicited, unsolicited, or serious adverse events between the placebo and vaccine groups, and no cases of anaphylaxis or vaccine-related deaths (Ella et al. 2021c).

ChAdOx1 nCoV-19 (also known as AZD1222) was developed by the University of Oxford in partnership with AstraZeneca and was supplied for use in more than 170 countries across the world, albeit with very limited deployment in the EU and none in the USA (Joe et al. 2022). ChAdOx1 nCoV-19 vaccine (AZD1222) consists of a replication-deficient chimpanzee adenoviral vector ChAdOx1, containing the codon-optimized SARS-CoV-2 structural surface glycoprotein antigen (spike protein; nCoV-19) gene. After technology transfer from Oxford University/AstraZeneca to Serum Institute of India Pvt. Ltd. (SIIPL), the ChAdOx1 nCoV-19 vaccine has been manufactured at SIIPL from the bulk stage (AstraZeneca 2020) and formulated at the commercial scale. The SIIPL-manufactured product is known as SII-ChAdOx1 nCoV-19 (Covishield™). More than 1 billion Covishield doses have been manufactured and distributed by SIIPL. While the neutralization titers were in the range of 40 in mice, in Rhesus monkeys, 14 days after an intramuscular boost immunization with 2.5×10^{10} viral particles of AZD1222, viral neutralization titers increased up to 160 (expressed as reciprocal value of the highest dilution of the serum that still inhibited virus replication). After challenge with a 50% tissue culture infective dose ($TCID_{50}$) of 2.6×10^6 of SARS-CoV-2 in both the upper and lower respiratory tracts of rhesus macaques, the average clinical score of control animals was higher compared with ChAdOx1 nCoV-19-vaccinated animals (van Doremalen et al. 2020). Following this, a phase 1/2, single-blind, randomized controlled trial was carried out in the UK with 5×10^{10} viral particles per dose. The median $PRNT_{50}$ titers went up to 218 in the vaccinated group 28 days post vaccination while the mean (rapid microneutralization assay) MNA_{80} titers were 136 upon boost immunization; pseudoviral neutralization titers were comparable to human convalescent sera (Folegatti et al. 2020). Vaccine efficacy for the prespecified primary analysis (combining dose groups) against the primary endpoint of COVID-19 occurring more than 14 days after the second dose was 70·4% reported for two of the four then ongoing trials from the UK and Brazil (Voysey et al. 2021; Knoll and Wonodi 2021).

A bridging study (Phase 2/3) was conducted in India to demonstrate non-inferiority of SII-ChAdOx1 nCoV-19 to AZD1222 in terms of geometric mean titer (GMT) ratio of anti-SARS-CoV-2 spike IgG antibodies 28 days after the second dose and to determine the incidence of serious adverse events (SAEs) causally related to SII-ChAdOx1 nCoV-19. Neutralizing antibody titers were < 6 international units (IU)/mL in both vaccine groups at baseline. These increased to >11 IU/mL after the first dose and > 56 IU/mL after the second dose in both groups. The rise in titers from baseline to post-second dose was >12-fold in both groups. The clinical symptoms reported for both groups were comparable and thus the non-inferiority of SII-ChAdOx1 nCoV-19 to AZD1222 was proven and subsequently, Covishield was approved by Indian regulatory authorities (Kulkarni et al. 2021).

Both Covaxin and Covishield have received full approval for manufacture for sale or for distribution in India (Central Drugs Standard Control Organization 2022). Humoral immune response of both ChAdOx1-nCOV (Covishield™) and BBV-152 (Covaxin™) vaccines in Indian healthcare workers (HCW) was assessed. This study reports that both seropositivity rate and GMT (95% CI) of anti-spike antibody were significantly higher in Covishield vs. Covaxin recipients (98.1 vs. 80.0%; 129.3 vs. 48.3 AU/mL; both $p < 0.001$) (Singh et al. 2021). A multi-centric, hospital-based, case-control study involving 1143 cases and 2541 control patients in India during the Delta wave (B.1.617 lineages) indicated that the vaccine effectiveness (VE) of complete vaccination was 85% with Covishield and 71% with Covaxin against severe COVID-19 among individuals aged ≥45 years (Bhatnagar et al. 2022). In other independent studies, Covishield was shown to be moderately effective (63–67% VE) against the Delta variant (Thiruvengadam et al. 2022; Lopez Bernal et al. 2021). The situation was different when the Omicron BA.1 variant surfaced. In a study conducted in the UK, primary immunization with two doses of ChAdOx1 nCov-19 provided limited protection against symptomatic disease caused by the Omicron BA.1 variant. Vaccine effectiveness was only 46.7% (95% CI, 34.3–56.7) against Omicron at 5–9 weeks even after the booster (third) dose, and is likely to be even lower against currently circulating variants such as XBB (WHO 2022a).

Leading Indian pharmaceutical companies, including Serum Institute of India, Gland Pharma, Hetero Biopharma, Panacea Biotec, Stelis Biopharma, Virchow Biotech, and Morepen initiated efforts to manufacture Sputnik V, a COVID-19 vaccine with two different adenoviral vectors (recombinant Ad26 [rAd26] and recombinant Ad5 [rAd5]), both carrying the gene for SARS-CoV-2 spike glycoprotein (rAd26-S and rAd5-S). A prime-boost regimen was followed with 10^{11} viral particles per dose for both recombinant adenoviruses in a Phase 1/2 study conducted in Russia. Both vaccine components were found to be safe, and volunteers who received the heterologous rAd26 and rAd5 vaccine elicited the same titer of SARS-CoV-2 neutralizing antibodies as did people who had recovered from COVID-19 (mean reciprocal neutralizing antibody titers of 40) (Logunov et al. 2020). Phase 3 data indicated vaccine efficacy of 91.6% (Logunov et al. 2021); however, manufacture of the rAd5 vector apparently had challenges and eventually Sputnik V Light with only rAd26-S as its component was released.

Jansen vaccine is the fourth vaccine to be produced in India. Originally developed by Johnson & Johnson–Janssen and manufactured in India by Biological E Limited, the Ad26.COV2.S vaccine is another recombinant, replication-incompetent human adenovirus type 26 (Ad26) vector encoding a prefusion stabilized full-length, membrane-bound SARS-CoV-2 spike protein. A single dose of Ad26.COV2.S provided 52.9% protection against moderate to severe–critical COVID-19 (Sadoff et al. 2022).

The vaccine developed by Novavax, named Covovax in India, is a subunit vaccine derived from the full-length spike (S) protein that is stable in the prefusion conformation. NVX-CoV2373 forms nanoparticles approximately 27 nm in size and is adjuvanted with the Saponin-based Matrix-M adjuvant. In preclinical studies, this vaccine induced high levels of functional immunity in mice and baboons, and protected mice expressing hACE2 receptors in a live SARS-CoV-2 challenge (Tian et al. 2021). In a Phase 1/2 study, in vaccinated individuals GMT neutralization titers were $> 10^3$, exceeded values seen in symptomatic COVID-19 outpatients and were of the magnitude seen in convalescent serum from hospitalized patients with COVID-19 (Keech et al. 2020). It was found to be safe and effective (vaccine efficacy 90.4%, 95%CI) in a clinical trial conducted in the USA, Mexico (Dunkle et al. 2022), and a few other locations.

Corbevax, produced in the methylotrophic yeast *Pichia pastoris*, consists of a version of the receptor binding domain (RBD) of the SARS-CoV-2 spike protein, together with the adjuvants aluminum hydroxide gel and CpG 1018. This has received emergency approval for use in children and adolescents in India. It is also a protein subunit vaccine (Pollet et al. 2021) developed by Texas Children's Hospital Center for Vaccine Development and Baylor College of Medicine in Houston, Texas, and Dynavax technologies based in Emeryville, California. It is licensed to the Indian biopharmaceutical firm Biological E Limited (BioE) for development and production. All doses tested were well tolerated in the Phase 1 study, and the optimum Corbevax formulation consisting of RBD antigen (25 μg) + aluminum hydroxide (750 μg) + CpG 1018 (750 μg) in a 0.5 mL volume (single dose for a human subject) was evaluated in the Phase 2/3 study. Pseudoviral neutralization titers measured in the Phase 2 study reached titers comparable to human convalescent sera for one of the groups by Day 42 following two immunizations (Thuluva et al. 2022a). In a Phase 3 study involving children (5–11 years) and adolescents (12–17 years), Corbevax was found to be safe and moderately immunogenic with post-vaccination nAb titers of ~ 1000 and 500 following two boosts for Ancestral and Delta strains of SARS-CoV-2, respectively. There were significant neutralization titers of ~50 at day 0, indicating prior exposure to the virus in most participants, which were non-inferior to adult cohorts. An efficacy evaluation for this vaccine is not yet published (Thuluva et al. 2022b).

The ZyCoV-D vaccine was developed by Cadila Healthcare Limited, Ahmedabad, India. The DNA vaccine candidate is comprised of a DNA plasmid (episomal) vector pVAX1 carrying a gene expressing the Wuhan strain spike-S protein of

SARS-CoV-2 and IgE signal peptide. Balb/C mice, guinea pigs, and rabbits showed good immunogenic responses, and all sera showed modest neutralization IC_{50} titers above 100 upon 3 immunizations with 100–500 μg of DNA vector on Day 42 (Dey et al. 2021). ZyCoV-D vaccine was well tolerated in 48 healthy vaccinated adults with no vaccine-related SAEs in a single-center, open-label, non-randomized, Phase 1 trial in India (Momin et al. 2021). The seroconversion rate based on neutralizing antibodies, measured by live virus neutralization PRNT assay, was seen in 50% (05/10) and 80% (08/10) of participants who received intradermally three doses of 2 mg vaccine either via needle and syringe or needle-free injection system (NFIS) device, respectively. In the interim analysis of a randomized Phase 3 study conducted in India, the ZyCoV-D vaccine efficacy was found to be 66·6% (95% CI 47·6–80·7). In this study, three doses of vaccine (2 mg per dose) or placebo, 28 days apart, were administered intradermally via a needle-free injection systems (Khobragade et al. 2022). This vaccine received emergency approval for use in individuals >12 years of age in India.

Levels of neutralizing antibodies in sera upon immunization are likely a correlate of protection (Khoury et al. 2021). The data from animal studies and human clinical trials for the above vaccines, wherever available, are tabulated below (Tables 6.2 and 6.3). However, as different assays are used by different groups, we encourage readers to refer to the relevant articles and their supplementary files for more details. Recently, data from a Phase 2/3 trial of a self-amplifying mRNA vaccine formulation GEMCOVAC, developed by the Indian company Gennova Biopharmaceuticals Ltd, have been published (Saraf et al 2024).

Newer technologies such as mRNA and intranasal vaccines are now being adopted by Indian biotechnology/manufacturing industries. HGC019 is an mRNA-based COVID-19 vaccine candidate being developed by Gennova Biopharmaceuticals, Pune, and HDT Bio, Seattle, that employs adsorption technology rather than encapsulation technology. Bharath Biotech has developed iNCOVACC, a recombinant replication-deficient adenovirus (ChAd36) vectored vaccine with a pre-fusion stabilized spike protein combined with technology licensed from Washington University in St. Louis. Efficacy of these vaccines is yet to be determined, though a recent Phase 1 clinical trial with an intranasally administered, adenovirus-vectored, COVID-19 vaccine gave disappointing results (Madhavan et al. 2022).

As discussed above, India has gained a lot of expertise in large-scale and rapid manufacturing of COVID-19 vaccines, use of which has helped control the spread of COVID-19 in the world, especially in poor and developing nations. This expertise is beneficial for future pandemics. However, all the vaccines that have been described above were developed in other countries except Covaxin, which is an inactivated whole-virus vaccine, and Zycov-D, the DNA vaccine developed by Cadila Healthcare. Research in newer technologies such as thermostable vaccines for last mile benefits, stable mRNA vaccines for quick efficacious delivery, and adjuvants for other parenteral routes of vaccine administration such as intranasal or patch technologies is limited to very few research labs in India.

Table 6.2 Preclinical neutralizing antibody titers reported for vaccines currently available in India. IP, IM, ID, and NFIS refer to intraperitoneal, intramuscular, intradermal, and needle-free injection system, respectively[a]

Vaccine	Dose	Route	Animal	Measurement day	nAb	Reference
BBV152B (Covaxin)	0.6 μg antigen with Algel-IMDG	IP	Balb/C mice	Day 21 after 3 immunizations	MNT_{50} = 10,000	Ganneru et al. (2021))
BBV152B (Covaxin)	6 μg antigen with Algel-IMDG	IM	New Zealand white rabbits	Day 21 after 3 immunizations	PRNT $_{90}$ = 1000	Ganneru et al. (2021))
BBV152B (Covaxin)	6 μg antigen with Algel-IMDG	Not mentioned	Hamsters	Day 48 after 3 immunizations	$PRNT_{50}$ = 25,000	Mohandas et al. (2021)
BBV152B (Covaxin)	6 μg antigen with Algel-IMDG	Not mentioned	Rhesus macaques	Day 26 after 2 immunizations	$PRNT_{50}$ = 2000	Yadav et al. (2021)
AZD1222 (Covishield)	6×10^9 virus particles	IM	Balb/C mice	Day 14 after 1 immunization	VNA = 40	van Doremalen et al. (2020)
AZD1222 (Covishield)	6×10^9 virus particles	IM	CD1 mice	Day 14 after 1 immunization	VNA = 80	van Doremalen et al. (2020)
AZD1222 (Covishield)	2.5×10^{10} virus particlesper animal	IM	Rhesus macaques	Day 42 after 2 immunizations	VNA = 500	van Doremalen et al. (2020)
NVXCoV2373 (Covavax)	10 μg antigen with MatrixM	IM	Balb/C mice	Day 21 after 2 immunizations	nAb (CPE_{100}) = 20,000	Tian et al. (2021))
NVXCoV2373 (Covavax)	5, 25 μg antigen with MatrixM	IM	Baboons	Day 35 after 2 immunizations	nAb (CPE_{100}) = 10,000	Tian et al. (2021)
RBD203-N1 (Corbevax)	7 μg antigen with Alhydrogel-CpG	IM	Balb/C mice	Day 35 after 2 immunizations	nAb (pseudovirus) = 1000	Chen et al. (2022)
Zycov-D	100 μg DNA	ID	Balb/C mice	Day 42 after 3 immunizations	Genscript neutralization assay (IC_{50}) = 168	Dey et al. (2021)
Zycov-D	100 μg DNA	ID	Guinea pigs	Day 42 after 3 immunizations	Genscript neutralization assay (IC_{50}) = 371	Dey et al. (2021)
Zycov-D	500 μg DNA	NFIS	New Zealand white rabbits	Day 42 after 3 immunizations	Genscript neutralization assay (IC_{50}) = 108	Dey et al. (2021)

[a]The vaccine trial data for the Gennova mRNA vaccine have recently been published in Nature Medicine (Saraf et al. 2024)

Table 6.3 Neutralizing antibody titers after the final immunization reported from Clinical Trials, for vaccines currently available in India. All vaccines elicit significant neutralization titers compared to pre-immune sera

Vaccine and dose	Immunization schedule	Neutralizing antibody titer	Reference
Covaxin - 6μgAg + Algel-IMDG	Prime and boost	GMT $PRNT_{50}$ = 197·0 [95% CI 155·6–249·4]	Ella et al. (2021b)
Covishield ChAdOx1 - 5x10^{10} particles	Prime only	PHE $PRNT_{50}$ GMT = 218 (IQR 122–395)	Folegatti et al. (2020)
Gam-COVID-Vac (sputnik V - 10^{11} particles)	Prime and boost	MNA GMT = 49·25	Logunov et al. (2020)
Covovax NVX-CoV2373 - 5 μg	Prime and boost	MNA (IC_{99}) GMT = 3906	Keech et al. (2020)
Corbevax - 25 μg ag +750 μg CpG adjuvant	Prime and boost	PNA NT_{50} GMT - 530	Thuluva et al. (2022a)
Zycov-D - 2 mg DNA	Prime, boost 1, boost 2	GMT $PRNT_{50}$ = 39.17 (15.43, 99.47)	Momin et al. (2021)

IISc and Mynvax Private Limited have co-developed a thermostable SARS-CoV-2 Spike RBD protein with vaccine potential (Ahmed et al. 2021). Using saturation suppressor mutagenesis (yeast surface display) technology, a triple mutant of SARS-CoV-2 RBD was identified that shows an increase in thermal melting temperature of ~7 °C with respect to the wild-type B.1 RBD. In mice and hamsters, squalene-in-water emulsion (SWE) adjuvanted formulations of the B.1-stabilized RBD were considerably more immunogenic than RBD lacking the stabilizing mutations and elicited antibodies that neutralized alpha, beta, and delta variants of concern with similar neutralization titers. The adjuvant used is equivalent to MF-59 that has been widely used in the context of human influenza vaccines and shown to be very safe. Lyophilized, stabilized proteins were resistant to high-temperature exposure and could be stored for over a month at 37 °C. One of the ways of improving immunogenicity is by stabilizing higher order structures, which was achieved by fusion of RBD to a heterologous, poorly immunogenic disulfide-linked trimerization domain derived from cartilage matrix protein (Malladi et al. 2021). Two intramuscular immunizations with an SWE-adjuvanted formulation of this trimeric-stabilized RBD elicited antibodies with pseudoviral neutralizing titers in guinea pigs and mice that were 25–250 fold higher than corresponding values in human convalescent sera and showed protection in Hamster challenge studies. When lyophilized, these proteins possessed remarkable functional stability to transient thermal stress of up to 100 °C and were stable over long-term storage of over 4 weeks at 37 °C. Vaccine formulations such as this that can be produced in large quantities and stored at room temperature are a useful modality to combat pandemics such as COVID-19, especially in remote and low-resource settings.

Recently, Omicron and its descendant sublineages including BA.2, BA.4, BA.5, XBB contain many immune evasion mutations in the RBD (Table 6.4), thus compromising vaccine effectiveness of pre-Omicron vaccines against these variants. Thus, many countries have authorized booster doses of available vaccines or updated vaccines that include closer matches with the circulating variant component in the

Table 6.4 RBD amino acid mutations in the most prevalent SARS-CoV-2 circulating variants

Position	339	346	368	371	373	375	376	405	408	417	440	444	445	446	*450*	452	460	477	478	484	486	490	493	496	498	501	505
B.1	G	R	L	S	S	S	T	D	**R**	K	N	K	V	G	N	L	N	S	T	**E**	F	F	**Q**	G	**Q**	N	Y
B.1.17																										Y	
B.1.351										N										K						Y	
P.1										T										K						Y	
B.1.617																R			K								
BA.1	D			L	P	F				N	K			S				N	K	A			R	S	R	Y	H
BA.2	D			F	P	F	A	N	s	N	K							N	K	A			R		R	Y	H
BA.4/5	D			F	P	F	A	N	S	N	K					R		N	K	A	V				R	Y	H
XBB	H	T	I	F	P	F	A	N	S	N	K		**p**	**s**			K	N	K	A	**s**	S			R	Y	**H**
XBB.1.5	**H**	T	**I**	F	P	F	A	**N**	**S**	**N**	**K**		**p**	**s**			**K**	**N**	**K**	A	**p**	**s**			R	Y	**H**
BQ.1	**D**			F	P	F	A	N	S	N	K	T				R	K	N	K	A	V				R	Y	H
BQ.1.1	D	T		F	P	F	A	N	S	N	K	T				R	K	N	K	A	V				R	Y	H
BF.7	D	T		F	P	F	A	N	S	N	K					R		N	K	A	V				R	Y	H

vaccine. An mRNA booster was initially effective in increasing serum breadth and reducing the risk of infection and hospitalization during Omicron BA.1 circulation (Petrie et al. 2023). However, upon emergence of BA.2 and BA.4/5 strains, a second boost of the original SARS-CoV-2 mRNA vaccine did not add any significant protection over the first booster against these variants (Petrie et al. 2023; Kislaya et al. 2022; McConeghy et al. 2022; Barda et al. 2022). The waning protection of the monovalent WT-based vaccine boosters and the recurrent emergence of new neutralization resistant variants led to the development of bivalent vaccines that contain a component from a recent circulating variant. The bivalent mRNA vaccine contained a component of the ancestral SARS-CoV-2 strain and an updated component of Omicron BA.1 or recent BA.5 variant. The bivalent mRNA booster was reported to protect against severe and symptomatic disease (Arbel et al. 2023) and reduce hospitalization cases related to COVID-19 (Surie et al. 2022; Tenforde et al. 2022; Lin et al. 2023; Kirsebom et al. 2023; Andersson et al. 2023a, b; Link-Gelles et al. 2023). Novavax has recently reported that immunization with an updated monovalent protein subunit vaccine based on the monovalent recombinant spike protein of XBB.1.5 induced robust and higher neutralization titers than the bivalent (prototype + XBB.1.5) vaccine against XBB variants in mice and NHPs (Patel et al. 2023). ZF2001, a homodimer-based prototype RBD protein vaccine, was also updated to a heterotypic chimeric RBD (Delta+BA.1) or (Delta+BA.2) vaccine that induced balanced neutralization response against the early and recent circulating variants, including XBB and its sublineages (Xu et al. 2022). Currently, WHO recommends that new vaccines should be based on XBB descendent sequence (World Health Organization 2023). The continual emergence of new SARS-CoV-2 variants that evade prior immune response thus necessitates the need to develop better vaccines with improved protective breadth.

6.3 Influenza Virus

The world had witnessed three major influenza, colloquially known as flu, pandemics in the twentieth century, caused by 1918 H1N1, 1957 H2N2, and 1968 H3N2 strains of influenza virus (Kilbourne 2006). A fourth pandemic was caused by swine origin influenza virus in 2009 (H1N1) (Wu and Wilson 2020). Both the 1957 and 1968 pandemics and most likely the 1918 pandemic were caused by avian origin viruses (Lycett et al. 2019). The 1918 pandemic was particularly severe with deaths likely in excess of 50 million, with an estimated death toll in India alone of 10–20 million (Chandra and Kassens-Noor 2014). Estimation of the influenza burden of disease (BoD) helps in strengthening preparedness, prevention, and control for future pandemics. The number of respiratory failure-related death cases due to seasonal influenza was estimated to be between 290,000 and 650,000 annually (Iuliano et al. 2018). According to the study, southeast Asian countries, including India, have the highest respiratory-associated death rate (25% of total deaths) across all age groups, and the deaths among children below 5 years in sub-Saharan Africa, southeast Asia, and the Eastern Mediterranean region account for 95% of global respiratory infection mortality in this age group (Iuliano et al. 2018). The precise disease burden in India is unknown due to insufficient surveillance, though a recent study (Narayan et al. 2020) estimated that ~130,000 deaths from influenza occur annually in India, and another estimated ~30,000 deaths in children under five in India in 2016 (Ram Purakayastha et al. 2018). This highlights the importance of implementing vaccination to reduce disease burden, especially for the high-risk groups including aged individuals, pregnant women, children, and healthcare workers. Influenza vaccination uptake in India is very poor. According to a study of a cohort of 64,715 adults over 45, the influenza vaccine uptake rate was 1.5% (Rizvi and Singh 2022). Establishment of an influenza preparedness and vaccination strategy should be considered to prevent future pandemics and to minimize the ongoing disease burden.

Influenza viruses are enveloped viruses of the Orthomyxoviridae family with segmented, negative-sense, single-strand RNA genomes. Three types of influenza have been identified so far: A, B, and C. Type A and B influenza viruses frequently infect humans and comprise eight RNA segments coding for eight proteins. The major surface glycoproteins are Hemagglutinin and Neuraminidase (HA and NA, respectively). The other proteins are the polymerase protein complex (PB1, PB2, and PA), matrix protein (M), non-structural protein (NS), and nucleocapsid protein (NP) (Blut et al. 2009). Influenza A viruses circulate in humans and domestic animals and to date, a total of 18 and 11 antigenically different HA and NA serotypes have been identified (Krammer et al. 2018). However, the influenza B type is restricted to humans with two circulating lineages identified: B/Victoria/2/1987-like and B/Yamagata/16/1988-like. Both the lineages along with the currently circulating strain of H1N1 and H3N2 subtypes are components of current annual influenza vaccines. Hemagglutinin (HA) is the major surface antigen of influenza virus, and it is expressed as a trimeric precursor, HA0, which is processed by host proteases and cleaved into HA1 and HA2 subunits (Wu and Wilson 2020). The majority of the neutralizing antibodies are targeted towards the immunodominant globular head domain (composed of HA1), which is less conserved between different subtypes than the stem domain, and harbors the conserved receptor binding site. In contrast, the stem region (composed of both HA1 and HA2) is more highly conserved between different influenza subtypes but is an immune-subdominant target compared to the head (Raymond et al. 2018).

6.3.1 Influenza Vaccine Strategies

Currently, influenza vaccines are available and can help prevent seasonal influenza infections (Nypaver et al. 2021). Influenza viruses undergo constant antigenic drift and shift that compromise and reduce vaccine effectiveness. Hence vaccine formulations need to be updated continuously to contain the most recent circulating viruses. All the currently approved vaccines are quadrivalent formulations that protect against four circulating influenza strains. The majority of influenza vaccines are derived from inactivated viruses (grown in eggs or mammalian cells). Recently, recombinant subunit vaccines (U.S. Food and Drug Administration 2022) have also become available. Currently available influenza vaccines have moderate efficacy of 40–60% and offer poor or no protection against heterologous strains (Jackson et al. 2017). Egg-based vaccines are the most common, because of their high yield, are inexpensive, and can be produced rapidly, but have several drawbacks. H3N2 vaccines produced in hen eggs have altered immunogenicity due to introduction of egg-adaptive mutations in HA protein which can reduce vaccine efficacy (Wu et al. 2019; Zost et al. 2017; Skowronski et al. 2014). In addition, these conventional egg-grown vaccines confer protection for relatively short periods, are vulnerable to antigenic shift and drift, and there is limited availability in many parts of the world (Wong and Webby 2013). There is a critical need for improved vaccines with higher efficiency and longer duration of protection against influenza infections (Belongia et al. 2016). Different novel vaccine modalities have been developed, and many of them are in clinical trials (Rockman et al. 2020).

6.3.1.1 HA Stem-Based Vaccines

The HA globular head is highly variable and is the immunodominant target of neutralizing antibodies. This leads to strain-specific responses, steering the immune response away from the highly conserved, immune-subdominant stem region (Kirkpatrick et al. 2018). The HA stem region is relatively conserved within influenza subtypes, compared to the globular head of HA, and antibodies elicited against this domain display cross reactivity against different HA subtypes. This suggests that the HA stem is an attractive target for developing a future universal vaccine (Kallewaard et al. 2016; Paules et al. 2017). Several studies proposed different approaches to focus on the immune response towards the conserved stem, such as use of a protein minimization approach combined with introducing destabilizing mutations for the HA2 post fusion conformation (Bommakanti et al. 2010, 2012), the design of headless well-folded stalk substructures, and introducing different stabilizing mutations and oligomerization motifs (Steel et al. 2010; Mallajosyula et al. 2013, 2014, 2015; Lu et al. 2014; Wohlbold et al. 2015; Yassine et al. 2015; Valkenburg et al. 2016; Impagliazzo et al. 2015; Wang et al. 2019). In these studies, most stem immunogens induced cross-reactive responses and provided protection against divergent strains. The results of these studies suggest that expanding the antibody response towards the conserved region of HA stem can provide a promising approach on the path of developing universal influenza vaccines. However, because responses to the HA head are more potent, albeit less broad than to the stem, it has been challenging to show the additional benefit of adding HA stem components to the existing influenza vaccine formulations.

6.3.1.2 Viral Vector-Based Vaccine

Viral vector vaccines are considered safe, because they are attenuated or deficient in replication. These viruses harbor the expression cassette for the desired antigen and upon vaccination and host cell infection, will express and deliver the antigen of interest for induction of an immune response. Some common viral vector systems used to generate influenza vaccines are based on adenovirus (Tang et al. 2002;

Holman et al. 2008; Weaver and Barry 2013) or modified vaccinia virus Ankara vectors (MVA) (Hessel et al. 2014). The results of these studies seem promising; these vaccines induced balanced titers of humoral and cellular immunity with a degree of cross reactivity and provide protection in immunized animals. One potential difficulty is elicitation of anti-vector immunity, which can potentially diminish efficacy in repeat immunizations.

6.3.1.3 HA-Based Nanoparticles

Nanoparticles provide a platform to multimerize and display repetitive pattern of the antigen, in order to focus and amplify the immune response towards that antigen. Ferritin nanoparticles typically can display 24 copies of HA monomers or 8 copies of HA trimer (Yassine et al. 2015; Georgiev et al. 2018; Kelly et al. 2020). These elicited high neutralizing antibody titers and conferred protection in vivo. A ferritin nanoparticle design is currently in clinical trials (Houser et al. 2022). The Ferritin-like protein from *M. smegmatis*, MsDps2, displaying the HA stem region was shown to be immunogenic and conferred protection in mice against homologous and heterologous challenge (Kar et al. 2022). Computationally designed two-component nanoparticles were also used to display 20 copies of the four HA proteins present in the quadrivalent vaccine strains (Boyoglu-Barnum et al. 2021). One disadvantage of the nanoparticles is the scaffold-specific titers, which are often quite substantial and may present problems upon repeated annual immunizations (Malladi et al. 2021; Kar et al. 2022).

6.3.1.4 Nucleic Acid Vaccines

DNA vaccines have many benefits: they are safe, non-infectious, and non-replicative, and have high stability and rapid production advantages over other approaches (McMillan et al. 2021). DNA vaccines encoding for HA protein (Robinson et al. 1993) or HA and NP (Robinson et al. 1997) or the matrix protein (Tompkins et al. 2007) have been developed, and one DNA vaccine candidate for H5N1-HA was tested in clinical trials (Ledgerwood et al. 2012). These studies demonstrated that DNA vaccines are robust, cheap, and stable; however immunogenicity in humans is lower than competing modalities, and special devices are required for delivery.

Recently, the mRNA vaccine platform has been extensively used to combat COVID-19 and proven to be highly effective (Jackson et al. 2020). mRNA was recently tested for use as an influenza vaccine platform (Bahl et al. 2017; McMahon et al. 2022), induced cellular and humoral immunity, and conferred protection against challenge with heterologous strains. A significant advantage of nucleic acid platforms is that no or little process development is required when a new target is encountered.

6.3.1.5 Chimeric HA and Non-HA Vaccines

Chimeric vaccines aim to direct and focus the immune response to the conserved region of the antigen. These employ sequential immunization of chimeric HAs consisting of a fixed, conserved stem domain fused to different HA heads from strains the host is naïve to. These vaccines induced high, long-lasting and cross-reactive antibody titers and protect animals in vivo (Nachbagauer et al. 2016; Krammer et al. 2013, 2014). These have been tested in Phase 1 human clinical trials, and while immunogenic, the lack of a known correlate of protection is a barrier to further clinical development (Nachbagauer et al. 2021).

While many studies focused on HA as the primary target, others have explored the use of other influenza proteins as vaccine candidates. The M2e fragment is part of the M2 protein that is exposed on virion surface and is highly conserved between influenza A subtypes, and is hence an attractive model for developing a universal vaccine (Wong and Webby 2013). Several studies showed that the M2e protein provides protection as a vaccine in animal models, either alone (Tompkins et al. 2007; Fan et al. 2004; Neirynck et al. 1999) or in combination with another influenza conserved protein, NP (Carignan et al. 2015). The NA protein was also explored as a vaccine candidate due to the fact that drift in HA and NA is uncorrelated, and elicited cross protection in immunized animals (Deroo et al. 1996; Johansson et al. 1998; Bosch et al. 2010; Strohmeier et al. 2021, 2022). However, the immunodominance of HA over NA makes it difficult to clearly demonstrate the incremental benefit of adding NA to existing influenza vaccine formulations.

6.3.2 Influenza and India

India, like many other countries, was affected by the pandemic H1N1 outbreak in 2009 and 2010, which resulted in 27,236 and 20,604 infections, and 981 and 1763 deaths in these 2 years, respectively (Bagchi 2015). Recently, in December 2020 and the beginning of 2021, outbreaks of H5N1 and H5N8 were reported in poultry in addition to a human case of H5N1 reported later in June 2021 (Potdar et al. 2022). In addition to H5, H7 subtypes (Li and Chen 2021) are also of concern though so far this has not been detected in India. Sporadic avian H9 and H10 infections have also been reported in humans, outside India (Hu et al. 2015). In addition, H1 and H3 subtypes commonly circulate in swine which are also potential reservoirs for viral recombination and sources of pandemic viruses. Influenza viruses of subtype H1 and H3 undergo rapid evolution by acquiring mutations and evading the immune system and in addition to H5, they can lead to potential pandemics in the future. As shown in Fig. 6.1, the majority of human influenza circulating strains in India are of H3 and to a lesser extent H1 and B subtypes; Fig. 6.2 shows the distribution of these subtypes in India from mid-2021 until now.

An influenza pandemic preparedness plan should focus on strengthening systematic surveillance to be able to detect novel strains as soon as they emerge and monitoring poultry and swine populations for any potential spill over. Another key component of the preparedness process is increased rates of vaccination, by implementing vaccination drives for all age groups as a measure of control and prevention. Increasing vaccine uptake rate can be achieved through increase in awareness of the risks of influenza as well as the efficacy of vaccines, and through including influenza vaccination in the government vaccination program, as well as by lowering the prices of influenza shots, thus making it affordable and accessible. Domestically manufactured vaccine formulations offer an ideal

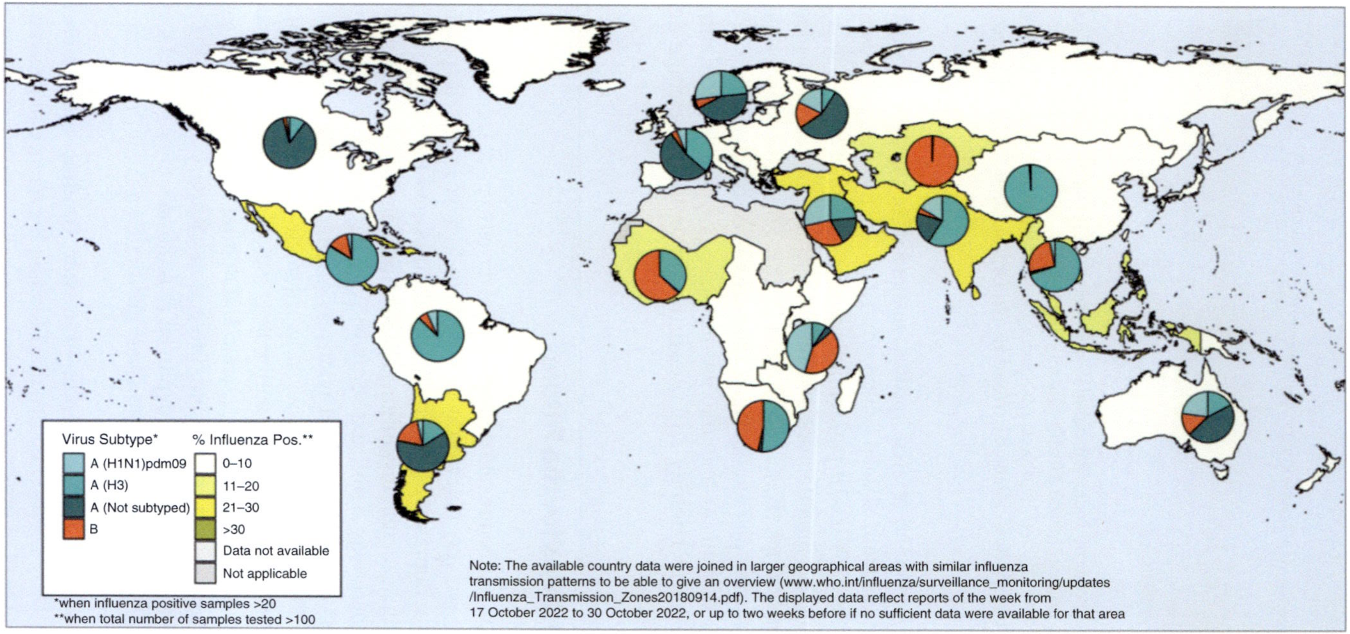

Fig. 6.1 Percentage of respiratory specimens that tested positive for influenza, by influenza transmission zone. The data reported to FluNet for the period from 17 October 2022 to 30 October 2022 from 105 countries. Map generated on 11 November 2022 (WHO 2022b). [Influenza Update N° 432]. [Geneva]: World Health Organization [2022]. Licence: CC BY-NC-SA 3.0 IGO

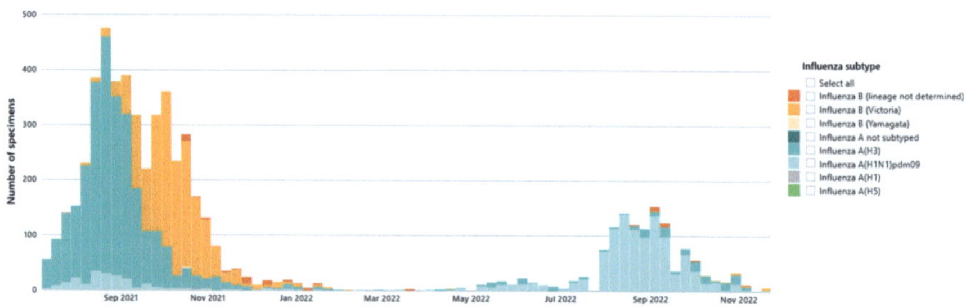

Fig. 6.2 Virus detection in India by subtype reported to FluNet between 10/07/2021 and 20/11/2022. (Image source: World Health Organization, WHO, 2022, and data source: FluNet: http://www.who.int/tools/flunet). [Influenza laboratory surveillance information, Virus detection by subtype reported to FluNet]. [Geneva]: World Health Organization [2022]. Licence: CC BY-NC-SA 3.0 IGO

path to overcome vaccine-related hurdles, instead of relying on expensive formulations imported from abroad which contribute to the delay in response to pandemics. As the COVID-19 pandemic has clearly demonstrated, it is important to have domestic manufacture and manufacturing processes in place well before a pandemic arises (Kant and Guleria 2018). The Government can facilitate this by sponsoring and encouraging the national research domain to design, and the industry domain to manufacture influenza vaccines that eliminate the reliance on foreign technology and industries, for influenza vaccine development and manufacture.

6.4 Suggestions for the Future

India's ability to manufacture and deliver vaccine doses to its citizens in a relatively short span of time is a very noteworthy achievement. Yet, we need to do much better.

Despite India's impressive manufacturing abilities and the array of expertise available in Indian academic institutions, by the time vaccines reached the bulk of the population, there was significant seropositivity, suggestive of prior exposure. It was also fortunate that we have a relatively young population with a current median age of 28.4 years (https://www.worldometers.info/world-population/), that vaccine-induced neutralizing antibodies, especially against the original SARS-CoV-2 strains, were easy to generate and that relatively low levels of neutralizing antibodies confer significant protection (Khoury et al. 2021). The Covishield vaccine that was widely deployed in India saw only limited use in the EU and none in the USA, which also likely facilitated availability within India. Below are some suggestions:

Academic research and Academia—Industry funding and interactions: Within academia, there needs to be sustained funding for science in general, so that there is a sufficient pool of expertise in diverse areas. The funding situation over the past two years has been very precarious with many sanctioned grants failing to receive funding for long periods. Vaccine projects have long timelines and sustained funding is required, especially in the case of vaccines for challenging pathogens. Current academic funding initiatives which demand translational outcomes are unlikely to succeed without significant early stage Industry participation, and some incentives for Industry. This is currently not possible in DBT-funded grants. BIRAC allows for this possibility, yet BIRAC grants are largely for late-stage innovations where Academia has a limited role; there is therefore an important gap to be filled in order to incentivize early stage Academia:Industry partnerships, which are essential to foster innovation and promote greater interactions. In addition funding and planning for potential pandemic threats needs to be much more proactive. During the COVID-19 pandemic, the first funding calls were only in April 2020, well after the begining of the pandemic. Currently, with the recent wide dispersal of avian origin H5 influenza virus into many mammalian species including dairy cattle, there is a real possibility of an H5 influenza pandemic. However, in India there has been little visible action or planning for this and no funding calls targeted to address this threat till date.

Reagents and infrastructure: For the future, it is imperative that India has the expertise to both rapidly develop and manufacture vaccines within its borders and secure all the necessary components including gene synthesis, cell growth and chromatography media, nucleotides, cell lines, adjuvants, etc. Non-human primates are a valuable resource for vaccine research. Despite this, there are only a few facilities where these are available, challenge studies are virtually impossible, and access needs to be facilitated for both Academia and Industry. There is currently very little expertise in working with and developing mRNA vaccines within India. Although the efforts by Gennova in developing an mRNA vaccine for COVID-19 are a welcome development yet, this is an important area that needs significant effort and funding.

Vaccine approvals and publication: While not directly related to pandemic preparedness, it is essential that data from preclinical and clinical studies on which vaccine approvals are based are in the public domain before vaccines are approved. At the present time, approvals for clinical trials can be given before preclinical data is in the public domain, and EUAs are given without any requirement for clinical trial data to be in the public domain. This has the potential to greatly undermine public trust and can lead to vaccine hesitancy. With the availability of fora such as biorxiv and medrxiv, there is no reason that this cannot be done.

Indian regulatory framework: The current regulatory framework is oriented largely towards "generic and biosimilars" low risk-low reward molecules/formulations for which there is existing efficacy data. These moieties are expected to be "exact copies" of innovator molecules that have surpassed their patent life and are available for production without a need to license the innovation. Such a framework is completely unsuited for innovator molecules, where the costs and the long time taken act as a significant barrier to enter clinical testing. Especially for novel vaccines where administration is limited to 1–2 immunizations with 10–50 μg doses, changes in manufacturing process between Phase 1 and Phase 3 are not expected to alter immunogenicity or safety.

There are currently no separate guidelines for innovator-developed novel molecules, and developers of novel first-in-human molecules are expected to follow the same regulatory guidelines as that of biosimilars. The system makes it especially onerous for start-ups that are making innovative molecules for the Indian and global market, and adds significantly to their cost of product development. It should be realized that such product development is high risk and high reward, and since many innovator molecules will fail in Phase 1/2 clinical trials, a simplified regulatory pathway up to these stages is essential.

In contrast to the situation in India, globally, for new molecules, the regulatory framework is broadly as follows:

1. Does not require the production of 3 consistency batches for running GLP safety toxicity studies.
2. Does not require permission from regulators to run the GLP safety toxicity studies; however such studies are required and the drug/vaccine can be produced in small laboratory-scale "shake flasks" or small-scale bioreactors.
3. GMP or GMP-like production in a shake flask or small-scale bioreactors including material generated from transient transfections and with molecules containing tags such as His-tags is accepted for running Phase 1 clinical trials.
4. In Australia, the Ethics Committees are empowered to approve trials; in the USA, an IND has to be filed with the USFDA for Phase 1 studies.
5. There is a single regulatory agency, namely the TGA in Australia or the USFDA in the USA and EMEA in Europe, unlike in India where multiple agencies are involved.

Adopting similar practices in India will ensure that Indian and international regulatory systems are harmonized; will encourage drug research and development in India and help Indian innovators generate data from India for their innovative molecules; result in the creation of significant IP within the country, and importantly create access to more efficacious drugs and pharmaceuticals for Indian citizens at much reduced costs. It will lessen the desire of the small number of Indian innovators and start-ups to seek advancement of their molecules overseas, thereby depriving the Indian market of much needed innovative molecules. The current policies of disallowing material from transient transfections, refusal to allow administration of His-tagged molecules, and the requirement for multiple batches prior to a GLP safety toxicity study add huge time and costs to development, and during pandemics, these policies lead to many vaccine-preventable deaths. In addition to these general points, a clearly articulated strategy for rapid approval of vaccines for the most vulnerable sections of the population during a pandemic is required. Similarly, the pathway for approval when naïve populations are no longer available to determine efficacy, and where a clear correlate of protection is not established, also needs to be clearly stated.

Acknowledgments We sincerely acknowledge funding from DBT-ENDFLU (BT/IN/EU-INF/15/RV/19-20) funded by European Union's Horizon 2020 research and innovation program under grant agreement No. 874650 and grants BT/NBM0010/017 and BT/COVID0093/01/20 from the Biotechnology Industry Research and Assistance Council, Government of India. RV is a JC Bose Fellow of DST and KK is supported by a Fellowship from MHRD, Government of India.

References

Ahmed S, Khan MS, Gayathri S, Singh R, Kumar S, Patel UR et al (2021) A stabilized, monomeric, receptor binding domain elicits high-titer neutralizing antibodies against all SARS-CoV-2 variants of concern. Front Immunol 12(December):765211. http://www.ncbi.nlm.nih.gov/pubmed/34956193

Andersson NW, Thiesson EM, Baum U, Pihlström N, Starrfelt J, Faksová K et al (2023a) Comparative effectiveness of bivalent BA.4-5 and BA.1 mRNA booster vaccines among adults aged ≥50 years in Nordic countries: nationwide cohort study. BMJ 382:e075286. https://www.bmj.com/content/382/bmj-2022-075286

Andersson NW, Thiesson EM, Hansen JV, Hviid A (2023b) Safety of BA.4-5 or BA.1 bivalent mRNA booster vaccines: nationwide cohort study. BMJ 382:e075015. https://www.bmj.com/content/382/bmj-2023-075015

Arbel R, Peretz A, Sergienko R, Friger M, Beckenstein T, Duskin-Bitan H et al (2023) Effectiveness of a bivalent mRNA vaccine booster dose to prevent severe COVID-19 outcomes: a retrospective cohort study. Lancet Infect Dis 23(8):914–921. /pmc/articles/PMC10156150/

AstraZeneca (2020) AstraZeneca takes next steps towards broad and equitable access to Oxford University's potential COVID-19 vaccine (Published: 4 June 2020) [Internet]. [cited 2022 Dec 3]. https://www.astrazeneca.com/media-centre/press-releases/2020/astrazeneca-takes-next-steps-towards-broad-and-equitable-access-to-oxford-universitys-covid-19-vaccine.html#!

Bagchi S (2015) India tackles H1N1 influenza outbreak. Lancet 385(9972):e21. https://doi.org/10.1016/S0140-6736(15)60501-3

Bahl K, Senn JJ, Yuzhakov O, Bulychev A, Brito LA, Hassett KJ et al (2017) Preclinical and clinical demonstration of immunogenicity by mRNA vaccines against H10N8 and H7N9 influenza viruses. Mol Ther 25(6):1316–1327. https://doi.org/10.1016/j.ymthe.2017.03.035

Barda N, Canetti M, Gilboa M, Indenboim V, Asraf K, Weiss-Ottolenghi Y et al (2022) Comparing immunogenicity and efficacy of two different mRNA-based COVID-19 vaccines as a fourth dose; six-month follow-up, Israel, 27 December 2021 to 24 July 2022. Eurosurveillance 27(39):2200701. /pmc/articles/PMC9524052/

Belongia EA, Simpson MD, King JP, Sundaram ME, Kelley NS, Osterholm MT et al (2016) Variable influenza vaccine effectiveness by subtype: a systematic review and meta-analysis of test-negative design studies. Lancet Infect Dis 16(8):942–951. https://doi.org/10.1016/S1473-3099(16)00129-8

Bhatnagar T, Chaudhuri S, Ponnaiah M, Yadav PD, Sabarinathan R, Sahay RR et al (2022) Effectiveness of BBV152/Covaxin and AZD1222/Covishield vaccines against severe COVID-19 and B.1.617.2/Delta variant in India, 2021: a multi-centric hospital-based case-control study. Int J Infect Dis 122(January):693–702. http://www.ncbi.nlm.nih.gov/pubmed/35843496

Blut A, Bewertung U, Krankheitserreger B (2009) Influenza virus. Transfus Med Hemotherapy 36(1):32–39. https://www.karger.com/Article/FullText/197314

Bommakanti G, Citron MP, Hepler RW, Callahan C, Heidecker GJ, Najar TA et al (2010) Design of an HA2-based *Escherichia coli* expressed influenza immunogen that protects mice from pathogenic challenge. Proc Natl Acad Sci 107(31):13701–13706. https://doi.org/10.1073/pnas.1007465107

Bommakanti G, Lu X, Citron MP, Najar TA, Heidecker GJ, ter Meulen J et al (2012) Design of Escherichia coli-expressed stalk domain immunogens of H1N1 hemagglutinin that protect mice from lethal challenge. J Virol 86(24):13434–13444

Bosch BJ, Bodewes R, de Vries RP, Kreijtz JHCM, Bartelink W, van Amerongen G et al (2010) Recombinant soluble, multimeric HA and NA exhibit distinctive types of protection against pandemic swine-origin 2009 a(H1N1) influenza virus infection in ferrets. J Virol 84(19):10366–10374

Boyoglu-Barnum S, Ellis D, Gillespie RA, Hutchinson GB, Park YJ, Moin SM et al (2021) Quadrivalent influenza nanoparticle vaccines induce broad protection. Nature 592(7855):623–628. https://doi.org/10.1038/s41586-021-03365-x

Carignan D, Thérien A, Rioux G, Paquet G, Gagné MÈL, Bolduc M et al (2015) Engineering of the PapMV vaccine platform with a shortened M2e peptide leads to an effective one dose influenza vaccine. Vaccine 33(51):7245–7253. https://doi.org/10.1016/j.vaccine.2015.10.123

Central Drugs Standard Control Organization (2022) Approved COVID-19 vaccines as on 04.10.2022. https://cdsco.gov.in/opencms/opencms/system/modules/CDSCO.WEB/elements/download_file_division.jsp?num_id=OTA4MQ==

Chandra S, Kassens-Noor E (2014) The evolution of pandemic influenza: evidence from India, 1918-19. BMC Infect Dis 14(1):1–10

Chen WH, Pollet J, Strych U, Lee J, Liu Z, Kundu RT et al (2022) Yeast-expressed recombinant SARS-CoV-2 receptor binding domain RBD203-N1 as a COVID-19 protein vaccine candidate. Protein Expr Purif 190(January):106003

Deroo T, Min Jou W, Fiers W (1996) Recombinant neuraminidase vaccine protects against lethal influenza. Vaccine 14(6):561–569

Dey A, Chozhavel Rajanathan TM, Chandra H, Pericherla HPR, Kumar S, Choonia HS et al (2021) Immunogenic potential of DNA vaccine candidate, ZyCoV-D against SARS-CoV-2 in animal models. Vaccine 39(30):4108–4116

Dinnon KH, Leist SR, Schäfer A, Edwards CE, Martinez DR, Montgomery SA et al (2020) A mouse-adapted model of SARS-CoV-2 to test COVID-19 countermeasures. Nature [Internet]. 586(7830):560–566. /pmc/articles/PMC8034761/

Dunkle LM, Kotloff KL, Gay CL, Áñez G, Adelglass JM, Barrat Hernández AQ et al (2022) Efficacy and safety of NVX-CoV2373 in adults in the United States and Mexico. N Engl J Med 386(6):531–543

Ella R, Vadrevu KM, Jogdand H, Prasad S, Reddy S, Sarangi V et al (2021a) Safety and immunogenicity of an inactivated SARS-CoV-2 vaccine, BBV152: a double-blind, randomised, phase 1 trial. Lancet Infect Dis 21(5):637–646. https://doi.org/10.1016/S1473-3099(20)30942-7

Ella R, Reddy S, Jogdand H, Sarangi V, Ganneru B, Prasad S et al (2021b) Safety and immunogenicity of an inactivated SARS-CoV-2 vaccine, BBV152: interim results from a double-blind, randomised, multicentre, phase 2 trial, and 3-month follow-up of a double-blind, randomised phase 1 trial. Lancet Infect Dis 21(7):950–961. https://doi.org/10.1016/S1473-3099(21)00070-0

Ella R, Reddy S, Blackwelder W, Potdar V, Yadav P, Sarangi V et al (2021c) Efficacy, safety, and lot-to-lot immunogenicity of an inactivated SARS-CoV-2 vaccine (BBV152): interim results of a randomised, double-blind, controlled, phase 3 trial. Lancet 398(10317):2173–2184. https://linkinghub.elsevier.com/retrieve/pii/S0140673621020006

Fan J, Liang X, Horton MS, Perry HC, Citron MP, Heidecker GJ et al (2004) Preclinical study of influenza virus a M2 peptide conjugate vaccines in mice, ferrets, and rhesus monkeys. Vaccine 22(23–24):2993–3003

Folegatti PM, Ewer KJ, Aley PK, Angus B, Becker S, Belij-Rammerstorfer S et al (2020) Safety and immunogenicity of the ChAdOx1 nCoV-19 vaccine against SARS-CoV-2: a preliminary report of a phase 1/2, single-blind, randomised controlled trial. Lancet (London, England) [Internet]. 396(10249):467–478. http://www.ncbi.nlm.nih.gov/pubmed/32702298

Ganneru B, Jogdand H, Daram VK, Das D, Molugu NR, Prasad SD et al (2021) Th1 skewed immune response of whole virion inactivated SARS CoV 2 vaccine and its safety evaluation. iScience 24(4):102298. https://doi.org/10.1016/j.isci.2021.102298

Georgiev IS, Joyce MG, Chen RE, Leung K, McKee K, Druz A et al (2018) Two-component ferritin nanoparticles for multimerization of diverse trimeric antigens. ACS Infect Dis. 4(5):788–796

Gu H, Chen Q, Yang G, He L, Fan H, Deng Y-Q et al (2020) Adaptation of SARS-CoV-2 in BALB/c mice for testing vaccine efficacy. Science 369(6511):1603–1607. /pmc/articles/PMC7574913/

Hessel A, Savidis-Dacho II, Coulibaly S, Portsmouth D, Kreil TR, Crowe BA et al (2014) MVA vectors expressing conserved influenza proteins protect mice against lethal challenge with H5N1, H9N2 and H7N1 viruses. PLoS One 9(2):e88340

Holman DH, Wang D, Raja NU, Luo M, Moore KM, Woraratanadharm J et al (2008) Multi-antigen vaccines based on complex adenovirus vectors induce protective immune responses against H5N1 avian influenza viruses. Vaccine 26(21):2627–2639

Houser KV, Chen GL, Carter C, Crank MC, Nguyen TA, Burgos Florez MC et al (2022) Safety and immunogenicity of a ferritin nanoparticle H2 influenza vaccine in healthy adults: a phase 1 trial. Nat Med 28(2):383–391

Hu M, Li X, Ni X, Wu J, Gao R, Xia W et al (2015) Coexistence of avian influenza virus H10 and H9 subtypes among chickens in live poultry markets during an outbreak of infection with a novel H10N8 virus in humans in Nanchang, China. Jpn J Infect Dis 68(5):364–369

Huang C, Wang Y, Li X, Ren L, Zhao J, Hu Y et al (2020) Clinical features of patients infected with 2019 novel coronavirus in Wuhan, China. Lancet (London, England) 395(10223):497–506. http://www.ncbi.nlm.nih.gov/pubmed/31986264

Huang K, Zhang Y, Hui X, Zhao Y, Gong W, Wang T et al (2021) Q493K and Q498H substitutions in Spike promote adaptation of SARS-CoV-2 in mice. EBioMedicine 67:103381. /pmc/articles/PMC8118724/

Imai M, Iwatsuki-Horimoto K, Hatta M, Loeber S, Halfmann PJ, Nakajima N et al (2020) Syrian hamsters as a small animal model for SARS-CoV-2 infection and countermeasure development. Proc Natl Acad Sci U S A 117(28):16587–16595. /pmc/articles/PMC7368255/

Impagliazzo A, Milder F, Kuipers H, Wagner MV, Zhu X, Hoffman RMB et al (2015) A stable trimeric influenza hemagglutinin stem as a broadly protective immunogen. Science 349(6254):1301–1306. https://doi.org/10.1126/science.aac7263

Iuliano AD, Roguski KM, Chang HH, Muscatello DJ, Palekar R, Tempia S et al (2018) Estimates of global seasonal influenza-associated respiratory mortality: a modelling study. Lancet 391(10127):1285–1300

Jackson ML, Chung JR, Jackson LA, Phillips CH, Benoit J, Monto AS et al (2017) Influenza Vaccine Effectiveness in the United States during the 2015–2016 Season. N Engl J Med 377(6):534. https://doi.org/10.1056/NEJMoa1700153

Jackson LA, Anderson EJ, Rouphael NG, Roberts PC, Makhene M, Coler RN et al (2020) An mRNA vaccine against SARS-CoV-2—preliminary report. N Engl J Med 383(20):1920–1931

Joe CCD, Jiang J, Linke T, Li Y, Fedosyuk S, Gupta G et al (2022) Manufacturing a chimpanzee adenovirus-vectored SARS-CoV-2 vaccine to meet global needs. Biotechnol Bioeng 119(1):48–58. https://doi.org/10.1002/bit.27945

Johansson BE, Matthews JT, Kilbourne ED (1998) Supplementation of conventional influenza a vaccine with purified viral neuraminidase results in a balanced and broadened immune response. Vaccine 16(9–10):1009–1015

Kallewaard NL, Corti D, Collins PJ, Neu U, McAuliffe JM, Benjamin E et al (2016) Structure and function analysis of an antibody recognizing all influenza a subtypes. Cell 166(3):596–608. https://doi.org/10.1016/j.cell.2016.05.073

Kant L, Guleria R (2018) Pandemic flu, 1918: after hundred years, India is as vulnerable. Indian J Med Res 147(March):221–224. http://www.ncbi.nlm.nih.gov/pubmed/23144490

Kar U, Khaleeq S, Garg P, Bhat M, Reddy P, Vignesh VS et al (2022) Comparative immunogenicity of bacterially expressed soluble trimers and nanoparticle displayed influenza hemagglutinin stem immunogens. Front Immunol 13(May):1–16

Keech C, Albert G, Cho I, Robertson A, Reed P, Neal S et al (2020) Phase 1–2 trial of a SARS-CoV-2 recombinant spike protein nanoparticle vaccine. N Engl J Med 383(24):2320–2332

Kelly HG, Tan HX, Juno JA, Esterbauer R, Ju Y, Jiang W et al (2020) Self-assembling influenza nanoparticle vaccines drive extended germinal center activity and memory B cell maturation. JCI Insight 5(10):1–15

Khobragade A, Bhate S, Ramaiah V, Deshpande S, Giri K, Phophle H et al (2022) Efficacy, safety, and immunogenicity of the DNA SARS-CoV-2 vaccine (ZyCoV-D): the interim efficacy results of a phase 3, randomised, double-blind, placebo-controlled study in India. Lancet 399(10332):1313–1321. http://www.ncbi.nlm.nih.gov/pubmed/35367003

Khoury DS, Cromer D, Reynaldi A, Schlub TE, Wheatley AK, Juno JA et al (2021) Neutralizing antibody levels are highly predictive of immune protection from symptomatic SARS-CoV-2 infection. Nat Med 27(7):1205–1211. https://doi.org/10.1038/s41591-021-01377-8

Kilbourne ED (2006) Influenza pandemics of the 20th century. Emerg Infect Dis 12(1):9–14. http://wwwnc.cdc.gov/eid/article/12/1/05-1254_article.htm

Kirkpatrick E, Qiu X, Wilson PC, Bahl J, Krammer F (2018) The influenza virus hemagglutinin head evolves faster than the stalk domain. Sci Rep 8(1):1–14

Kirsebom FCM, Andrews N, Stowe J, Ramsay M, Lopez BJ (2023) Duration of protection of ancestral-strain monovalent vaccines and effectiveness of bivalent BA.1 boosters against COVID-19 hospitalisation in England: a test-negative case-control study. Lancet Infect Dis 23:1235. http://www.thelancet.com/article/S1473309923003651/fulltext

Kislaya I, Machado A, Magalhães S, Rodrigues AP, Franco R, Leite PP et al (2022) COVID-19 mRNA vaccine effectiveness (second and first booster dose) against hospitalisation and death during omicron BA.5 circulation: cohort study based on electronic health records, Portugal, May to July 2022. Eurosurveillance 27(37):2200697. /pmc/articles/PMC9479470/

Knoll MD, Wonodi C (2021) Oxford-AstraZeneca COVID-19 vaccine efficacy. Lancet (London, England) 397(10269):72–74. http://www.ncbi.nlm.nih.gov/pubmed/33306990

Krammer F, Pica N, Hai R, Margine I, Palese P (2013) Chimeric hemagglutinin influenza virus vaccine constructs elicit broadly protective stalk-specific antibodies. J Virol 87(12):6542–6550

Krammer F, Margine I, Hai R, Flood A, Hirsh A, Tsvetnitsky V et al (2014) H3 stalk-based chimeric hemagglutinin influenza virus constructs protect mice from H7N9 challenge. J Virol 88(4):2340–2343

Krammer F, Smith GJD, Fouchier RAM, Peiris M, Kedzierska K, Doherty PC et al (2018) Influenza. Nat Rev Dis Prim 4(1):3. https://www.nature.com/articles/s41572-018-0002-y#citeas

Kulkarni PS, Padmapriyadarsini C, Vekemans J, Bavdekar A, Gupta M, Kulkarni P et al (2021) A phase 2/3, participant-blind, observer-blind, randomised, controlled study to assess the safety and immunogenicity of SII-ChAdOx1 nCoV-19 (COVID-19 vaccine) in adults in India. eClinicalMedicine [Internet] 42:101218. https://doi.org/10.1016/j.eclinm.2021.101218

Ledgerwood JE, Hu Z, Gordon IJ, Yamshchikov G, Enama ME, Plummer S et al (2012) Influenza virus H5 DNA vaccination is immunogenic by intramuscular and intradermal routes in humans. Clin Vaccine Immunol 19(11):1792–1797

Li C, Chen H (2021) H7n9 influenza virus in China. Cold Spring Harb Perspect Med 11(8):1–18

Li C, Ye Z, Zhang AJX, Chan JFW, Song W, Liu F et al (2022) Severe acute respiratory syndrome coronavirus 2 (SARS-CoV-2) infection by intranasal or Intratesticular route induces testicular damage. Clin Infect Dis 75(1):e974–e990. /pmc/articles/PMC8903466/

Lin D-Y, Xu Y, Gu Y, Zeng D, Wheeler B, Young H et al (2023) Effectiveness of bivalent boosters against severe omicron infection. N Engl J Med 388(8):764–766. /pmc/articles/PMC9933929/

Link-Gelles R, Ciesla AA, Roper LE, Scobie HM, Ali AR, Miller JD et al (2023) Early Estimates of Bivalent mRNA booster dose vaccine effectiveness in preventing symptomatic SARS-CoV-2 infection attributable to omicron BA.5– and XBB/XBB.1.5–related sublineages among immunocompetent adults—increasing community access to testing program. MMWR Morb Mortal Wkly Rep 72(5):119–124. https://www.cdc.gov/mmwr/volumes/72/wr/mm7205e1.htm

Logunov DY, Dolzhikova IV, Zubkova OV, Tukhvatullin AI, Shcheblyakov DV, Dzharullaeva AS et al (2020) Safety and immunogenicity of an rAd26 and rAd5 vector-based heterologous prime-boost COVID-19 vaccine in two formulations: two open, non-randomised phase 1/2 studies from Russia. Lancet 396(10255):887–897

Logunov DY, Dolzhikova IV, Shcheblyakov DV, Tukhvatulin AI, Zubkova OV, Dzharullaeva AS et al (2021) Safety and efficacy of an rAd26 and rAd5 vector-based heterologous prime-boost COVID-19 vaccine: an interim analysis of a randomised controlled phase 3 trial in Russia. Lancet (London, England) 397(10275):671–681. http://www.ncbi.nlm.nih.gov/pubmed/33545094

Lopez Bernal J, Andrews N, Gower C, Gallagher E, Simmons R, Thelwall S et al (2021) Effectiveness of covid-19 vaccines against the B.1.617.2 (Delta) variant. N Engl J Med 385(7):585–594

Lu Y, Welsh JP, Swartz JR (2014) Production and stabilization of the trimeric influenza hemagglutinin stem domain for potentially broadly protective influenza vaccines. Proc Natl Acad Sci USA 111(1):125–130

Lycett SJ, Duchatel F, Digard P (2019) A brief history of bird flu. Philos Trans R Soc B Biol Sci 374(1775):0–3

Madhavan M, Ritchie AJ, Aboagye J, Jenkin D, Provstgaad-Morys S, Tarbet I et al (2022) Tolerability and immunogenicity of an intranasally-administered adenovirus-vectored COVID-19 vaccine: an open-label partially-randomised ascending dose phase I trial. eBioMedicine 85:104298. https://doi.org/10.1016/j.ebiom.2022.104298

Malladi SK, Patel UR, Rajmani RS, Singh R, Pandey S, Kumar S et al (2021) Immunogenicity and protective efficacy of a highly thermotolerant, trimeric SARS-CoV-2 receptor binding domain derivative. ACS Infect Dis 7(8):2546–2564

Mallajosyula VVA, Citron M, Lu X, Ter MJ, Varadarajan R, Liang X (2013) In vitro and in vivo characterization of designed immunogens derived from the CD-helix of the stem of influenza hemagglutinin. Proteins 81(10):1759–1775

Mallajosyula VVA, Citron M, Ferrara F, Lu X, Callahan C, Heidecker GJ et al (2014) Influenza hemagglutinin stem-fragment immunogen elicits broadly neutralizing antibodies and confers heterologous protection. Proc Natl Acad Sci USA 111(25):E2514

Mallajosyula VVA, Citron M, Ferrara F, Temperton NJ, Liang X, Flynn JA et al (2015) Hemagglutinin sequence conservation guided stem immunogen design from influenza a H3 subtype. Front Immunol 6(JUN):1–12

McConeghy KW, White EM, Blackman C, Santostefano CM, Lee Y, Rudolph JL et al (2022) Effectiveness of a second COVID-19 vaccine booster dose against infection, hospitalization, or death among nursing home residents—19 States, March 29–July 25, 2022. MMWR Morb Mortal Wkly Rep 71(39):1235–1238. /pmc/articles/PMC9533729/

McMahon M, O'Dell G, Tan J, Sárközy A, Vadovics M, Carreño JM et al (2022) Assessment of a quadrivalent nucleoside-modified mRNA vaccine that protects against group 2 influenza viruses. Proc Natl Acad Sci USA 119(45):1–11

McMillan CLD, Young PR, Watterson D, Chappell KJ (2021) The next generation of influenza vaccines: towards a universal solution. Vaccine 9(1):1–20

Mittal N, Kumar S, Rajmani RS, Singh R, Lemoine C, Jakob V et al (2023) Enhanced protective efficacy of a novel, thermostable, RBD-S2 vaccine formulation against SARS-CoV-2 and its variants. bioRxiv 2023:03.19.533338. https://doi.org/10.1101/2023.03.19.533338v2

Mohandas S, Yadav PD, Shete-Aich A, Abraham P, Vadrevu KM, Sapkal G et al (2021) Immunogenicity and protective efficacy of BBV152, whole virion inactivated SARS-CoV-2 vaccine candidates in the Syrian hamster model. iScience 24(2):102054. https://doi.org/10.1016/j.isci.2021.102054

Momin T, Kansagra K, Patel H, Sharma S, Sharma B, Patel J et al (2021) Safety and Immunogenicity of a DNA SARS-CoV-2 vaccine (ZyCoV-D): results of an open-label, non-randomized phase I part of phase I/II clinical study by intradermal route in healthy subjects in India. eClinicalMedicine 38:101020. https://linkinghub.elsevier.com/retrieve/pii/S258953702100300X

Moreau GB, Burgess SL, Sturek JM, Donlan AN, Petri WA, Mann BJ (2020) Evaluation of K18-hACE2 mice as a model of SARS-CoV-2 infection. Am J Trop Med Hyg 103(3):1215–1219. /pmc/articles/PMC7470527/

Nachbagauer R, Kinzler D, Choi A, Hirsh A, Beaulieu E, Lecrenier N et al (2016) A chimeric haemagglutinin-based influenza split virion vaccine adjuvanted with AS03 induces protective stalk-reactive antibodies in mice. NPJ Vaccines 1(July):1–10. https://doi.org/10.1038/npjvaccines.2016.15

Nachbagauer R, Feser J, Naficy A, Bernstein DI, Guptill J, Walter EB et al (2021) A chimeric hemagglutinin-based universal influenza virus vaccine approach induces broad and long-lasting immunity in a randomized, placebo-controlled phase I trial. Nat Med [Internet]. 27(1):106–114. https://doi.org/10.1038/s41591-020-1118-7

Narayan VV, Iuliano AD, Roguski K, Bhardwaj R, Chadha M, Saha S et al (2020) Burden of influenza-associated respiratory and circulatory mortality in India, 2010-2013. J Glob Health 10(1):1–11

Neirynck S, Deroo T, Saelens X, Vanlandschoot P, Jou WM, Fiers W (1999) A universal influenza a vaccine based on the extracellular domain of the M2 protein. Nat Med 5(10):1157–1163

Nypaver C, Dehlinger C, Carter C (2021) Influenza and influenza vaccine: a review. J Midwifery Women's Heal 66(1):45–53

Patel N, Trost JF, Guebre-Xabier M, Zhou H, Norton J, Jiang D et al (2023) XBB.1.5 spike protein COVID-19 vaccine induces broadly neutralizing and cellular immune responses against EG.5.1 and emerging XBB variants. bioRxiv 2023:08.30.554497. https://doi.org/10.1101/2023.08.30.554497v1

Paules CI, Lakdawala S, McAuliffe JM, Paskel M, Vogel L, Kallewaard NL et al (2017) The hemagglutinin a stem antibody MEDI8852 prevents and controls disease and limits transmission of pandemic influenza viruses. J Infect Dis 216(3):356–365

Petrie JG, King JP, McClure DL, Rolfes MA, Meece JK, Pattinson D et al (2023) Effectiveness of first and second COVID-19 mRNA vaccine monovalent booster doses during a period of circulation of Omicron variant sublineages: December 2021–July 2022. Influenza Other Respi Viruses 17(3):e13104. /pmc/articles/PMC9975792/

Piret J, Boivin G (2021) Pandemics throughout history. Front Microbiol 11(January):631736. https://doi.org/10.3389/fmicb.2020.631736/full

Pollet J, Chen WH, Versteeg L, Keegan B, Zhan B, Wei J et al (2021) SARS-CoV-2 RBD219-N1C1: a yeast-expressed SARS-CoV-2 recombinant receptor-binding domain candidate vaccine stimulates virus neutralizing antibodies and T-cell immunity in mice. Hum Vaccines Immunother 17(8):2356–2366. https://doi.org/10.1080/21645515.2021.1901545

Potdar V, Brijwal M, Lodha R, Yadav P, Jadhav S, Choudhary ML et al (2022) Identification of human case of avian influenza a(H5N1) infection, India. Emerg Infect Dis 28(6):1269–1273

Ram Purakayastha D, Vishnubhatla S, Rai SK, Broor S, Krishnan A (2018) Estimation of burden of influenza among under-five children in India: a meta-analysis. J Trop Pediatr 64(5):441–453

Raymond DD, Bajic G, Ferdman J, Suphaphiphat P, Settembre EC, Moody MA et al (2018) Conserved epitope on influenza-virus hemagglutinin head defined by a vaccine-induced antibody. Proc Natl Acad Sci USA 115(1):168–173

Rizvi AA, Singh A (2022) Vaccination coverage among older adults: a population-based study in India. Bull World Health Organ 100(6):375–384

Rizvi ZA, Dalal R, Sadhu S, Binayke A, Dandotiya J, Kumar Y et al (2022) Golden Syrian hamster as a model to study cardiovascular complications associated with SARS-CoV-2 infection. Elife 11:e73522. https://pubmed.ncbi.nlm.nih.gov/35014610/

Roberts DL, Rossman JS, Jarić I (2021) Dating first cases of COVID-19. PLoS Pathog 17(6):1–10

Robinson HL, Hunt LA, Webster RG (1993) Protection against a lethal influenza virus challenge by immunization with a haemagglutinin-expressing plasmid DNA. Vaccine 11(9):957–960

Robinson HL, Boyle CA, Feltquate DM, Morin MJ, Santoro JC, Webster RG (1997) DNA immunization for influenza virus: studies using hemagglutinin-and nucleoprotein-expressing DNAs. J Infect Dis 176(SUPPL. 1):50–55

Rockman S, Laurie KL, Parkes S, Wheatley A, Barr IG (2020) New technologies for influenza vaccines. Microorganisms 8(11):1–20

Sadoff J, Gray G, Vandebosch A, Cárdenas V, Shukarev G, Grinsztejn B et al (2022) Final analysis of efficacy and safety of single-dose Ad26.COV2.S. N Engl J Med 386(9):847–860

Saraf A, Gurjar R, Kaviraj S, Kulkarni A, Kumar D, Kulkarni R, et al (2024) An Omicron-specific, self-amplifying mRNA booster vaccine for COVID-19: a phase 2/3 randomized trial. Nat Med 30(5):1363–1372. https://doi.org/10.1038/s41591-024-02955-2

Sia SF, Yan L-M, Chin AWH, Fung K, Choy K-T, Wong AYL et al (2020) Pathogenesis and transmission of SARS-CoV-2 in golden hamsters. Nature 583(7818):834–838. /pmc/articles/PMC7394720/

Singh AK, Phatak SR, Singh R, Bhattacharjee K, Singh NK, Gupta A et al (2021) Antibody response after first and second-dose of ChAdOx1-nCOV (CovishieldTM®) and BBV-152 (CovaxinTM®) among health care workers in India: the final results of cross-sectional coronavirus vaccine-induced antibody titre (COVAT) study. Vaccine [Internet]. 39(44):6492–6509. http://www.ncbi.nlm.nih.gov/pubmed/34600747

Skowronski DM, Janjua NZ, De Serres G, Sabaiduc S, Eshaghi A, Dickinson JA et al (2014) Low 2012–13 influenza vaccine effectiveness associated with mutation in the egg-adapted H3N2 vaccine strain not antigenic drift in circulating viruses. PLoS One 9(3):e92153. https://doi.org/10.1371/journal.pone.0092153

Steel J, Lowen AC, Wang TT, Yondola M, Gao Q, Haye K et al (2010) Influenza virus vaccine based on the conserved hemagglutinin stalk domain. mBio 1(1):e00018. https://doi.org/10.1128/mBio.00018-10

Strohmeier S, Amanat F, Zhu X, McMahon M, Deming ME, Pasetti MF et al (2021) A novel recombinant influenza virus neuraminidase vaccine candidate stabilized by a measles virus phosphoprotein Tetramerization domain provides robust protection from virus challenge in the mouse model. MBio 12(6):e0224121

Strohmeier S, Amanat F, Campbell JD, Traquina P, Coffman RL, Krammer F (2022) A CpG 1018 adjuvanted neuraminidase vaccine provides robust protection from influenza virus challenge in mice. NPJ Vaccines 7(1):1–13

Surie D, DeCuir J, Zhu Y, Gaglani M, Ginde AA, Douin DJ et al (2022) Early estimates of bivalent mRNA vaccine effectiveness in preventing COVID-19–Associated Hospitalization Among Immunocompetent Adults Aged ≥65 Years — IVY Network, 18 States, September 8–November 30, 2022. MMWR Morb Mortal Wkly Rep 71(5152):1625–1630. /pmc/articles/PMC9812444/

Tang M, Harp JA, Wesley RD (2002) Recombinant adenovirus encoding the HA gene from swine H3N2 influenza virus partially protects mice from challenge with heterologous virus: a/HK/1/68 (H3N2). Arch Virol 147(11):2125–2141

Tenforde MW, Weber ZA, Natarajan K, Klein NP, Kharbanda AB, Stenehjem E et al (2022) Early estimates of bivalent mRNA vaccine effectiveness in preventing COVID-19–Associated Emergency Department or Urgent Care Encounters and Hospitalizations Among Immunocompetent Adults—VISION network, Nine States, September–November 2022. Morb Mortal Wkly Rep 71(5152):1616. /pmc/articles/PMC9812442/

Thiruvengadam R, Awasthi A, Medigeshi G, Bhattacharya S, Mani S, Sivasubbu S et al (2022) Effectiveness of ChAdOx1 nCoV-19 vaccine against SARS-CoV-2 infection during the delta (B.1.617.2) variant surge in India: a test-negative, case-control study and a mechanistic study of post-vaccination immune responses. Lancet Infect Dis 22(4):473–482. http://www.ncbi.nlm.nih.gov/pubmed/34838183

Thuluva S, Paradkar V, Gunneri SR, Yerroju V, Mogulla R, Turaga K et al (2022a) Evaluation of safety and immunogenicity of receptor-binding domain-based COVID-19 vaccine (Corbevax) to select the optimum formulation in open-label, multicentre, and randomised phase-1/2 and phase-2 clinical trials. eBioMedicine 83:104217. https://doi.org/10.1016/j.ebiom.2022.104217

Thuluva S, Paradkar V, Gunneri SR, Yerroju V, Mogulla RR, Suneetha PV et al (2022b) Safety, tolerability and immunogenicity of biological E's CORBEVAX™ vaccine in children and adolescents: a prospective, randomised, double-blind, placebo controlled, phase-2/3 study. Vaccine 40(49):7130–7140. https://doi.org/10.1016/j.vaccine.2022.10.045

Tian JH, Patel N, Haupt R, Zhou H, Weston S, Hammond H et al (2021) SARS-CoV-2 spike glycoprotein vaccine candidate NVX-CoV2373 immunogenicity in baboons and protection in mice. Nat Commun 12(1):372

Tompkins SM, Zhao ZS, Lo CY, Misplon JA, Liu T, Ye Z et al (2007) Matrix protein 2 vaccination and protection against influenza viruses, including subtype H5N1. Emerg Infect Dis 13(3):426–435

U.S. Food and Drug Administration (2022) Influenza Vaccine for the 2022–2023 Season. [cited 2022 Nov 17]. https://www.fda.gov/vaccines-blood-biologics/lot-release/influenza-vaccine-2022-2023-season

Valkenburg SA, Mallajosyula VVA, Li OTW, Chin AWH, Carnell G, Temperton N et al (2016) Stalking influenza by vaccination with pre-fusion headless HA mini-stem. Sci Rep 6:1–11

van Doremalen N, Lambe T, Spencer A, Belij-Rammerstorfer S, Purushotham JN, Port JR et al (2020) ChAdOx1 nCoV-19 vaccine prevents SARS-CoV-2 pneumonia in rhesus macaques. Nature 586(7830):578–582. http://www.ncbi.nlm.nih.gov/pubmed/32731258

Voysey M, Clemens SAC, Madhi SA, Weckx LY, Folegatti PM, Aley PK et al (2021) Safety and efficacy of the ChAdOx1 nCoV-19 vaccine (AZD1222) against SARS-CoV-2: an interim analysis of four randomised controlled trials in Brazil, South Africa, and the UK. Lancet 397(10269):99–111

Wang SC, Liao HY, Zhang JY, Cheng TJR, Wong CH (2019) Development of a universal influenza vaccine using hemagglutinin stem protein produced from Pichia pastoris. Virology 526:125–137. https://doi.org/10.1016/j.virol.2018.10.005

Wang J, Shuai L, Wang C, Liu R, He X, Zhang X et al (2020) Mouse-adapted SARS-CoV-2 replicates efficiently in the upper and lower respiratory tract of BALB/c and C57BL/6J mice. Protein Cell 11(10):776–782. /pmc/articles/PMC7401472/

Weaver EA, Barry MA (2013) Low Seroprevalent species D adenovirus vectors as influenza vaccines. PLoS One 8(8):1–14

WHO (2022a) TAG-VE statement on Omicron sublineages BQ.1 and XBB. https://www.who.int/news/item/27-10-2022-tag-ve-statement-on-omicron-sublineages-bq.1-and-xbb

WHO (2022b) Influenza Update N° 432 [Internet] [cited 2022 Dec 2]. https://cdn.who.int/media/docs/default-source/influenza/influenza-updates/2022/2022_11_14_surveillance_update_432.pdf?sfvrsn=19cda949_1&download=true

Wohlbold TJ, Nachbagauer R, Margine I, Tan GS, Hirsh A, Krammer F (2015) Vaccination with soluble headless hemagglutinin protects mice from challenge with divergent influenza viruses. Vaccine 33(29):3314–3321

Wong SS, Webby RJ (2013) Traditional and new influenza vaccines. Clin Microbiol Rev 26(3):476–492

World Health Organization (2023) Statement on the antigen composition of COVID-19 vaccines. WHO, Geneva. https://www.who.int/news/item/18-05-2023-statement-on-the-antigen-composition-of-covid-19-vaccines

Wu NC, Wilson IA (2020) Structural biology of influenza hemagglutinin: an amaranthine adventure. Viruses 12(9):1–18. https://www.ncbi.nlm.nih.gov/pmc/articles/PMC7551194/pdf/viruses-12-01053.pdf

Wu NC, Lv H, Thompson AJ, Paulson JC, Mok CKP, Wilson IA et al (2019) Preventing an antigenically disruptive mutation in egg-based H3N2 seasonal influenza vaccines by mutational incompatibility article preventing an antigenically disruptive mutation in egg-based H3N2 seasonal influenza vaccines by mutational incompatibility. Cell Host Microbe 25(6):836–844.e5. https://doi.org/10.1016/j.chom.2019.04.013

Xu K, Gao P, Liu S, Lu S, Lei W, Zheng T et al (2022) Protective prototype-beta and delta-omicron chimeric RBD-dimer vaccines against SARS-CoV-2. Cell 185(13):2265–2278.e14. https://linkinghub.elsevier.com/retrieve/pii/S009286742200527X

Yadav PD, Ella R, Kumar S, Patil DR, Mohandas S, Shete AM et al (2021) Immunogenicity and protective efficacy of inactivated SARS-CoV-2 vaccine candidate, BBV152 in rhesus macaques. Nat Commun 12(1):1–11. https://doi.org/10.1038/s41467-021-21639-w

Yassine HM, Boyington JC, McTamney PM, Wei CJ, Kanekiyo M, Kong WP et al (2015) Hemagglutinin-stem nanoparticles generate heterosubtypic influenza protection. Nat Med 21(9):1065–1070

Zost SJ, Parkhouse K, Gumina ME, Kim K, Diaz Perez S, Wilson PC et al (2017) Contemporary H3N2 influenza viruses have a glycosylation site that alters binding of antibodies elicited by egg-adapted vaccine strains. Proc Natl Acad Sci 114(47):12578–12583. https://doi.org/10.1073/pnas.1712377114

Development of Genetically Engineered Vaccine and Post-Exposure Therapy Against Anthrax

7

Himanshu Gogoi, Divya Kandari, and Rakesh Bhatnagar

Abstract

Anthrax, caused by the spore forming bacterium *Bacillus anthracis,* is a severe infectious disease in both animals and humans. Due to its potential to serve as a biological weapon, research and development of effective vaccines and post-exposure therapeutics has been crucial. Currently available anthrax vaccines such as Anthrax Vaccine Adsorbed (AVA) and Anthrax Vaccine Precipitate (AVP) are tightly regulated and not commercially available. FDA approved prophylactic approaches against the disease, Raxibacumab and Obiltoxaximab, can neutralize protective antigen (PA) which provides protective immunity and are effective against inhalational anthrax only. In an attempt to design an effective vaccine against anthrax, we successfully constructed and expressed recombinant PA in a heterologous host. The vaccine could offer 67–100% protection against anthrax challenge in pre-clinical animal models. We have also elucidated the role of amino acid residues of PA that are crucial for its oligomerization (F552, I562, L566, and I574) and translocation (L352 and W346) into the host cytosol, and the residues I207, P205, L203, F202 which are crucial for binding to lethal factor (LF). Further structural validation of LF and edema factor (EF) disclosed a homologous sequence consisting of the amino acid residues (valine, tyrosine, glutamic acid, isoleucine, glycine, and lysine) which are responsible for binding to PA. To design a post-prophylactic approach against the disease, we success-

H. Gogoi
Laboratory of Molecular Biology and Genetic Engineering, Jawaharlal Nehru University, New Delhi, India

Translational Health Science and Technology Institute, Faridabad, Haryana, India

D. Kandari · R. Bhatnagar (✉)
Laboratory of Molecular Biology and Genetic Engineering, Jawaharlal Nehru University, New Delhi, India
e-mail: nsc-rakesh@mail.jnu.ac.in

G. John, A. K. Panda (eds.), *Biotechnology in India - Reworking A Strategy*, https://doi.org/10.1007/978-981-97-0896-3_7

fully created a monoclonal antibody which can successfully bind to the common N-terminal of LF and EF and effectively neutralize the toxin components as well as provide 60% protection against a murine model of anthrax. This antibody in combination with ciprofloxacin provided 100% protection in mouse model. Toxin-antitoxin (TA) components in *Bacillus anthracis* assists the bacteria to cope up with adverse environmental conditions as well as negatively impact host molecular and cellular machinery. We demonstrated that blocking TA system (PemK–PemI) of *B. anthracis* results in decreased virulence. Most recently, by implementing computational as well as wet lab approaches, we demonstrated that blocking of the sites in PA crucial for binding to LF by quercetin dihydrate (QDH) and 1,3,6,-Trigalloyl-β-D-Glucose (TGG), results in reduced cytotoxicity.

7.1 Introduction

Anthrax, caused by exposure to the Gram-positive bacteria *Bacillus anthracis*, primarily affects livestock, but it is notorious for causing pathogenesis in humans (Bakhteeva and Timofeev 2022). It is an obligate pathogen within the genus bacillus and is classified within the *Bacillus cereus* group comprising of *B. cereus, B. mycoides, B. pseudomycoides, B. weihenstephanensis,* and *B. thuringiensis* (Turnbull 1999). Discrimination between the bacillus group has only been possible by multiple locus variable number tandem repeat analysis and amplified fragment length polymorphism (Keim et al. 1997, 2000). *B. anthracis* exhibits very low genetic diversity unlike the other members in the group. It is a spore forming, facultative anaerobic bacteria. However, spores do not form within the host unless exposed to air and under nutrient deprived conditions. Resistant spores can survive in the soil for decades and germinate in a nutrient rich milieu like blood or tissue of a human or animal host. Germination of spores to vegetative cells occur after ingestion by macrophages at the site of entry, followed by extracellular multiplication (Doganay et al. 2023). Vegetative bacteria can spread through blood and lymphatics where they can cause severe septicemia apart from the secretion of exotoxins which inhibits host defenses and leads to the formation of characteristic locally destructive lesions and can spread to systemic circulation and other organs via lymphatics (Friedlander 1999). Anthrax is endemic in India and commonly affects the communities living near forest areas. However, multiple cases of anthrax outbreak have been frequently reported from the states of Odisha, Chhattisgarh, Andhra Pradesh, Jammu and Kashmir, Tamil Nadu, and Karnataka (Bhattacharya et al. 2021; Lalitha and Kumar 1996).Anthrax poses a major public health concern as a potential biological weapon. World War I and II saw rapid research on biological agents by countries like Germany, France, Japan, England, and Russia (Riedel 2004). The British biological defense program deployed spores of a highly virulent *B. anthracis* strain as bombs on Gruinard island, near Scotland in 1942 (Manchee et al. 1981). The spores were reported viable till the 1980s after which the entire island was set on fire in 1986 and soaked with formaldehyde for an entire year (Manchee et al.

1983). In another instance, accidental release of *Bacillus anthracis* spores in a Soviet military settlement led to the death of 64 people (Meselson et al. 1994). In 1970, a WHO expert committee stated that release of anthrax spores over an urban population of five million would lead to 250,000 casualties, out of whom 95,000 are expected to die without treatment while severely affecting 125,000 people (Consultants, ROAWGO 1970). However, *B. anthracis* gained notoriety during the 2001 anthrax attacks in Florida, USA, where spores of *B. anthracis* Ames strain were deliberately distributed through the postal system causing 22 cases of anthrax including five deaths (Edmonds et al. 2016). Moreover, the recent experience of COVID-19 pandemic has made it imperative to be prepared for a looming biological threat (Tin et al. 2022).

Bacillus anthracis can present itself to humans in three clinical presentations: cutaneous, gastrointestinal, and inhalational. Spores of *B. anthracis* can gain entry into subepidermal tissue through a cut or abrasion. This form of infection initiates as a papule and progresses relatively to form a vesicle and then ulcerates. Sometimes massive edema surrounding the lesions develop into a characteristic black eschar. Humans are relatively resilient to cutaneous anthrax, but any abrasion in the skin can allow the organism to enter the skin. Timely intervention with antibiotics have reported a mortality of less than 2%, but if left unchecked the mortality rate can go up to 20% (Goel 2015). Accumulation of *B. anthracis* spores within the alveoli results in the inhalational form of anthrax. The spores germinate in blood vessels and lymphatics where they are phagocytosed by the immune cells and transported to lung draining lymph nodes. Within these lymph nodes, the spores germinate to vegetative form and multiply while producing toxins. Progression of the disease can lead to bloodstream infection leading to septic shock. Inhaled spores have been reported to reach the peribronchial lymph nodes within 15 minutes of inhalation (Weiner et al. 2012). Clinical symptoms of the infection includes severe sore throat or formation of ulcer in oropharyngeal cavity associated with neck swelling, fever, toxicity, and dysphagia (Friedlander 1999). Gastrointestinal anthrax results from ingestion of contaminated or undercooked meat or ingestion of spores. M cells have been suggested to play a crucial role in the uptake of anthrax bacilli due to the formation of localized lesions in Peyer's patches (World Health Organization 2008). Symptoms of this form of anthrax consists of nausea, vomiting, anorexia, and abdominal pain (Spencer 2003). Current antibiotic regimen against the disease consists of ciprofloxacin and doxycycline (Bell et al. 2002).

The virulence of *B. anthracis* is attributed to two extrachromosomal plasmids, a 183 kb pXO1 and 96 kb pXO2. pXO1 carries a pathogenicity island which encodes for protective antigen (PA), lethal factor (LF), and edema factor (EF) while pXO2 carries the operon *capBCADE* which encodes for poly- γ- D- glutamic acid capsule that inhibits host phagocytosis of the vegetative form of *B. anthracis* (Makino et al. 1989). Loss of either plasmid results in attenuation of virulence of the strain (Hanna 1998). The mechanism behind regulation and replication of both pXO1 and pXO2 has been well documented (Pomerantsev et al. 2009; Uchida et al. 1993; Drysdale et al. 2004; Guignot et al. 1997; Cryan et al. 2013). The anthrax toxins comprise three polypeptides that come together to form two toxins, lethal toxin (LT) and edema toxin (ET). Protective antigen (PA) serves as a common receptor binding component and facilitates the entry of the catalytic components, lethal factor (LF) and edema factor (EF) to the cytosol. PA is a 83 kDa protein and contains four domains (Petosa et al. 1997). The N terminus of the first domain (1–249 amino acids) contains the sequence RKKR (193–196 amino acids) which provides the cleavage site for cell surface proteases after binding to host cell receptors tumor endothelial marker 8 (TEM8) or capillary morphogenesis gene 2 (CMG2) (Petosa et al. 1997; van der Goot and Young 2009). Domain 2 (250–487 amino acids) aids in cargo translocation by the formation of a β-barrel core (Krantz et al. 2005). Domain 3 (488–594 amino acids) contains hydrophobic regions that assist in protein–protein interactions within the oligomer (Mogridge et al. 2001). Domain 4(595–735 amino acids) is involved in receptor binding and has been demonstrated to offer protection against anthrax challenge in murine models (Petosa et al. 1997; Manish et al. 2013). Both TEM8 and CMG2 receptors have a metal ion-dependent adhesion site (MIDAS) within a von Willebrand domain which is typical of integrins and is crucial for binding of PA (Bradley et al. 2003; Liu et al. 2007). Clipping off of a 20 kDa fragment from PA by cell surface furin like proteases exposes an endocytosis competent 63 kDa fragment to oligomerize into heptamer or octamer (Kintzer et al. 2009; Milne et al. 1994). This oligomer facilitates the binding of LF and/or EF and translocation into the cytosol via endocytosis (Cunningham et al. 2002; Mogridge et al. 2002). Oligomerization has been reported to induce rapid movement of the receptor–toxin complex to lipid rafts, inducing *src* like kinase activation which drives clathrin-dependent endocytosis (Abrami et al. 2010a, b). Palmitoylation and ubiquitination of receptors have also been implicated in endocytic process (Abrami et al. 2006). A low pH within the endocytic compartment induces conformational change within PA from a prepore structure to a SDS and heat resistant pore structure (Liu and Leppla 2003; Miller et al. 1999). Translocation of LF and EF occur in the late endosome which occurs through back fusion of intraluminal vesicles with limiting membranes (Abrami et al. 2004). LF contains four domains. N-terminal domain 1 of LF shares high homology with EF domain 1 which facilitates binding of LF/EF to PA oligomer (Arora and Leppla 1993). Domain 2 and 3 are crucial for substrate binding, while domain 4 contains a zinc-binding catalytic domain (Anthrax toxin lethal factor contains a zinc metalloprotease consensus sequence which is required for lethal toxin activity) (Kandari et al. 2019). LF has been demonstrated to cleave and inactivate members of the mitogen-activated protein kinase kinase (MAPKK) family (MEK and MAPKKs) (Duesbery et al. 1998; Vitale et al. 2000, 1999) and prevents activation of Erk1/2, p38, and JNK pathways (Lowe and Glomski 2012). LF has also been demonstrated to activate rodent NLRP1 leading to the secretion of pro-inflammatory cytokines like IL-18 and IL-1β and induction of a caspase 1 dependent pyroptosis (Muehlbauer et al. 2007; Wickliffe et al. 2008). Studies using mouse strains have also demonstrated the arrest of neutrophil recruitment (Weiner et al. 2014). EF, an 89 kDa protein is a calmodulin dependent adenylate cyclase and rapidly converts ATP to 3′- 5′ cyclic AMP up to 20,000 molecules per second (Leppla 1982, 1984). EF consists of three domains, an N-terminal PA binding domain, an adenylyl cyclase activity domain, and a helical domain (Tang and Guo 2009). High levels of cAMP result in activation of protein kinase A (PKA), exchange protein activated by cAMP (Epac), and guanine exchange factor Rap1 which regulates cellular functions like cytoskeletal changes and inhibition of chemotaxis (Hong et al. 2007). ET has also been reported to inhibit neutrophil chemotaxis by reducing F-actin formation in vitro and reduced expression of CD11b in neutrophils which is necessary for neutrophil adhesion and extravasation (Szarowicz et al. 2009).

7.2 Vaccines and Therapeutics

Almost all the anthrax vaccines currently under development or in use rely on PA as the primary immunogen and anti-PA antibodies have been reported to be protective against anthrax spore challenge (Ivins et al. 1998; Turnbull et al. 1986). The currently licensed vaccines US BioThrax and anthrax vaccine precipitate (AVP) in the UK are composed of culture filtrates of non-encapsulated strain of *B. anthracis* adsorbed or precipitated onto alum. These vaccines are, however, not available for civilian use (Spencer 2003). Moreover, they require frequent booster doses and show reactogenicity due to the presence of toxin components apart from PA. An improved vaccine should be immunogenic, well tolerated, requiring less doses, and offering long term immunity.

7.2.1 Development of Recombinant Vaccine Against Anthrax

In an attempt to empower India with a vaccine against anthrax, our laboratory at the School of Biotechnology, Jawaharlal Nehru University overexpressed and purified a recombinant PA in *Escherichia coli* under the transcriptional regulation of the T5 promoter (Gupta et al. 1999). Gene encoding for PA was amplified by polymerase chain reaction with pXO1 as a template and ligated onto pQE30 vector with BamHI and KpnI as the restriction sites. The ligated product was transformed *into E. coli* SG13009 (pREP4) competent cells. Purified recombinant PA (rPA) was obtained from transformed *E. coli* culture at a concentration of 2 mg/L. The rPA was demonstrated to be biologically and functionally active. rPA proteolytically cleaved with trypsin could bind to LF and the mobility of PA_{63} and LF complex was retarded on a nondenaturing gel. rPA (EC_{50} = 50 ng/mL) and LF (1 μg/mL) complex were able to lyse macrophage like cell line J774A.1 demonstrating the biological activity of rPA. As a part of pre-clinical evaluation of the immunogenicity of rPA, non-human primates like Rhesus monkeys and NZW rabbits were intramuscularly (I.M) immunized with a varying dose of rPA (5–100 μg) on day 0 and followed by a booster on day 28. Functional attributes of the anti-rPA antibodies were analyzed in vitro by a toxin neutralization assay in J774A.1 cell line and in vivo spore challenge. Immunized and control groups were challenged with 1000 LD_{50} spores of a virulent strain of *B. anthracis* (IVRI). Interestingly, even 5 μg rPA could offer 80% protection against the live spore challenge, and 25–100 μg rPA offering 100% protection

(Chawla et al. 2009). This system was designed where the PA gene is under the control of phage T5 promoter, which can be recognized by any *E. coli* host strain RNA polymerase. Upscaling the system in a 5 L Biostat B fermenter in a fed batch mode was executed. Carbon sources were optimized for enhanced yield of biomass and protein, and glycerol was found to be the optimal choice. A final OD of 17.75 units was achieved with a fourfold increase in the biomass compared to a shake flask. A single run of the fed batch culture in a 5 L volume could yield rPA equivalent to one million doses (Chauhan et al. 2001). This PA expression system was transferred to a Biopharma Company for validation and clinical studies. The vaccine has successfully completed Phase I/II evaluation for safety and immunogenicity.

7.2.1.1 Identification of Amino Acid Residues in PA Crucial for Binding to LF and Lethal Toxin (LT) Activity

Binding of LF to PA during oligomerization is a pre-requisite for LT to demonstrate its deleterious activity. Binding of PA to LF and/or EF occurs via domain 1 and furin cleavage of PA releases PA_{20} and exposes a large hydrophobic patch in PA_{63} (residues 168–249; designated as domain 1b) (Petosa et al. 1997; Ezzell Jr. and Abshire 1992). The hydrophobic patch consists of residues P184, L187, F202, L203, P205, I207, I210, W226, and F236 which are usually buried in PA_{83} but become exposed due to conformational changes brought about by the cleavage of PA_{83} to $PA_{63.}$ It has been hypothesized that these residues might play a crucial role in the binding of PA to LF. Side directed mutagenesis approach was implemented to construct nine mutants and each amino acid residue was replaced with alanine. All the mutants could bind to cell surface receptor present on RAW 264.7 cells and proteolytic cleavage of PA_{83} to $PA_{63,}$ this process was unhindered in all the mutants. However, binding of mutants I207A, P205A, L203A, and F202A to LF was completely hindered in solution as well as in RAW 264. 7 cells and PA_{63} -LF complex were not detected in these mutants. These four mutants were nontoxic in combination with LF to RAW 264.7 cell lines depicting that this small stretch of amino acids in PA is critical for binding to LF (Chauhan and Bhatnagar 2002).

7.2.1.2 Identification of the Amino Acids of PA Involved in Membrane Insertion and Translocation

Domain 2 of PA have been implicated in cell membrane insertion after oligomerization and translocation of LF and/ or EF (Petosa et al. 1997). Domain 2 has a β-barrel core structure and contains a flexible loop between $2\beta_2$ and $2\beta_3$ and a pH sensitive loop with three hydrophobic residues: W346, M350, and L352. Crystal structure of PA at a neutral pH and acidic pH (pH 6) have demonstrated that these three residues which otherwise are buried within the core of the β-barrel at neutral pH become exposed at pH 6 and induce conformational changes. Mutation of the amino acid residues at 346, 350, and 352 to alanine did not hinder binding to cell surface receptors, proteolytic cleavage of PA by trypsin, oligomerization, and binding to LF. However, substitution of W346 and L352 with alanine altered the cytotoxic profile of PA and the mutants were nontoxic to J774A.1 macrophages in combination with LF up to concentrations as high as 2.5 μg/mL. These residues thereby seemed to play an important role in membrane insertion and translocation of LF and/or EF into the cytosol (Batra et al. 2001).

7.2.1.3 Identification of Residues Involved in the Oligomerization of PA

During cell entry, the PA binds to host cell receptors TEM8 and/or CMG2, gets proteolytically cleaved, and oligomerizes to form a heptamer/octamer which assists in the binding and translocation of LF/EF. Domain 3 (residues 488–594) has a solvent exposed "hydrophobic patch," whose role was not elucidated. The hydrophobic residues Phe552, Phe554, Ile562, Leu566, and Ile574 are contained in a triangular strand-helix loop structure. In order to determine the potential role of the amino acids in this "exposed hydrophobic patch", we studied the impact of these five residues on in vitro toxicity, binding to LF and EF and oligomerization potential of PA. Site-directed mutagenesis approach was utilized to create five different constructs with single mutations where the amino acid of each residue was replaced with alanine (F552A, F554A, I562A, L566A, and I574A). Alanine substitution did not hamper the binding of the PA mutants to cell receptors on J774A.1 and proteolytic cleavage by trypsin to nick off a 20 kDa fragment. The cellular toxicity of mutant PA on J774A.1 and CHO cell line demonstrated a complete loss of activity in F552A and I574A and a substantial loss in mutant F554A. However, mutants I562A and L566A were highly toxic as compared to WT PA. A complete loss or defect in the toxicity indicated that mutant PA were defective in mediating the entry of LF/EF into the host cell. Proteolytically cleaved PA_{63} oligomerizes with LF and EF which are high molecular weight complexes having restricted mobility in non-denaturing PAGE. Moreover, WT PA_{63} oligomers are SDS and heat resistant. Except mutant F554A, other mutants failed to form this high molecular weight complex and were heat and SDS labile. Furthermore, a mutant with all the residues as alanine that failed to oligomerize was heat and SDS sensitive and formed high molecular weight complex with LF/ EF thereby demonstrating that residues F552, I562, L566, and I574 of domain 3 are crucial for oligomerization of PA (Ahuja et al. 2001).

7.2.1.4 Delineating the Homologous Sequence of LF and EF Crucial for Binding to PA

Internalization of PA_{63} oligomer and LF/EF complex by receptor mediated endocytosis into endosome results in acidification and translocation of LF/EF into the cytosol (Woodberry et al. 2009). LF is a zinc metalloprotease and acts as an endopeptidase which cleaves the amino terminal of mitogen and activates protein kinase kinases (MAPKK) 1–4, 6, 7 thereby inhibiting MAPKK signaling. MAPKK cleavage inhibits phosphorylation of the downstream components p38, ERK, and JNK which inhibits MAPK signaling (Duesbery et al. 1998, 2001; Vitale et al. 1999). MAPK pathway serves in proliferation, survival, and inflammation in cell types and inactivation of this pathway in macrophage and dendritic cells has been shown to down-regulate co-stimulatory molecules like CD80, CD86, cause ineffective T-cell priming and inefficient pro-inflammatory cytokine production (Agrawal and

Pulendran 2004). EF (90 kDa) has been known to elevate cAMP levels in susceptible cells resulting in changes in membrane permeability leading to edema. The rise in cellular cAMP levels reduces phagocytic ability of neutrophils, and blocks TNF-α production by monocytes thereby impairing their microbicidal activity (Wade et al. 1985; O'Brien et al. 1985; Hoover et al. 1994). EF is a bacterial adenylate cyclase and upon activation by its eukaryotic co-factor, calmodulin, mediates the rapid increase of cAMP levels. This drastic increase of cAMP has been attributed to two mechanisms; association of adenylyl cyclase with protein kinase A (PKA) and the exchange protein activation by cAMP (Epac). PKA consists of two regulatory and two catalytic subunits and exists as an inactive tetramer. Binding of the regulatory subunits to cAMP results in their dissociation from the catalytic subunits which are activated by ATP binding. Adenylyl cyclase like EF and cholera toxin activates cystic fibrosis transmembrane conductance regulator (CFTR) which is an ATP binding cassette transporter-class chloride ion channel found in epithelial cells of liver, digestive tract and skin. Activation of PKA in intestinal epithelial cells has been shown to result in phosphorylation of CFTR leading to ATP mediated efflux of Cl^- ions into the intestinal lumen followed by the secretion of Na^+. This increases total electrolyte concentration in the intestinal lumen resulting in the movement of water out of enterocytes and into intestinal lumen by osmosis (Muanprasat and Chatsudthipong 2013). Epac was identified in 1998 while unraveling the mechanism of cAMP dependent activation of the small GTPase Rap1 that was independent of PKA signaling (Schmidt et al. 2013). Epac, a downstream effector for cAMP regulates a wide range of biological responses like calcium homeostasis, muscle contraction, metabolism, and gene transcription. Epac regulated cytoskeletal changes, inhibition of VEGF-stimulated chemotaxis of human endothelial cells. However, edema formation is likely to be a synergistic effect of PKA and Epac activation (Hong et al. 2007).

LF and EF binds competitively to proteolytically cleaved $PA_{63.}$ A comparison of LF $_{1-250}$ and EF revealed a stretch of seven amino acids valine, tyrosine, tyrosine, glutamic acid, isoleucine, glycine, lysine, which is the same in both LF and EF. This stretch was present as 147VYYEIGK153 in LF and 136VYYEIGK142 in EF. The seven residues in LF were substituted with alanine and seven mutants were generated. LF mutants Y148A, Y149A, I151A, and K153A were defective in binding to PA while V147A, E150A, and G152A were able to bind to PA like WT LF. Toxicity studies on J774A.1 cells demonstrated that mutants Y148A and Y149A were nontoxic even at a concentration of 10 μg/mL. Mutants I151A and K153A were slightly cytotoxic at higher concentrations, meanwhile V147A, E150A, and G152A were cytotoxic like WT LF. This study revealed that residues Y148, Y149, I151, and K153played a crucial role in the process of binding of LF to PA_{63} (Gupta et al. 2001). A similar study using EF mutants demonstrated that mutants Y137A, Y138A, I140A, and K142A were defective in binding to PA, failed to elicit the elongation of CHO cell line in combination with PA and could not increase cellular cAMP levels (Kumar et al. 2001).

7.2.1.5 Designing Novel Vaccination Strategies Against Anthrax

Oral vaccines can stimulate immune response at mucosal sites and provide an essential line of defense. We successfully integrated PA gene in the nuclear genome of tobacco and tomato plants by *Agrobacterium tumifaciens* mediated leaf-disc transformation method. PA gene was cloned in a binary vector pBINAR and mobilized into *A. tumifaciens* (LBA4404) and leaf disc transformation of tobacco plants (*Nicotiana tabacum*) was performed. A high concentration of kanamycin (300 mg/L) was used for selection. Total soluble protein (TSP) from leaves of transformed plants were extracted in liquid nitrogen and PA could be detected in an immunoblot using anti- PA antibodies at the desired molecular weight. Biological activity of the TSP along with LF demonstrated killing of RAW 264.7 cell lines, however, the cytotoxic percentage varied between 26% to 98% owing to different expression levels in different plants (Aziz et al. 2002). In vivo immunization using 10 μg PA from transgenic plants could elicit appreciable anti-PA antibody titer, prevent killing of RAW 264.7 macrophage, by LT and offered up to 80% protection against anthrax challenge (Aziz et al. 2005; Gorantala et al. 2014).

In an attempt to create a DNA vaccine against anthrax, structural gene for PA_{63} was cloned as chimera containing targeting sequences toward various sub-cellular locations. The targeting sequences were lysosome-associated membrane protein 1 (LAMP1), tissue plasminogen activator (TPA), and ubiquitin. The construct pTPA-PA_{63} encoded a cell-secreted version of PA_{63}, while chimera pPA_{63}-LAMP1 encoded a lysosome-associated PA_{63}. Chimera pTPA-PA_{63}-LAMP1 encoded a lysosome-associated form of PA_{63} along with a cell-secreted form. pUQ-PA_{63} encoded a solely cell-associated PA_{63}. In vivo challenge in immunized mice against LeTx (50 μg PA and 22 μg LF) displayed a protection up to 88% in pTPA-PA_{63}-LAMP1 immunized group, while pTA-PA_{63} and pPA_{63}-LAMP1 immunized mice displayed protective efficacy up to 75%. This study showed that targeting PA_{63} to lysosomal compartments by DNA vaccination generated high end-point IgG titers which were maintained till 24-weeks post-immunization (Midha and Bhatnagar 2009) (Table 7.1).

Induction of neutralizing anti-PA antibodies is considered to be protective against anthrax (McBride et al. 1998). A small loop between 679 and 693 a.a residues of domain 4 of PA contains the host cell receptor binding region (Varughese et al. 1999). In an attempt to identify B-cell epitopes within domain 4, we used B-cell epitope prediction software's BCPred, BcePRED, and Emboss to identify three epitopes (ID-1: 604–622, ID- II: 626–676, and ID- III: 707–723) which were synthesized commercially. These epitopes elicited potent B-cell response with high IgG antibody titer with a predominantly IgG1/IgG2a response. However, serum from ID-II immunized mice offered significant protection (86.57%) against the cytotoxic

Table 7.1 depicts the immunological parameters induced by the vaccine candidates

S. No	Vaccine formulation	Mean antibody titer	% protection
1	Recombinant PA	≥50,000	66%
2	pTPA- PA_{63}-LAMP1 (DNA vaccine)	≥10,000	88%
3	Edible PA vaccine	≈60,000	80%

effects of LT in J774A.1 cells and offered 66% protection against 2 × LD_{50} dose of LT. This study demonstrated the potential of B-cell epitopes to act as vaccine candidate against anthrax (Kaur et al. 2009).

Inactivated whole spore vaccine can enhance protective efficacy against anthrax challenge. However, this can result in unintended infection and also show possible variations in preparation.

The exosporium components of spores of *B. anthracis* provide the initial point of interaction between the bacilli and host immune system during the initial stages of infection (Oliva et al. 2009). The spore components (ExsF and p5303) have been evaluated for their ability to augment the protective efficacy of PA based vaccines. In an attempt to exploit the immunogenic potential of exosporium components, we targeted a major glycoprotein; Bacillus collagen like antigen (BclA). This glycoprotein has been reported as a prominent vaccine target and the antibody response has been reported to be generated against the protein component of BclA and not the carbohydrate component. Domain 4 of PA is involved in host cell receptor binding and generates protective immune response (Williams et al. 2009). We generated a fusion protein, rPB, comprising of domain 4 of PA and C-terminal domain of BclA. Anti-rPB antibody titers were raised up to 1:64,000 after 42 days of immunization. In vitro biological efficacy demonstrated that, anti-rPB antibodies promoted phagocytosis of *B. anthracis* spores by RAW 264.7 cells and significantly inhibited germination of BA spores. The rPB vaccine was able to provide 100% protection against 10 x LD_{50} spores of BA. This study demonstrated the successful conjugating of the C-terminal of BclA with PAD-IV and generation of a successful vaccine candidate that can target both spore component as well as receptor binding component of PA (Majumder et al. 2018).

Nanoparticles can enhance vaccine efficacy by protecting antigens against premature proteolytic degradation, facilitate antigen uptake and processing by APCs, and controlled release (Pati et al. 2018). To enhance the bioavailability and efficacy of PAD-IV, polylactic-co-glycolic acid (PLGA) nanoparticles, non-ionic surfactant vesicles encapsulating PAD-IV and amorphous and crystalline aluminum hydroxide nanoparticles were prepared and evaluated for immune response (Manish et al. 2013; Gogoi et al. 2018, 2020). Nanoparticle mediated delivery of antigens elicited a balanced Th1/ Th2 response as compared to a biased Th2 response elicited by naked PAD-IV and further increased the protective efficacy of the vaccine formulation. Interestingly, we also demonstrated that nanoparticle crystallinity is an important factor in binding to an antigen which plays a crucial role in eliciting an immune response (Gogoi et al. 2020).

7.2.1.6 Post-Exposure Strategy Against Anthrax

Toxin-antitoxin (TA) loci are organized as an operon comprising of two adjacent open reading frames, where the first cistron encodes a small (65–85 aa) labile antitoxin and the second cistron encodes a large stable toxin (95–135 aa). Anti-toxins are vulnerable to proteolysis and require a constant de novo synthesis to maintain a steady-state level to sequester the toxin and circumvent its deleterious effects. TAs play a central role that is advantageous for its survival like switching into a dormant, drug resistant state to withstand high levels of antibiotic stress (Page and Peti 2016). Typically, TA toxins act on host crucial cellular processes like translation, replication, membrane integrity, cell wall biosynthesis, and cytoskeleton formation. Several families of TA system have been studied and reported in different bacterial chromosomes, relBE, higBA, mazEF, ccdAB, vapBC, parDE, yoeB-yefM, and phd-doc. CCdB and ParE function as gyrase inhibitors and block DNA replication, whereas MazF, YoeB, and VapC are potent ribonucleases. The genetic organization and chromosomal TA locus and gene products of BA have been observed to be similar with programmed cell death prevalent in other prokaryotes like *Leptospira interrogans*. We have identified and functionally characterized a TA system pemKI in BA and have found it to be a type II TA system. PemK protein, when overexpressed in *E. coli* and BA could inhibit the growth of cells. However, upon co-expression of PemK and PemI, normal growth of the cells were observed (Agarwal et al. 2007). Functional characterization of the PemKI TA system described PemK as a potent ribonuclease with affinity toward pyrimidines, and residues His-59 and Glu-78 play a crucial role in mediating this activity. PemI could inhibit ribonuclease activity at a stoichiometry of 1:1. Transcriptional levels of PemKI were enhanced during stress conditions of the microorganism like nutrition deprivation, high temperature, antibiotics, and UV irradiation. Very interestingly, this TA system proved to be a very potent target for targeting BA and synthetic peptides designed to target the PemK-PemI interacting sites and could partially (in 38%) inhibit PemK-PemI interaction (Agarwal et al. 2010).

Most recently, we have identified two natural compounds, namely quercetin dihydrate (QDH) and 1,3,6,–Trigalloyl-β-D-Glucose (TGG). Computational analysis has shown these molecules to bind with PA in the residues crucial for interaction with LF. Physico-chemical studies using surface plasmon resonance (SPR) demonstrated a significant reduction in the interaction between PA and LF in the presence of these molecules. A significant reduction of LeTx mediated cytotoxicity in RAW264.7 cells was observed in the presence of both these molecules. These molecules can act as potential post-prophylactic candidates; however, this claim must be further supported by in vivo studies.

B. anthracis toxin proteins, LF and EF share a common stretch of seven amino acid residues (Val, Tyr, Tyr, Glu, Ile, Gly, and Lys) at the N-terminal that binds with PA. A monoclonal antibody bi-specific toward LF and EF can neutralize the toxin components during post-infection and reduce disease severity (Kandari and Bhatnagar 2023). Utilizing hybridoma technology, a monoclonal antibody targeting the N-terminal of LF, "H10 mAb" was created and selected based on cross-reactivity toward both LF and EF. H10 mAb could recognize both LF and EF in immunoblots but could not recognize PA or a non-specific protein BSA. In vitro and in vivo assays with H10 mAb could obliterate adenylate cyclase activity of ET as detected by insignificant cAMP production in CHO.K1 cells compared to control cells, and reduce ET induced edema in murine footpads. H10 mAb could also neutralize the cytolytic activity of LT in J774.A1 cells as well as protect mice from LT as well as *B. anthracis* which otherwise resulted in 100% mortality within 48 hours of infection. This development of bi-specific monoclonal antibody which offered protection against both ET and LT in mice has resulted in an advancement and opened up new avenues for the development of humanized monoclonal antibodies against anthrax (Kulshreshtha and Bhatnagar 2011).

7.2.2 The Road Ahead for Anthrax Detection, Diagnosis, and Intervention

The Department of Biotechnology which was created in 1986 has been a huge catalyst in promoting this revolutionary field in India by encouraging research and development, technology validation and transfer, human resource development, setting up state-of-the-art research centers, biological repositories, and promoting industry–academia interactions. Biotechnology has been a huge boon for India and successfully implemented toward the betterment of the society. Considering the recent SARS-CoV2 pandemic, the development of indigenous vaccines, "Covaxin," "iNCOVACC," and global distribution of these vaccines demonstrated India's advances in Biotechnology research and production. More instances of vaccine development that can be listed are the indigenous Hib vaccine against *Hemophilus influenzae* type B, rotavac vaccine against rotavirus, and Jenvac (Japanese encephalitis virus vaccine). Apart from advances in vaccine development, research on genomics, stem cell, gene therapy (The most recent inclusion of the indigenous CAR-T-cell therapy "NexCAR19" developed by "ImmunoACT" to treat B-cell lymphoma and leukemia), genetically modified crops designed to withstand stress conditions, have showcased India's advances in the field of Biotechnology.

Our group at Jawaharlal Nehru University has focused on implementing fundamental understanding of the toxin components of *Bacillus anthracis* to conduct translational research and devise interventional strategies. Our strategy to successfully clone and express recombinant PA successfully led to a technology transfer more than two decades ago when the concept of academia-industry collaboration was naïve in the Indian subcontinent. This technology transfer was well acknowledged by both Indian as well as US media. The clones developed at our laboratory have been distributed across the National laboratories for research and development purposes. However, there is still a need to develop an effective detection and intervention strategy against the pathogen. Currently, we are in a process of developing humanized therapeutic vaccine which may eventually be used as an emergency intervention for the anthrax. Cases of anthrax are regularly reported by various states in India. An efficient and sensitive detection tool to diagnose spores of *Bacillus anthracis,* effective vaccines, and monoclonal antibodies against the toxin components that can be used for mass administration in times of an epidemic are of utmost importance. Apart from interventional strategies such as vaccination of livestock, strategic measures like strengthening of surveillance system for early detection and reporting of anthrax cases, categorizing "risk zones" via GIS mapping of cases for establishment of dedicated anthrax diagnosis and treatment facility in endemic zones, standardizing operational protocols for anthrax outbreaks, and creating awareness about food habits can be crucial in containing the dreaded disease.

In summary, our work has shed light on the crucial role of the amino acid residues crucial for anthrax toxin activity, delineated the common stretch of LF and EF crucial for binding to PA and designed a bi-specific monoclonal antibody which can bind to LF and EF and neutralize their deleterious effects. Furthermore, we successfully cloned and expressed a recombinant protective antigen against anthrax and transferred the technology for further clinical trials. Our current endeavor focuses on developing a humanized antibody bi-specific for LF and EF which could successfully neutralize both the toxin components and act as therapeutic intervention against anthrax.

Acknowledgement We acknowledge the works of Pankaj Gupta, S. Mohsin Waheed, Anil Chawla, Suchi Midha, Vibha Chauhan, Samer Singh, Nidhi Ahuja, Praveen Kumar, Smriti Batra, Aparna Singh, Mohd. Azhar Aziz, P Anand Kumar, Deepa Sikriwal, Sridhar Jarugula, Manpreet Kaur, Saugata Majumder, Shreya Das, Vikas Somani, Shivakiran S. Makam, Kingston J. Joseph, Shivangi Agarwal, Shivani Agarwal, Neeraj Kumar Mishra, Sonika Bhatnagar, Parul Kulshrestha, Manish Manish, Amit Rahi, Rajesh Mani, Anshu Malik, Soumya Aggarwal and Manoj Munde which has immensely contributed toward the preparation of this article. We would also like to acknowledge Department of Biotechnology (DBT), Science and Engineering Research Board-National Science Chair (SERB- NSC), Pennsylvania State University-USA, and SERB- Ramanujan fellowship (RJF/2022/000040) for funding.

References

Abrami L, Lindsay M, Parton RG, Leppla SH, van der Goot FG (2004) Membrane insertion of anthrax protective antigen and cytoplasmic delivery of lethal factor occur at different stages of the endocytic pathway. J Cell Biol 166:645–651. https://doi.org/10.1083/jcb.200312072

Abrami L, Leppla SH, van der Goot FG (2006) Receptor palmitoylation and ubiquitination regulate anthrax toxin endocytosis. J Cell Biol 172:309–320. https://doi.org/10.1083/jcb.200507067

Abrami L, Bischofberger M, Kunz B, Groux R, van der Goot FG (2010a) Endocytosis of the anthrax toxin is mediated by clathrin, actin and unconventional adaptors. PLoS Pathog 6:e1000792. https://doi.org/10.1371/journal.ppat.1000792

Abrami L, Kunz B, van der Goot FG (2010b) Anthrax toxin triggers the activation of src-like kinases to mediate its own uptake. Proc Natl Acad Sci USA 107:1420–1424. https://doi.org/10.1073/pnas.0910782107

Agarwal S, Agarwal S, Bhatnagar R (2007) Identification and characterization of a novel toxin-antitoxin module from *Bacillus anthracis*. FEBS Lett 581:1727–1734. https://doi.org/10.1016/j.febslet.2007.03.051

Agarwal S, Mishra NK, Bhatnagar S, Bhatnagar R (2010) PemK toxin of *Bacillus anthracis* is a ribonuclease: an insight into its active site, structure, and function. J Biol Chem 285:7254–7270. https://doi.org/10.1074/jbc.M109.073387

Agrawal A, Pulendran B (2004) Anthrax lethal toxin: a weapon of multisystem destruction. Cell Mol Life Sci 61:2859–2865. https://doi.org/10.1007/s00018-004-4251-4

Ahuja N, Kumar P, Bhatnagar R (2001) Hydrophobic residues Phe552, Phe554, Ile562, Leu566, and Ile574 are required for oligomerization of anthrax protective antigen. Biochem Biophys Res Commun 287:542–549. https://doi.org/10.1006/bbrc.2001.5613

Arora N, Leppla SH (1993) Residues 1-254 of anthrax toxin lethal factor are sufficient to cause cellular uptake of fused polypeptides. J Biol Chem 268:3334–3341

Aziz MA, Singh S, Anand Kumar P, Bhatnagar R (2002) Expression of protective antigen in transgenic plants: a step towards edible vaccine against anthrax. Biochem Biophys Res Commun 299:345–351. https://doi.org/10.1016/s0006-291x(02)02625-6

Aziz MA et al (2005) Transformation of an edible crop with the pagA gene of *Bacillus anthracis*. FASEB J 19:1501–1503. https://doi.org/10.1096/fj.04-3215fje

Bakhteeva I, Timofeev V (2022) Some peculiarities of anthrax epidemiology in herbivorous and carnivorous animals. Life 12(6):870

Batra S, Gupta P, Chauhan V, Singh A, Bhatnagar R (2001) Trp 346 and Leu 352 residues in protective antigen are required for the expression of anthrax lethal toxin activity. Biochem Biophys Res Commun 281:186–192. https://doi.org/10.1006/bbrc.2001.4320

Bell DM, Kozarsky PE, Stephens DS (2002) Clinical issues in the prophylaxis, diagnosis, and treatment of anthrax. Emerg Infect Dis 8:222–225. https://doi.org/10.3201/eid0802.01-0521

Bhattacharya D et al (2021) One health approach for elimination of human anthrax in a tribal district of Odisha: study protocol. PLoS One 16:e0251041. https://doi.org/10.1371/journal.pone.0251041

Bradley KA et al (2003) Binding of anthrax toxin to its receptor is similar to alpha integrin-ligand interactions. J Biol Chem 278:49342–49347. https://doi.org/10.1074/jbc.M307900200

Chauhan V, Bhatnagar R (2002) Identification of amino acid residues of anthrax protective antigen involved in binding with lethal factor. Infect Immun 70:4477–4484. https://doi.org/10.1128/iai.70.8.4477-4484.2002

Chauhan V, Singh A, Waheed SM, Singh S, Bhatnagar R (2001) Constitutive expression of protective antigen gene of *Bacillus anthracis* in *Escherichia coli*. Biochem Biophys Res Commun 283:308–315. https://doi.org/10.1006/bbrc.2001.4777

Chawla A, Midha S, Bhatnagar R (2009) Efficacy of recombinant anthrax vaccine against *Bacillus anthracis* aerosol spore challenge: preclinical evaluation in rabbits and rhesus monkeys. Biotechnol J 4:391–399. https://doi.org/10.1002/biot.200800213

Consultants, ROAWGO (1970) Health Aspects of chemical and biological weapons. World Health Organization, Geneva

Cryan LM et al (2013) Identification of small molecules that inhibit the interaction of TEM8 with anthrax protective antigen using a FRET assay. J Biomol Screen 18:714–725. https://doi.org/10.1177/1087057113478655

Cunningham K, Lacy DB, Mogridge J, Collier RJ (2002) Mapping the lethal factor and edema factor binding sites on oligomeric anthrax protective antigen. Proc Natl Acad Sci USA 99:7049–7053. https://doi.org/10.1073/pnas.062160399

Doganay M, Dinc G, Kutmanova A, Baillie L (2023) Human anthrax: update of the diagnosis and treatment. Diagnostics 13(6):1056

Drysdale M, Bourgogne A, Hilsenbeck SG, Koehler TM (2004) atxA controls *Bacillus anthracis* capsule synthesis via acpA and a newly discovered regulator, acpB. J Bacteriol 186:307–315. https://doi.org/10.1128/jb.186.2.307-315.2004

Duesbery NS et al (1998) Proteolytic inactivation of MAP-kinase-kinase by anthrax lethal factor. Science 280:734–737. https://doi.org/10.1126/science.280.5364.734

Duesbery NS et al (2001) Suppression of ras-mediated transformation and inhibition of tumor growth and angiogenesis by anthrax lethal factor, a proteolytic inhibitor of multiple MEK pathways. Proc Natl Acad Sci USA 98:4089–4094. https://doi.org/10.1073/pnas.061031898

Edmonds J, Lindquist HA, Sabol J, Martinez K, Shadomy S, Cymet T, Emanuel P (2016) Multigeneration cross-contamination of mail with bacillus anthracis spores. PLoS One 11(4):e0152225

Ezzell JW Jr, Abshire TG (1992) Serum protease cleavage of *Bacillus anthracis* protective antigen. J Gen Microbiol 138:543–549. https://doi.org/10.1099/00221287-138-3-543

Friedlander AM (1999) Clinical aspects, diagnosis and treatment of anthrax. J Appl Microbiol 87:303. https://doi.org/10.1046/j.1365-2672.1999.00896.x

Goel AK (2015) Anthrax: a disease of biowarfare and public health importance. World J Clin Cases 3:20–33. https://doi.org/10.12998/wjcc.v3.i1.20

Gogoi H, Mani R, Bhatnagar R (2018) A niosome formulation modulates the Th1/Th2 bias immune response in mice and also provides protection against anthrax spore challenge. Int J Nanomedicine 13:7427

Gogoi H et al (2020) Crystalline and amorphous preparation of aluminum hydroxide nanoparticles enhances protective antigen domain 4 specific immunogenicity and provides protection against anthrax. Int J Nanomedicine 15:239

Gorantala J et al (2014) Generation of protective immune response against anthrax by oral immunization with protective antigen plant-based vaccine. J Biotechnol 176:1–10
Guignot J, Mock M, Fouet A (1997) AtxA activates the transcription of genes harbored by both *Bacillus anthracis* virulence plasmids. FEMS Microbiol Lett 147:203–207. https://doi.org/10.1111/j.1574-6968.1997.tb10242.x
Gupta P, Waheed SM, Bhatnagar R (1999) Expression and purification of the recombinant protective antigen of *Bacillus anthracis*. Protein Expr Purif 16:369–376. https://doi.org/10.1006/prep.1999.1066
Gupta P, Singh A, Chauhan V, Bhatnagar R (2001) Involvement of residues 147VYYEIGK153 in binding of lethal factor to protective antigen of *Bacillus anthracis*. Biochem Biophys Res Commun 280:158–163. https://doi.org/10.1006/bbrc.2000.4099
Hanna P (1998) Anthrax pathogenesis and host response. Curr Top Microbiol Immunol 225:13–35. https://doi.org/10.1007/978-3-642-80451-9_2
Hong J et al (2007) Anthrax edema toxin inhibits endothelial cell chemotaxis via Epac and Rap1. J Biol Chem 282:19781–19787. https://doi.org/10.1074/jbc.M700128200
Hoover DL et al (1994) Anthrax edema toxin differentially regulates lipopolysaccharide-induced monocyte production of tumor necrosis factor alpha and interleukin-6 by increasing intracellular cyclic AMP. Infect Immun 62:4432–4439. https://doi.org/10.1128/iai.62.10.4432-4439.1994
Ivins BE et al (1998) Comparative efficacy of experimental anthrax vaccine candidates against inhalation anthrax in rhesus macaques. Vaccine 16:1141–1148. https://doi.org/10.1016/s0264-410x(98)80112-6
Kandari D, Bhatnagar R (2023) Antibody engineering and its therapeutic applications. Int Rev Immunol 42(2):156–183
Kandari D, Gopalani M, Gupta M, Joshi H, Bhatnagar S, Bhatnagar R (2019) Identification, functional characterization, and regulon prediction of the zinc uptake regulator (zur) of *Bacillus anthracis*–an insight into the zinc homeostasis of the pathogen. Front Microbiol 9:3314
Kaur M et al (2009) Identification and characterization of immunodominant B-cell epitope of the C-terminus of protective antigen of *Bacillus anthracis*. Mol Immunol 46:2107–2115. https://doi.org/10.1016/j.molimm.2008.12.031
Keim P et al (1997) Molecular evolution and diversity in *Bacillus anthracis* as detected by amplified fragment length polymorphism markers. J Bacteriol 179:818–824. https://doi.org/10.1128/jb.179.3.818-824.1997
Keim P et al (2000) Multiple-locus variable-number tandem repeat analysis reveals genetic relationships within *Bacillus anthracis*. J Bacteriol 182:2928–2936. https://doi.org/10.1128/jb.182.10.2928-2936.2000
Kintzer AF et al (2009) The protective antigen component of anthrax toxin forms functional octameric complexes. J Mol Biol 392:614–629. https://doi.org/10.1016/j.jmb.2009.07.037
Krantz BA et al (2005) A phenylalanine clamp catalyzes protein translocation through the anthrax toxin pore. Science 309:777–781. https://doi.org/10.1126/science.1113380
Kulshreshtha P, Bhatnagar R (2011) Inhibition of anthrax toxins with a bispecific monoclonal antibody that cross reacts with edema factor as well as lethal factor of *Bacillus anthracis*. Mol Immunol 48:1958–1965. https://doi.org/10.1016/j.molimm.2011.05.024
Kumar P, Ahuja N, Bhatnagar R (2001) Purification of anthrax edema factor from *Escherichia coli* and identification of residues required for binding to anthrax protective antigen. Infect Immun 69:6532–6536. https://doi.org/10.1128/iai.69.10.6532-6536.2001
Lalitha MK, Kumar A (1996) Anthrax in southern India. Lancet (London, England) 348:553–554. https://doi.org/10.1016/s0140-6736(05)64718-6
Leppla SH (1982) Anthrax toxin edema factor: a bacterial adenylate cyclase that increases cyclic AMP concentrations of eukaryotic cells. Proc Natl Acad Sci USA 79:3162–3166. https://doi.org/10.1073/pnas.79.10.3162
Leppla SH (1984) Bacillus anthracis calmodulin-dependent adenylate cyclase: chemical and enzymatic properties and interactions with eucaryotic cells. Adv Cyclic Nucleotide Protein Phosphorylation Res 17:189–198
Liu S, Leppla SH (2003) Cell surface tumor endothelium marker 8 cytoplasmic tail-independent anthrax toxin binding, proteolytic processing, oligomer formation, and internalization. J Biol Chem 278:5227–5234. https://doi.org/10.1074/jbc.M210321200

Liu S, Leung HJ, Leppla SH (2007) Characterization of the interaction between anthrax toxin and its cellular receptors. Cell Microbiol 9:977–987. https://doi.org/10.1111/j.1462-5822.2006.00845.x
Lowe DE, Glomski IJ (2012) Cellular and physiological effects of anthrax exotoxin and its relevance to disease. Front Cell Infect Microbiol 2:76. https://doi.org/10.3389/fcimb.2012.00076
Majumder S et al (2018) A bivalent protein r-PB, comprising PA and BclA immunodominant regions for comprehensive protection against *Bacillus anthracis*. Sci Rep 8:7242. https://doi.org/10.1038/s41598-018-25502-9
Makino S, Uchida I, Terakado N, Sasakawa C, Yoshikawa M (1989) Molecular characterization and protein analysis of the cap region, which is essential for encapsulation in *Bacillus anthracis*. J Bacteriol 171:722–730. https://doi.org/10.1128/jb.171.2.722-730.1989
Manchee RJ, Broster MG, Melling J, Henstridge RM, Stagg AJ (1981) *Bacillus anthracis* on Gruinard Island. Nature 294:254–255. https://doi.org/10.1038/294254a0
Manchee RJ, Broster MG, Anderson IS, Henstridge RM, Melling J (1983) Decontamination of *Bacillus anthracis* on Gruinard Island? Nature 303:239–240. https://doi.org/10.1038/303239a0
Manish M, Rahi A, Kaur M, Bhatnagar R, Singh S (2013) A single-dose PLGA encapsulated protective antigen domain 4 nanoformulation protects mice against *Bacillus anthracis* spore challenge. PLoS One 8:e61885. https://doi.org/10.1371/journal.pone.0061885
McBride BW et al (1998) Protective efficacy of a recombinant protective antigen against *Bacillus anthracis* challenge and assessment of immunological markers. Vaccine 16:810–817. https://doi.org/10.1016/s0264-410x(97)00268-5
Meselson M et al (1994) The Sverdlovsk anthrax outbreak of 1979. Science (New York, N.Y.) 266:1202–1208. https://doi.org/10.1126/science.7973702
Midha S, Bhatnagar R (2009) Anthrax protective antigen administered by DNA vaccination to distinct subcellular locations potentiates humoral and cellular immune responses. Eur J Immunol 39:159–177. https://doi.org/10.1002/eji.200838058
Miller CJ, Elliott JL, Collier RJ (1999) Anthrax protective antigen: prepore-to-pore conversion. Biochemistry 38:10432–10441. https://doi.org/10.1021/bi990792d
Milne JC, Furlong D, Hanna PC, Wall JS, Collier RJ (1994) Anthrax protective antigen forms oligomers during intoxication of mammalian cells. J Biol Chem 269:20607–20612
Mogridge J, Mourez M, Collier RJ (2001) Involvement of domain 3 in oligomerization by the protective antigen moiety of anthrax toxin. J Bacteriol 183:2111–2116. https://doi.org/10.1128/jb.183.6.2111-2116.2001
Mogridge J, Cunningham K, Lacy DB, Mourez M, Collier RJ (2002) The lethal and edema factors of anthrax toxin bind only to oligomeric forms of the protective antigen. Proc Natl Acad Sci USA 99:7045–7048. https://doi.org/10.1073/pnas.052160199
Muanprasat C, Chatsudthipong V (2013) Cholera: pathophysiology and emerging therapeutic targets. Future Med Chem 5:781–798. https://doi.org/10.4155/fmc.13.42
Muehlbauer SM et al (2007) Anthrax lethal toxin kills macrophages in a strain-specific manner by apoptosis or caspase-1-mediated necrosis. Cell Cycle (Georgetown, Tex) 6:758–766. https://doi.org/10.4161/cc.6.6.3991
O'Brien J, Friedlander A, Dreier T, Ezzell J, Leppla S (1985) Effects of anthrax toxin components on human neutrophils. Infect Immun 47:306–310. https://doi.org/10.1128/iai.47.1.306-310.1985
Oliva C, Turnbough CL Jr, Kearney JF (2009) CD14-mac-1 interactions in *Bacillus anthracis* spore internalization by macrophages. Proc Natl Acad Sci USA 106:13957–13962. https://doi.org/10.1073/pnas.0902392106
Page R, Peti W (2016) Toxin-antitoxin systems in bacterial growth arrest and persistence. Nat Chem Biol 12:208–214. https://doi.org/10.1038/nchembio.2044
Pati R, Shevtsov M, Sonawane A (2018) Nanoparticle vaccines against infectious diseases. Front Immunol 9:2224. https://doi.org/10.3389/fimmu.2018.02224
Petosa C, Collier RJ, Klimpel KR, Leppla SH, Liddington RC (1997) Crystal structure of the anthrax toxin protective antigen. Nature 385:833–838. https://doi.org/10.1038/385833a0
Pomerantsev AP, Camp A, Leppla SH (2009) A new minimal replicon of *Bacillus anthracis* plasmid pXO1. J Bacteriol 191:5134–5146. https://doi.org/10.1128/jb.00422-09
Riedel S (2004) Biological warfare and bioterrorism: a historical review. Proc (Bayl Univ Med Cent) 17:400–406. https://doi.org/10.1080/08998280.2004.11928002

Schmidt M, Dekker FJ, Maarsingh H (2013) Exchange protein directly activated by cAMP (epac): a multidomain cAMP mediator in the regulation of diverse biological functions. Pharmacol Rev 65:670–709. https://doi.org/10.1124/pr.110.003707

Spencer RC (2003) *Bacillus anthracis*. J Clin Pathol 56:182–187. https://doi.org/10.1136/jcp.56.3.182

Szarowicz SE et al (2009) *Bacillus anthracis* edema toxin impairs neutrophil actin-based motility. Infect Immun 77:2455–2464. https://doi.org/10.1128/iai.00839-08

Tang WJ, Guo Q (2009) The adenylyl cyclase activity of anthrax edema factor. Mol Asp Med 30:423–430. https://doi.org/10.1016/j.mam.2009.06.001

Tin D, Sabeti P, Ciottone GR (2022) Bioterrorism: an analysis of biological agents used in terrorist events. Am J Emerg Med 54:117–121

Turnbull PC (1999) Definitive identification of *Bacillus anthracis*—a review. J Appl Microbiol 87:237–240. https://doi.org/10.1046/j.1365-2672.1999.00876.x

Turnbull PC, Broster MG, Carman JA, Manchee RJ, Melling J (1986) Development of antibodies to protective antigen and lethal factor components of anthrax toxin in humans and Guinea pigs and their relevance to protective immunity. Infect Immun 52:356–363. https://doi.org/10.1128/iai.52.2.356-363.1986

Uchida I, Hornung JM, Thorne CB, Klimpel KR, Leppla SH (1993) Cloning and characterization of a gene whose product is a trans-activator of anthrax toxin synthesis. J Bacteriol 175:5329–5338. https://doi.org/10.1128/jb.175.17.5329-5338.1993

van der Goot G, Young JA (2009) Receptors of anthrax toxin and cell entry. Mol Asp Med 30:406–412. https://doi.org/10.1016/j.mam.2009.08.007

Varughese M, Teixeira AV, Liu S, Leppla SH (1999) Identification of a receptor-binding region within domain 4 of the protective antigen component of anthrax toxin. Infect Immun 67:1860–1865. https://doi.org/10.1128/iai.67.4.1860-1865.1999

Vitale G et al (1999) Anthrax lethal factor cleaves the N-terminus of MAPKKS and induces tyrosine/threonine phosphorylation of MAPKS in cultured macrophages. J Appl Microbiol 87:288. https://doi.org/10.1046/j.1365-2672.1999.00893.x

Vitale G, Bernardi L, Napolitani G, Mock M, Montecucco C (2000) Susceptibility of mitogen-activated protein kinase kinase family members to proteolysis by anthrax lethal factor. Biochem J 352(Pt 3):739–745

Wade BH, Wright GG, Hewlett EL, Leppla SH, Mandell GL (1985) Anthrax toxin components stimulate chemotaxis of human polymorphonuclear neutrophils. Proc Soc Exp Biol Med 179:159–162. https://doi.org/10.3181/00379727-179-42078

Weiner ZP, Glomski IJ (2012) Updating perspectives on the initiation of *Bacillus anthracis* growth and dissemination through its host. Infect Immun 80(5):1626–1633

Weiner ZP et al (2014) Circulating lethal toxin decreases the ability of neutrophils to respond to *Bacillus anthracis*. Cell Microbiol 16:504–518. https://doi.org/10.1111/cmi.12232

Wickliffe KE, Leppla SH, Moayeri M (2008) Anthrax lethal toxin-induced inflammasome formation and caspase-1 activation are late events dependent on ion fluxes and the proteasome. Cell Microbiol 10:332–343. https://doi.org/10.1111/j.1462-5822.2007.01044.x

Williams AS, Lovell S, Anbanandam A, El-Chami R, Bann JG (2009) Domain 4 of the anthrax protective antigen maintains structure and binding to the host receptor CMG2 at low pH. Protein Sci 18:2277–2286. https://doi.org/10.1002/pro.238

Woodberry MW et al (2009) ATP depletion via mitochondrial F1F0 complex by lethal factor is an early event in B Anthracis-induced sudden cell death. J Cell Death 2:25–39. https://doi.org/10.4137/jcd.s2811

World Health Organization (2008) Anthrax in Humans and Animals Copyright ©. World Health Organization, Geneva

Artificial Intelligence in Healthcare: India Case Study

8

Raji Susan Mathew, Naveen Paluru, and Phaneendra K. Yalavarthy

Abstract

The field of artificial intelligence (AI) has shown a profound impact on the healthcare sector. The healthcare sector has been actively exploring the potential of AI to enable autonomous treatment options. One of the key factors driving the adoption of AI in healthcare, particularly in countries like India, is the shortage of trained medical professionals. With a large population and limited resources, AI offers a promising solution to bridge the gap between healthcare demand and supply. The COVID-19 pandemic has further highlighted the need for advanced technologies like AI to support healthcare systems overwhelmed by crisis. This chapter aims to delve into the advancements of AI in healthcare, with a specific focus on radiology and medical image analysis. This chapter will provide an overview of machine learning and deep learning methods used in radiology, along with challenges and potential solutions for improving AI-based healthcare. Further, this chapter will highlight the existing AI products and solutions. These products encompass various applications, such as computer-aided diagnosis, automated tumor detection, and predictive analytics for patient outcomes. The discussion will showcase the real-world impact of AI in healthcare and how it is transforming the delivery of medical services. Finally, this chapter will provide a futuristic outlook for AI in healthcare for India.

Raji Susan Mathew and Naveen Paluru contributed equally with all other contributors.

R. S. Mathew
School of Data Science, Indian Institute of Science Education and Research, Thiruvananthapuram, India

N. Paluru · P. K. Yalavarthy (✉)
Department of Computational and Data Sciences, Indian Institute of Science, Bengaluru, India
e-mail: yalavarthy@iisc.ac.in

G. John, A. K. Panda (eds.), *Biotechnology in India - Reworking A Strategy*, https://doi.org/10.1007/978-981-97-0896-3_8

8.1 Introduction

Artificial intelligence (AI) has revolutionized the interpretation of sensory information by allowing machines to mimic human intelligence to better represent and interpret complex data. With this capability, AI has been emerging as an irreplaceable tool in various applications ranging from web search and self-driving vehicles to natural language processing and computer vision. AI enables task-dependent tailoring of prior information that is particularly suited for the relevant task under consideration. Further, it decouples the computationally intense training step from the testing step. Inspired by its success, the healthcare sector has been making strides to make autonomous treatment a possibility.

The primary motive behind the development of AI-based medical imaging tools is to achieve greater efficiency in clinical care. The increased proportion of radiological imaging data led to a large-scale increase in radiologists workload. As shown (McDonald et al. 2015), to meet the workload demands, on average radiologists must interpret one image every 3–4 s in an 8-h workday. Under such constrained conditions, which involve a high amount of visual perception as well as decision-making, the chances of committing errors are inevitable (Fitzgerald 2001). Given the shortage of trained clinicians and with COVID-19 pandemic, AI utility has become a necessity for India.

There exist a myriad of tasks in radiology where AI can be utilized to achieve improved accuracy in the diagnosis and assessing the treatment response. Typically, the imaging pipeline comprises data acquisition/scanning, reconstruction of data to obtain the underlying image and preprocessing of the image, image analysis (image-based clinical tasks), generation of reports, and the integration of patient information from multiple data sources (Hosny et al. 2018). With recent advances in AI, it has the capability to be utilized in all the steps involved in the imaging pipeline, starting right from efficient data acquisition to the systematic interpretation of patient data from different sources, thereby significantly improving diagnostic accuracy. A schematic representation of the overview of clinical workflow in radiology is shown in Fig. 8.1. In the data acquisition part, the main objectives are to enable faster acquisition to avoid patient discomfort as well as to predetermine the data samples to be collected for recovering a clinical-quality image. In the subsequent step, the image reconstruction, the task is to enable the recovery of high-quality images from the acquired limited data samples such as magnetic resonance imaging (MRI) and computed tomography (CT) or low signal-to-noise ratio (SNR) data such as optical coherence tomography (OCT). The image-based clinical tasks involve the quantification of features to enable the detection of abnormalities by the characterization of objects in images, diagnosis and staging, and the monitoring of objects for assessment of treatment response. This is the step that utilizes most of the AI-driven techniques. The following step is the auto-generation of reports using the current observations and the previous history of the patient. And the final step involves the systematic interpretation of the patient data from different sources.

This chapter is a comprehensive study detailing the workflow in radiology, the challenges involved, and providing potential solutions for improving/accelerating the image-based tasks in radiology. Specifically, the discussion in this chapter is on

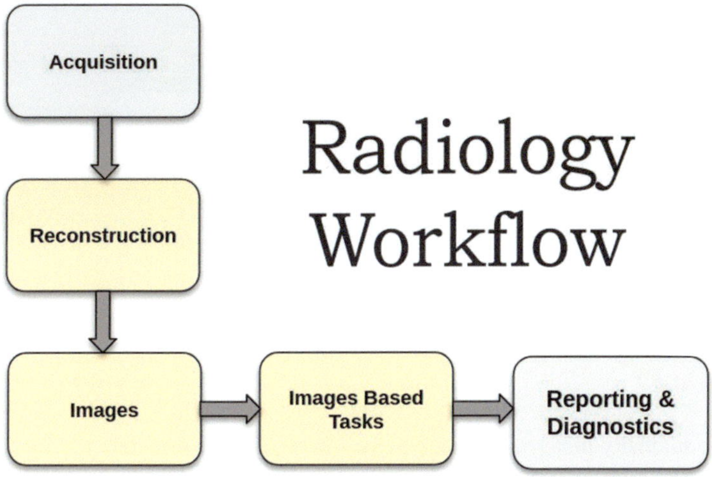

Fig. 8.1 Overview of clinical workflow in radiology. The image is motivated from (Hosny et al. 2018). Major applications of AI are in image reconstruction and image-based tasks like detection, characterization, and monitoring

the diagnostic medical imaging (radiology) aspects of healthcare, with a focus on image-based clinical tasks specific to India. Section 8.2 provides a brief overview of feature engineering (machine learning) and feature learning (deep learning) methods. Image-based clinical tasks in radiology have been discussed in Sect. 8.3. Section 8.4 introduces various challenges involved in developing deep learning based automated medical image analysis and provides potential solutions for these challenges. Section 8.5 discusses the role of AI in healthcare, specific to the Indian context along with current AI-based initiatives, challenges, and potential solutions. Finally, a concluding summary is provided in Sect. 8.6.

8.2 Feature Representations

The image-based clinical tasks are automated by learning meaningful representations from the imaging data. Learning such representations can be broadly categorized into (1) feature engineering and (2) feature learning. Specifically, feature engineering methods facilitate the machine learning models to perform the tasks. Conversely, deep learning techniques automate the feature learning process for performing imaging-based tasks. The fundamental difference between machine learning and deep learning methods is shown in Fig. 8.2. These methods are further explained below.

8.2.1 Machine Learning Methods

As discussed earlier, handcrafted feature representations drive machine learning methods. For example, the most popular handcrafted features for imaging-based

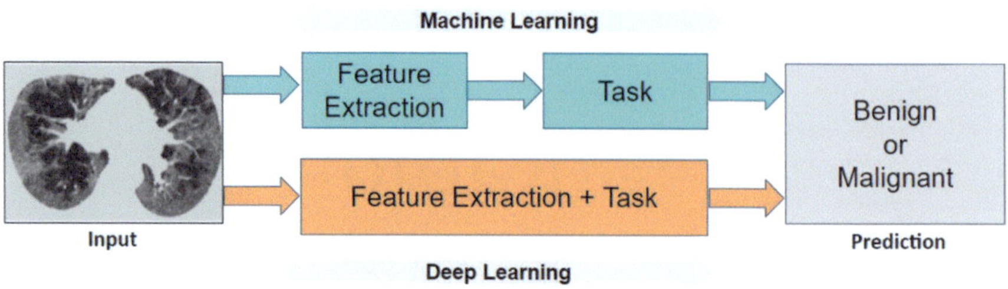

Fig. 8.2 Fundamental difference between machine learning and deep learning methods. Example of chest CT image (Ref. (Anon 2020)) classification (benign or malignant) is shown in the case of feature engineering (machine learning) and feature learning (deep learning). More details are provided in Sect. 8.2

tasks are intensity histograms, edge maps, histograms of oriented gradients (HoG) (Dalal and Triggs 2005), low dimensional embeddings (principal components (Mackiewicz and Ratajczak 1993)), and scale-invariant feature transforms (SIFT) (Lowe 2004), etc. These representations capture low- to medium-level semantics of the imaging data for driving machine learning methods. Given such feature representations, unsupervised techniques like clustering (Likas et al. 2003) and supervised techniques like support vector machine (SVM) (Noble 2006) are most popular among other machine learning based imaging tasks. However, the performance of these methods is sub-optimal with the increase in the data/task variability. The advancement of feature learning methods like convolutional neural networks (CNNs) leads to the enormous growth of automated imaging analysis.

8.2.2 Deep Learning Methods

Recently, deep learning frameworks have shown promising results in performing some kind of analysis like segmentation of medical images, enhancements like sinogram super-resolution, reconstruction of medical images, and dose estimation for radiotherapy. The capability of the deep learning models to predict the label on a given test sample is highly dependent on the training data. This section introduces different training strategies for building a deep learning model on a given dataset. Figure 8.3 shows the process of training deep learning models under various data constrained settings. As shown in Fig. 8.3, training schemes for building deep learning models can be broadly categorized into local data training, centralized training, and federated training. Let f represent a deep model characterized by a set of parameters (or weights), θ be the cost function, x, y be the data pairs, n be the number of training data samples for a given data center, and N be the total number of data centers. The training schemes for building deep models are explained as shown below:

1. **Local Data Training**: In local data training, each data center has access only to its own data. The deep learning models are trained locally as shown in Eq. (8.1). This kind of training has poor generalization while evaluating on the data

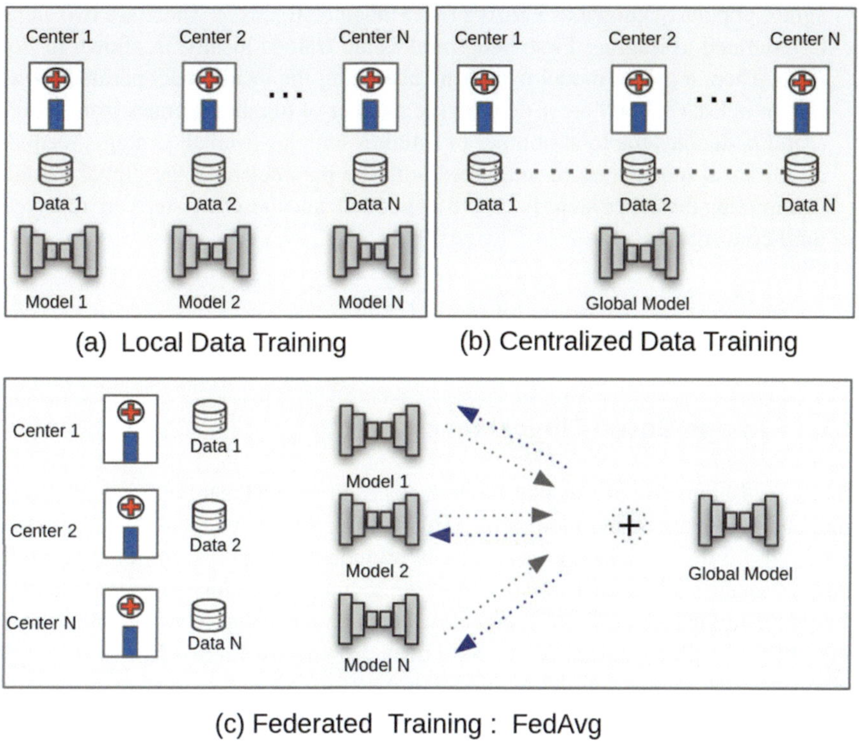

Fig. 8.3 Training deep learning models under different data constrained settings. (**a**) Local data training (data privacy is maintained but poor generalization), (**b**) centralized data training (no data privacy but good generalization), and (**c**) federated data training (good generalization with data privacy)

obtained from a different center. In Eq. (8.1), i represents the index of the data center and j represents the index of training samples.

$$\underset{\theta_i}{\text{argmin}} \frac{1}{n_i} \sum_{j=1}^{n_i} \mathcal{L}\left(f\left(x_{ij}, \theta_i\right), y_{ij}\right) \tag{8.1}$$

2. **Centralized Training:** In centralized training, a central server has access to the data samples from all the data centers and the deep learning model is trained on the centralized data as shown in Eq. (8.2), where n_t represents the total number of training samples collected from all the data centers. The deep model trained in this fashion has the highest capacity to generalize across the data samples from various centers.

$$\underset{\theta}{\text{argmin}} \frac{1}{n_t} \sum_{j=1}^{n_t} \mathcal{L}\left(f\left(x_j, \theta\right), y_j\right) \tag{8.2}$$

3. **Federated Training:** In federated training, the deep model utilizes the data from all the data centers by maintaining data privacy. Several advancements in federated training have shown that the performance can match the centralized training setup. The well-known baseline method in federated learning is federated aver-

aging, popularly known as FedAvg (McMahan et al. 2017). There are two steps in federated averaging: First, deep models are trained locally, as shown in Eq. (8.1). Then, a global model is built by averaging the local model parameters as shown in Eq. (8.3), where n_i denotes the number of training samples from center i, and n_t denotes the total number of training samples from all centers. Second, all the local models are re-initialized with the parameters of the global model. Completing these two steps is known as a round, and these two steps are repeated until convergence.

$$\theta_{global} = \sum_{i=1}^{N} \frac{n_i}{n_t} \theta_i \tag{8.3}$$

8.3 Image-Based Clinical Tasks

The typical hierarchy of imaging-based tasks used in healthcare consists of detection, characterization, and monitoring (Hosny et al. 2018). A schematic representation of the different image-based tasks is provided in Fig. 8.4. Further, example medical image analysis tasks (reconstruction, segmentation, denoising, and classification) spanning MRI, CT, OCT, and ultrasound from existing literature, along with reference to implementation details have been provided in Table 8.1.

8.3.1 Detection

In a manual detection workflow, clinicians visually examine the stacks of images with periodic adjustments of viewing planes, window width, and level settings. This

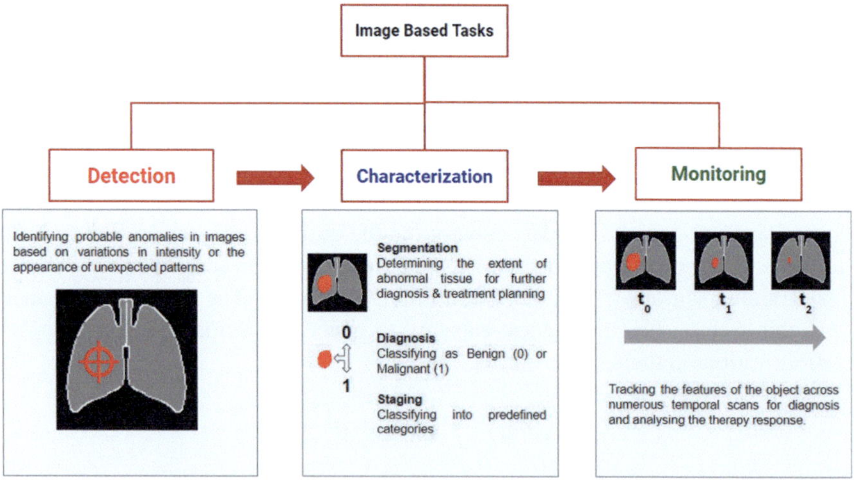

Fig. 8.4 Popular imaging-based tasks in radiology workflow: detection, characterization, and monitoring (Hosny et al. 2018). Each of these tasks has been explained in detail in Sect. 8.3. (The image is adapted and modified from (Hosny et al. 2018))

Table 8.1 Description of example medical image analysis tasks (reconstruction, segmentation, denoising, and classification) spanning MRI, CT, OCT, and ultrasound from existing literature, along with reference to implementation details

Modality	Study	Task	Organ (region)	Diagnostic information	References
MRI	Rastogi et al.	Reconstruction	Brain	Anatomical images	Rastogi and Yalavarthy (2021)
	Cheng et al.	Segmentation	Abdomen	Detecting organs	Ouyang et al. (2022)
	Lopez et al.	Denoising	Brain	Quality images	Lopez et al. (2021)
	Korolev et al.	Classification	Brain	Alzheimer's diseases	Korolev et al. (2017)
CT	Kang et al.	Reconstruction	Abdomen	Anatomical images	Kang et al. (2017)
	Paluru et al.	Segmentation	Lungs	COVID-19 anomalies	Paluru and Dayal (2021)
	Yang et al.	Denoising	Abdomen	Quality images	Yang et al. (2018)
	Aniello et al.	Classification	Lungs	COVID-19 anomalies	Castiglione et al. (2021)
OCT	Hao et al.	Reconstruction	Eyes	Anatomical images	Hao et al. (2020)
	Mukherjee et al.	Segmentation	Eyes	Retinal layers	Mukherjee et al. (2022)
	Nilesh et al.	Denoising	Eyes	Quality images	Kande et al. (2020)
	Paluru et al.	Classification	Eyes	Retinal diseases	Paluru et al. (2023)
Ultrasound	Yoon et al.	Reconstruction	Abdomen	Anatomical images	Yoon et al. (2018)
	Yang et al.	Segmentation	Wrist	Median nerve	Horng et al. (2020)
	Onur et al.	Denoising	Fetal head	Quality images	Karaoglu et al. (2022)
	Awasthi et al.	Classification	Lungs	COVID-19 anomalies	Awasthi et al. (2021)

sort of visual inspection relies on manual perceptual skills to identify potential abnormalities, followed by cognitive skills to confirm or refute the findings. This totally depends on their education, experience, and training to recognize abnormalities based on changes in imaging intensities or the emergence of unexpected patterns. These criteria, along with many others, are contained in a rather arbitrary decision matrix that permits the detection of anomalies in various human tissues.

Given the requirement to assess large amounts of radiographic images, it has become necessary to use computer-aided (or assisted) detection (CAD) to assist clinicians in segmentations that are specific to the application, imaging modality, and type of body part to be studied (Castellino 2005; Srinivasan Nagaraj et al. 2010). CAD is a pattern recognition software that detects suspicious features on the image and brings them to the attention of the radiologist that might otherwise be missed. As a result, the radiologist performs a second evaluation using the CAD software, before preparing the report. The most important objectives of automated segmentation procedures are (Sharma et al. 2010): (1) Handle a large number of radiographic images with the same accuracy so that the results are not impaired by fatigue, data overload, or skipped manual steps. (2) Generate fast and precise results using the latest high-speed computing resources. (3) Support faster communication to deliver medical care to patients from remote areas using information technology.

However, there are two key problems with the CAD system's effectiveness and dependability. First, the accuracy of conventional CAD-based systems remains questionable and can result in false positives. Second, the CAD system makes the whole procedure labor-intensive by requiring further examination by the radiologists to determine whether a generated annotation needs further assessment or not. For example, in the mammogram examination, it has been reported that the use of CAD did not result in any further improvements compared to that without using CAD (Cole et al. 2014; Lehman et al. 2015).

With the recent advances in deep learning, many efforts have been put forward to explore the deep learning based CAD to detect different anomalies including pulmonary nodules in CT (Huang et al. 2017), prostate cancer in multiparametric MRI (Tsehay et al. 2017; Wildeboer et al. 2020; De Vente et al. 2020), and lesions in mammograms (Kooi et al. 2017; Kleppe et al. 2021; Houssein et al. 2021). Studies demonstrate that the deep learning outperforms traditional CAD systems and performs on par with the radiologist's assessment. These facts indicate that deep learning can serve as an efficient tool in developing highly robust and accurate CAD software.

8.3.2 Characterization

In image-based clinical tasks, the term "characterization" refers to a variety of tasks including segmentation, diagnosis (classification), and disease staging (Hosny et al. 2018). These tasks are accomplished by calculating the size, extent, and interior texture of the abnormality from its radiological properties. In medical imaging, segmentation is the process of localizing (identifying) and delineating (separating) the anatomical region of interest from the rest. In essence, the medical image segmentation subdivides the objects in an image to identify the anatomical structure, particularly the region of interest (ROI), i.e., locate the tumor, lesion, and other abnormalities, measure tissue volume to analyze the spread of the tumor and help in treatment planning prior to radiation therapy such as radiation dose calculation (Sharma et al. 2010).

The segmentation of normal organs is comparatively easier, whereas identifying the extent of diseased tissue is more difficult. The typical standard tumor segmentation procedures in clinical radiology only include high-level parameters like the biggest in-plane diameter. A higher level of specificity and precision, however, is crucial in some other clinical conditions. For example, in clinical radiation oncology, it is necessary to precisely segment the extent of both tumor and non-tumor tissues in order to arrange radiation therapy. Under-segmentation can cause parts of the malignant tissue to remain after the therapy. Conversely, excessive segmentation might reduce the functionality of the organ rather than preserving it. Therefore, it is crucial to correctly identify the malignant tissue and formulate a treatment strategy. Deep learning architectures utilize CNNs to automatically extract features that are optimized for the task at hand. Architectural designs such as the UNet (Ronneberger et al. 2015) have been specifically designed for the segmentation of medical images (Yousef et al. 2023). CNNs have also been effectively used to segment medical images of knee cartilage (Prasoon et al. 2013), brain regions (Moeskops et al. 2016a; de Brebisson and Montana 2015; Zhou and Zhu 2023), the pancreas (Roth et al. 2015; Kumar et al. 2019; Lim et al. 2022), and coronary artery (Wolterink et al. 2016; Tian et al. 2021; Song et al. 2022). A single deep learning system has been shown to be capable of carrying out a variety of segmentation tasks across numerous modalities and tissue types, such as brain MRI, breast MRI, and cardiac CT angiography (CTA), without the need for task-specific training (Moeskops et al. 2016b; Wang et al. 2019; Zhao et al. 2022).

Various radiographic characteristics are evaluated as part of the succeeding diagnosis task. For instance, radiographic features of a tumor can include its size, maximal diameter, sphericity, internal texture, and margin definition. The parameters used for diagnosis are frequently arbitrary, allowing the differentiation of benign from malignant tissues. Some traditional CAD techniques are unable to generalize to various objects. For instance, despite the fact that tracking growth rates over time is thought to be an important aspect in determining risk, pulmonary nodule CAD systems built around this standard sometimes struggle to correctly identify unique nodules like cavity nodules (El-Baz et al. 2013). For the precise detection and diagnosis of such nodules (Edey and Hansell 2009), this necessitates the use of additional descriptors, which results in a variety of methods that are specifically designed for individual instances and have little generalizability. The deep learning based CAD, however, has the ability to automatically pick up on these ill-defined discriminative tasks, enabling it to capture both common and unique cases.

The information obtained from the segmentation and diagnosis is used to categorize the patients into various predetermined categories using the staging systems. This makes it possible to make an informed decision about the type of treatment and helps to determine the likelihood and prognosis of survival. Because staging relies on qualitative descriptions, which are frequently challenging and difficult to quantify, there has typically been little to no automation in the staging systems. Compared to traditional machine learning that depends on ensemble approaches, where numerous separate models are integrated, deep learning based approaches can learn joint data representations simultaneously, which make it ideally suited for such complex categorization challenges (Mirsadraee et al. 2012).

8.3.3 Monitoring

Monitoring is a crucial step for both disease diagnosis and therapy response assessment. This primarily includes an image registration preprocess in which the diseased tissue is aligned over several scans, followed by an assessment of straightforward metrics on them using preset protocols. A straightforward data comparison methodology is then utilized to measure change. For example, these procedures specify information about tumor size in oncology (Carl Jaffe et al. 2006). Although some characteristics such as significant differences in size, shape, and cavitation are immediately detectable by humans, other properties, including subtle variations in texture and heterogeneity within the object, are more difficult to identify. In addition to this, monitoring efficiency is also affected by unavoidable interobserver variability and poor image registration (Thiesse et al. 1997). The application of computer-aided change analysis is regarded as a comparatively younger field and has not yet attained as much of mainstream adoption as CAD in detection and characterization (Khorasani et al. 2006). The earlier attempts to automate change analysis operations depended on automatically registering numerous images, subtracting one from another, and then highlighting and presenting changed pixels to the reader. Other, more advanced techniques use pixel-by-pixel classification based on specified discriminative features to identify altered locations and then create a map of change (Patriarche and Erickson 2007).

A multistep process incorporating various feature sets is necessary since the predefined features used for registration and the subsequent change analysis are two different feature sets. As it becomes highly sensitive to registration errors, the change analysis stage might be compromised. Deep learning based computer-assisted change analysis does not rely on feature engineering and allows for the learning of a unified data representation. Hence, they are expected to find a wide range of applications in monitoring jobs.

8.4 Development to Deployment

The end-to-end pipeline describing the process flow from development to deployment is shown in Fig. 8.5. The development stage involves data preprocessing, deep learning model design, training, and validation. Deployment deals with generating lite versions of the trained models and designing user interface on the edge device. For example, popular deep learning libraries that are being used for AI model development and deployment have been shown in Tables 8.2 and 8.3, respectively. This

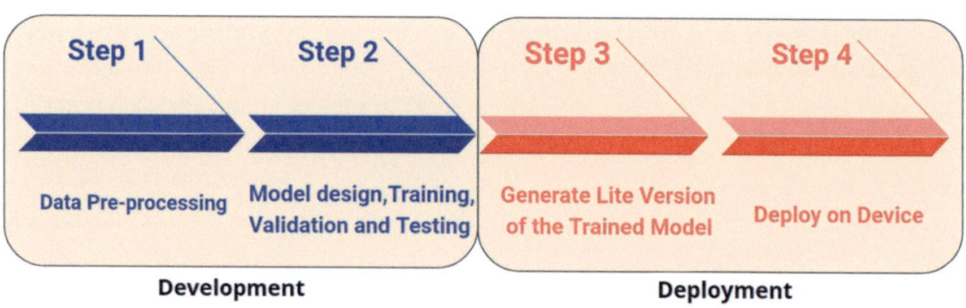

Fig. 8.5 Major steps involved from development to deployment stage of a deep learning model. Popular deep learning libraries that are being used for AI model development and deployment are listed in Tables 8.2 and 8.3, respectively

Table 8.2 Examples of popular deep learning libraries that are being used for AI model development

Library	Description
PyTorch	PyTorch is an open-source deep learning library developed by Facebook®. It provides a dynamic and intuitive programming interface, making it popular among researchers and practitioners. PyTorch allows for dynamic graph construction, making it easy to experiment and iterate [https://pytorch.org]
TensorFlow	TensorFlow is an open-source deep learning library developed by Google®. It provides a comprehensive ecosystem for building and deploying machine learning models, with a focus on neural networks. TensorFlow offers a flexible architecture, allowing users to define and train models using TensorFlow's lower-level operations [https://www.tensorflow.org/]
Keras	Keras is a high-level deep learning library written in Python that runs on top of other frameworks such as TensorFlow and Theano. It emphasizes ease of use, providing a user-friendly application programming interface (API) for defining and training neural networks. Keras abstracts away the complexities of lower-level operations, making it accessible to beginners while still being powerful enough for advanced use cases [https://keras.io/]
MXNet	MXNet is a flexible deep learning library designed for both efficiency and scalability. It offers a symbolic programming interface, allowing users to define and train models. MXNet supports distributed training across multiple devices and machines, making it suitable for large-scale deployments [https://mxnet.apache.org/]

Table 8.3 Examples of popular deep learning libraries that are being used for AI model deployment

Library	Description
ONNX Runtime	Open Neural Network Exchange (ONNX) Runtime is an open-source framework developed by Microsoft® that enables efficient deployment of machine learning models across different platforms. It supports models from various frameworks, including TensorFlow, PyTorch, and more and provides high-performance execution with hardware acceleration support [https://onnxruntime.ai/]
OpenVINO	OpenVINO (Open Visual Inference and Neural network Optimization) is an optimization toolkit developed by Intel®. It helps deploy deep learning models efficiently on Intel CPUs, GPUs, FPGAs, and other hardware accelerators [https://docs.openvino.ai/]
TFLite Micro	TFLite Micro is a specialized version of TensorFlow Lite developed by Google®. TFLite Micro is designed for deploying models on microcontrollers and other resource-constrained edge devices. It offers a small runtime footprint and optimized execution capabilities, making it suitable for IoT applications and edge computing [https://www.tensorflow.org/lite/]
Core ML	Core ML is a framework provided by Apple® for deploying machine learning models on Apple devices, including iPhones, iPads, and Macs. It supports models from various frameworks, and its optimized runtime provides efficient execution with hardware acceleration on Apple's A-series chips [https://apple.coreml.org/]

section introduces various challenges involved in developing deep learning based automated medical image analysis. We also discuss and highlight solutions for these challenges occurring at different stages of the process. On a broader note, these challenges can be classified into five categories: (1) data efficiency, (2) annotation efficiency, (3) generalizability, (4) deployability, and (5) data privacy. These are further explained below:

8.4.1 Data Efficiency

Data plays an important role in training deep learning models. The quantity and quality of available training data have a significant impact on the performance of any deep learning model. Deep learning models become more generalizable as more training samples are available, which is why they are often referred to as "data-hungry."

Large-scale datasets, however, may not always be accessible in practice due to propriety or privacy concerns, or because of the cost of data generation, annotation, and transmission. However, the availability of pre-trained deep learning models and data augmentation can handle the problem of data availability to a certain extent. Pre-trained deep learning models can generate data impressions which can be used to generate auxiliary datasets. Data impression techniques (Nayak et al. 2021) have been successful in training deep models in the absence of training data. However, similar experiments in the context of medical image analysis are yet to be explored. Further, training deep learning models with limited training data can be improved by transfer learning (Raghu et al. 2019; Zhuang et al. 2020; Niu et al. 2021). The details of example medical imaging datasets of Indian origin are summarized in Table 8.4.

8.4.2 Annotation Efficiency

Annotation-efficient deep learning refers to building high-performance deep learning models without requiring highly annotated data. Annotation is a tedious process, and having high-quality annotations, especially in tasks like segmentation, is very difficult. Further, it is challenging to get accurate annotations from medical experts, even for small-scale datasets, given their busy schedules (Tajbakhsh et al. 2021). However, there are few exciting solutions in literature dealing with annotation efficiency. The methods like unsupervised or semi-supervised learning utilize partially annotated data/noisy or weak annotations for training deep models (Huang et al. 2021). Conversely, techniques like active learning can identify the highly uncertain samples to be annotated from a given large database (Nath et al. 2020). Further, the techniques like zero-short and few-shot learning can train deep models with limited annotations (Al Chanti et al. 2021). Training the deep models for self-supervised pre-tasks (like rotation and inpainting, etc.) leverages learning high-quality embeddings; such deep models can be fine-tuned for a specific task with limited expert annotations (Chen et al. 2019). Finally, the highly celebrated method "domain adaptation" can generalize a deep model from a source dataset to an unlabeled target dataset (Tomar et al. 2021).

Table 8.4 Description of example medical imaging datasets of Indian origin

Study	Description
Sivaswamy et al. (2019)	The Indian Brain Segmentation Dataset (IBSD) comprises high-quality 1.5T T1w MRI data from 114 subjects, captured using a fixed imaging protocol. Expert radiologists have manually annotated the data, providing annotations for 14 sub-cortical structures. The dataset includes an approximately equal number of male and female subjects from a young age group (20–30 years). With its ample quantity of high-quality data, this dataset can be leveraged for various tasks, including segmenting structures of interest, aligning/registering images, and exploring both traditional methods and deep learning approaches
Porwal et al. (2018)	The Indian Diabetic Retinopathy Image Dataset (IDRiD) stands as the first database representative of the Indian population, offering a comprehensive collection of typical diabetic retinopathy lesions and pixel-level annotations of normal retinal structures. This dataset includes information on disease severity, encompassing diabetic retinopathy and diabetic macular edema, for each image. Consequently, IDRiD serves as an ideal resource for the development and evaluation of image analysis algorithms aimed at the early detection of diabetic retinopathy
Mandal et al. (2023)	The DBT-National Brain Research Centre (DBT-NBRC) has introduced the SWADESH project, an innovative brain initiative aimed at providing certified neuroimaging, neurochemical, and neuropsychological data, along with advanced analytics, to researchers for the management of brain disorders. SWADESH stands as the first large-scale multimodal neuroimaging database designed specifically for the Indian population. This initiative enables researchers to access valuable resources for in-depth exploration and understanding of brain disorders
Project MIDAS (2023)	Indian Council of Medical Research (ICMR) and Indian Institute of Science (IISc), Bengaluru, have collaborated and developed a Centre for Health Data, backed by ARTPARK, to create and offer India's healthcare Gold Standard Datasets as part of the national push for health research data. The goal of Project MIDAS is to capture the real diversity of India's healthcare spectrum in order to accelerate medical imaging AI research and innovation for inclusive Digital Health Programmes in India. This will accelerate the development, deployment, and effectiveness of medical imaging AI solutions centered on automatic disease screening, second opinion systems, and telemedicine programs

8.4.3 Generalizability

Deep learning models are highly biased toward the training data distribution. As a result, the performance of deep learning models drops significantly when evaluated on the data obtained from a different distribution. Given the variability in scanning protocols, acquisition devices, and the expertise of the device technician, it is tough to have a unified way of modeling the data distribution. For example, the lack of generalizability of deep models using MRI scans can be attributed to variability in the MR image acquisition; specifically, variations in the magnetic field strength, TE (time to echo), TR (repetition time), and spatial resolution. On a similar note, the source for the lack of generalizability of deep models in retinal diagnosis using OCT images can be attributed to the speckle noise present in the OCT images

(Schmitt et al. 1999; Karamata 2005). The methods focusing on improving the generalizability of deep models come under domain generalization. Domain generalization aims at improving the accuracy/performance of the deep models on test cases from the same or related domain. Techniques like domain alignment, domain adaptation, meta learning, ensemble learning, and regularization strategies collectively fall under domain generalization methods (Zhou et al. 2022; Guan and Liu 2021; Kouw and Loog 2019; Liu et al. 2022). However, specific to medical image analysis, developing modality or task-specific regularizing priors using the physics of image acquisition has more potential to promote the generalizability of deep learning models.

8.4.4 Deployability

One of the primary and essential problems associated with deep learning models is the specialized hardware requirement with high-end computational capabilities. Deep models having parameters in orders of tens of millions cannot give inferences in a point-of-care setting. Lightweight models having on-par performance with the existing heavy state-of-the-art models can be easily deployed in a point-of-care setting on resource-constrained settings using embedded/edge computing devices like NVIDIA Jetson, Raspberry Pi, etc. Note that one can also train the lightweight models on NVIDIA Jetson. Further, lightweight models have minimal training time and thus can be utilized to train site-specific models in a point-of-care setting. The computer vision community has elaboratively studied the design and development of lightweight deep learning methods. The techniques like knowledge distillation (KD) (Hinton et al. 2015; Wang and Yoon 2021), network pruning (Blalock et al. 2020; Zhuang et al. 2018), network quantization (Hubara et al. 2016; Liang et al. 2021), and lightweight designs (Sandler et al. 2018; Ma et al. 2018; Tan and Le 2019) are detailed to a great extent in the literature. However, task-specific deep models for medical image analysis in a point-of-care setting (Paluru and Dayal 2021; Awasthi et al. 2021) still need to be well explored.

8.4.5 Data Privacy

Data privacy related to developing deep learning models for medical image analysis can be studied in two stages: (1) during training and (2) during inference. Again, as detailed earlier, deep models' performance highly depends on the amount of heterogeneous data used for training. However, accessing such heterogeneous data on a large scale is extremely difficult. Techniques like federated learning (McMahan et al. 2017; Antunes et al. 2022; Rieke et al. 2020) have been proposed to utilize data from different centers without breaking data privacy. Nevertheless, there are also methods in the literature detailing the data leakage in the context of federated learning (Li et al. 2020). Particularly in the healthcare industry, hospitals or diagnostic facilities frequently refuse to allow data transfer to locations outside their facilities; this also applies to the collaborators. One potential solution for such cases is establishing an AI lab within the hospital or diagnostic center premises and then developing site-specific deep models for imaging-based tasks. On the inference side, data leakage is possible if the AI models run on some cloud-based centers. Also, such cloud-based AI solutions often have delayed responses because of round trip time; they require network connectivity, and the maintenance cost is high. The solution to such challenges is to design lightweight deep models that can provide promising results on edge devices like NVIDIA Jetson, Raspberry Pi, etc. (Paluru and Dayal 2021; Awasthi et al. 2021).

8.5 AI in Healthcare: Indian Scenario

The development of AI in India is still in its early stages. The Government of India recently launched a number of initiatives connected to AI, including the creation of an AI task force and the preparation of the NITI Aayog's National Strategy for Artificial Intelligence (Bajpai and Wadhwa 2021). Additionally, a few state governments in India have taken steps to build equivalent institutions.

One of the main benefits of AI in healthcare is that it may be extremely beneficial in locations with a shortage of human resources, particularly in rural and distant areas. With about one million new instances of cancer reported each year, the incidence of cancer is rising in India. However, India scarcely possesses more than a few thousand skilled oncopathologists. In this case, a pathologist can benefit from AI technology to make accurate diagnoses. NITI Aayog is in its final phase to create a national repository of annotated and curated pathology images as part of an endeavor to make high-quality pathology datasets accessible (Bajpai and Wadhwa 2021).

In August 2019, the Central TB Division under the Ministry of Health and Family Welfare (MoHFW) signed a Memorandum of Understanding (MoU) with Wadhwani Institute for Artificial Intelligence with the aim of developing and deploying AI-based solutions for tuberculosis (TB) (Ministry of Health and Family Welfare 2019). In addition to assisting the Revised National TB Control Programme (RNTCP) in adopting other AI technologies, it would support the program in vulnerability and hotspot mapping, modeling novel techniques of screening and diagnostics, and enabling decision assistance for clinicians.

In addition, NITI Aayog is collaborating with Microsoft and Forus Health on a pilot study to investigate the application of AI for the early identification of diabetic retinopathy (DR) (NITI Aayog 2020). The goal is to create an AI system for DR sign detection in diabetic patients and to scientifically validate the algorithm for use as a primary care screening tool.

Chronic Obstructive Pulmonary Disease Diagnosis (COPD) can also be aided by AI-based technologies (Bajpai and Wadhwa 2021). An AI-based early warning system in particular employs a sophisticated spirometer and cutting-edge analytics to help patients identify COPD triggers, symptoms, trends, and other specific insights. AI-assisted inhaler-based medication adherence solutions track patient's compliance with the recommended drug regimen and monitor the effectiveness of drug delivery strategy. Pulmonologists can visualize the structural and functional aspects of the lungs with the aid of modern computational fluid dynamics (CFD) techniques, high-resolution CT scans, and AI. Dealing with the COVID-19 crisis in India has proven to be a success, thanks to AI and its applications (Mohapatra and Giri 2020). With the help of this new technology, the pandemic can be tracked, COVID-19 cases screened, the virus contained, contacts traced, quarantines and social seclusion enforced, and suspects tracked. COVID-19 patients can also be treated and remotely monitored, all aiding in vaccine and drug development. According to a global survey by PricewaterhouseCoopers (PwC) India, during COVID-19, India saw the biggest increase in the application of AI (PWC 2020). According to the survey, 73% of pharmaceutical and healthcare organizations used AI.

8.5.1 Challenges

The absence of high-quality datasets, cultural biases in the datasets that are accessible, and the fact that many health records are handwritten in regional languages make the process much more difficult in the implementation of AI-driven healthcare in India (Khalid et al. 2020). Additionally, the healthcare data is still dispersed among numerous institutions, including hospitals, clinics, pharmacies, testing labs, etc. The lack of open (publicly accessible) medical datasets in India is another barrier, forcing start-ups to employ data from other nations, resulting in AI algorithms that do not take Indian demographics into account.

The lack of suitable infrastructure in India is one of the main obstacles for implementing AI in healthcare. Since most cloud computing infrastructure is found on servers outside of India, many Indian start-ups have established themselves abroad (Haider 2020). Another obstacle for using AI in healthcare in India is a lack of professionals with the necessary training (Pradhan et al. 2021). The labor that is readily available lacks the expertise needed to effectively employ AI technologies (Shaw et al. 2019; DonHee Lee and Seong No Yoon 2021; Rathod et al. 2024). A significant barrier to India's adoption of AI in the healthcare sector is the need for staff that is properly trained to handle sensitive health information carefully, defend against data theft, and use AI systems optimally. Another main challenge for a low-resource setting like India is the initial investment needed to implement AI in healthcare. The Indian government continues to make only modest investments in fields like AI and research (Paul et al. 2018).

The acceptability of the outcomes produced by AI algorithms is another issue with the use of AI in healthcare in India (Paul et al. 2018). Particularly in the Indian context, where the doctor–patient connection is accorded complete trust, the decisions taken by doctors based on AI solutions must be explainable. Additionally, there is currently a lack of knowledge about AI, its potential uses, and its advantages among both the general public and medical and healthcare experts.

Generalizability is another issue that needs to be addressed. For instance, the under representation of minority groups in data used to create AI algorithms may lead to biased findings (Panch et al. 2019). Due to skewed data (such as belonging to a particular ethnicity, gender, age, or religion), AI algorithms may produce findings that are unjust and more appropriate for some Indian demographics than others (Pinninti and Rajappa 2020).

Additionally, India lacks any legislative framework or organization to direct the creation of AI solutions and guarantee their security, privacy, and quality. This is another big challenge that needs to be addressed for India to adopt AI-based healthcare on a large scale.

Yet, another major challenge is regarding the privacy of information (Jagdev and Singh 2015). Cybersecurity is a significant problem as well. The online availability of confidential health information across the cloud computing environment poses a risk to data security. One such instance is the 2016 disclosure of medical records of more than 35,000 patients from a diagnostic lab in Maharashtra, India (Bajpai and Wadhwa 2021).

8.5.2 Discussion

Artificial intelligence is on the verge of revolutionizing our understanding and application of knowledge across various domains. In particular, the impact of AI in biotechnology, medical sciences, and related fields is expected to be unprecedented. With the power to process and analyze large amounts of data, AI can unlock hidden patterns, discover novel insights, and accelerate scientific discoveries. In biotechnology, AI-driven algorithms can aid in the design of more effective drugs, predicting their efficacy, potential side effects, and facilitating the development of targeted therapies. Furthermore, AI's ability to interpret complex biological data, such as genomic information, can lead to personalized medicine approaches, where treatments are tailored to individual patients based on their unique genetic profiles. In medical sciences, AI-powered diagnostic tools can learn from databases of medical records, imaging data, and clinical observations to aid in disease detection, risk assessment, and prognosis. Artificial intelligence based medical image analysis improves the accuracy and efficiency of radiology interpretations, enabling early detection of diseases, resulting in improved patient outcomes.

By automating routine tasks, AI can also alleviate the burden on healthcare professionals, allowing them to focus on more complex decision-making and patient care. The potential impact is not limited to research and healthcare delivery, but extends to areas such as disease prevention, population health management, and the optimization of healthcare systems. As AI's influence continues to grow, it is essential for stakeholders, including researchers, policymakers, and healthcare professionals, to understand and navigate the opportunities and challenges. Robust collaborations between AI experts and domain specialists will be crucial in harnessing the full potential of these technologies while addressing ethical considerations, privacy concerns, and ensuring the responsible use of AI in healthcare. By embracing AI in biotechnology, medical sciences, and related fields, there is immense opportunity to redefine the boundaries of knowledge and transform healthcare outcomes for the better.

Current adaptation of AI in healthcare is limited by seamless integration into the existing clinical workflows. Specific to India, this will be a paramount challenge given the diversity and lack of resources (personnel/computational). With 5G connectivity, one could expect AI to penetrate to tier II and III cities. Moreover, Ayushman Bharat Digital Mission (ABDM) being in place and digital health being seen as the twenty first century development, there are significant incentives being proposed by the National Health Authority of India (DHIS 2023). Thus, AI will have a big impact in the near future and change the current clinical practice. Nevertheless, end-to-end deployment of AI tools is still in the very nascent stage in India and is expected to take atleast another 5 years to make a difference.

The smooth integration of AI in healthcare can be speeded up by encouraging public–private partnerships, enacting laws and legislation related to AI and Health, such as those that address concerns about confidentiality and privacy in AI-driven healthcare, and establish certification systems for AI-based healthcare solutions. Additionally, it is crucial to train the workforce in AI so that they can properly manage delicate health information, safeguard data from theft, and utilize AI systems. Further, any healthcare decisions made using AI technologies should be justifiable and explainable.

8.6 Conclusion

Artificial intelligence (AI) is a rapidly expanding field that has great potential in all aspects of healthcare, particularly radiology. This chapter discussed currently used AI approaches, challenges, and potential applications in the future. Although high diagnostic accuracy has been reported in most of the AI-based studies, it is important to note that these findings are not always generalizable across clinical sites and settings, and reported outcomes overestimate the efficacy of these algorithms. Improvements in the application of AI can only be achieved with standardized study design and reporting which would further expand the clinical utility. End-to-end deployment of these AI tools is still a major hurdle for clinicians to see an impact across the globe. In India, digitization of the hospitals is on the rise and one could expect that the adaptation of AI tools will emerge as one of the most important part of this digitization.

References

Al Chanti D, Duque VG, Crouzier M, Nordez A, Lacourpaille L, Mateus D (2021) Ifss-net: interactive few-shot Siamese network for faster muscle segmentation and propagation in volumetric ultrasound. IEEE Trans Med Imaging 40(10):2615–2628

Anon (2020) Covid-19 CT segmentation dataset. http://medicalsegmentation.com/covid19/. Accessed 29 May 2020.

Antunes RS, da Costa CA, Kuderle A, Yari IA, Eskofier B (2022) Federated learning for healthcare: systematic review and architecture proposal. ACM Trans Intell Syst Technol 13(4):1–23
Awasthi N, Dayal A, Cenkeramaddi LR, Yalavarthy PK (2021) Mini-covidnet: efficient lightweight deep neural network for ultrasound based point-of-care detection of covid-19. IEEE Trans Ultrason Ferroelectr Freq Control 68(6):2023–2037
Nirupam Bajpai and Manisha Wadhwa (2021) Artificial intelligence and healthcare in India. Technical report, ICT India Working Paper 43
Blalock D, Ortiz JJG, Frankle J, Guttag J (2020) What is the state of neural network pruning? Proc Mach Learn Syst 2:129–146
Carl Jaffe C et al (2006) Measures of response: RECIST, WHO, and new alternatives. J Clin Oncol 24(20):3245–3251
Castellino RA (2005) Computer aided detection (CAD): an overview. Cancer Imaging 5(1):17
Castiglione A, Vijayakumar P, Nappi M, Sadiq S, Umer M (2021) Covid-19: automatic detection of the novel coronavirus disease from CT images using an optimized convolutional neural network. IEEE Trans Industr Inform 17(9):6480–6488
Chen L, Bentley P, Mori K, Misawa K, Fujiwara M, Rueckert D (2019) Self-supervised learning for medical image analysis using image context restoration. Med Image Anal 58:101539
Cole EB, Zhang Z, Marques HS, Edward Hendrick R, Yaffe MJ, Pisano ED (2014) Impact of computer-aided detection systems on radiologist accuracy with digital mammography. AJR Am J Roentgenol 203(4):909
Dalal N, Triggs B (2005) Histograms of oriented gradients for human detection. In: In 2005 IEEE computer society conference on computer vision and pattern recognition (CVPR'05), vol 1. IEEE, London, pp 886–893
de Brebisson A, Montana G (2015) Deep neural networks for anatomical brain segmentation. In: Proceedings of the IEEE conference on computer vision and pattern recognition workshops. Cornell University, Ithaca, pp 20–28
De Vente C, Vos P, Hosseinzadeh M, Pluim J, Veta M (2020) Deep learning regression for prostate cancer detection and grading in bi-parametric MRI. IEEE Trans Biomed Eng 68(2):374–383
DHIS (2023) Digital Health Incentive Scheme (DHIS) for ABDM adoption. https://abdm.gov.in/DHIS
Edey AJ, Hansell DM (2009) Incidentally detected small pulmonary nodules on CT. Clin Radiol 64(9):872–884
El-Baz A, Beache GM, Gimel'farb G, Suzuki K, Okada K, Elnakib A, Soliman A, Abdollahi B (2013) Computer-aided diagnosis systems for lung cancer: challenges and methodologies. Int J Biomed Imaging 2013:1
Fitzgerald R (2001) Error in radiology. Clin Radiol 56(12):938–946
Guan H, Liu M (2021) Domain adaptation for medical image analysis: a survey. IEEE Trans Biomed Eng 69(3):1173–1185
Haider H (2020) Barriers to the adoption of artificial intelligence in healthcare in India. K4D Helpdesk Report 780. Institute of Development Studies, Brighton
Hao Q, Zhou K, Yang J, Yan H, Chai Z, Ma Y, Liu G, Zhao Y, Gao S, Liu J (2020) High signal-to-noise ratio reconstruction of low bit-depth optical coherence tomography using deep learning. J Biomed Opt 25(12):123702–123702
Hinton G, Vinyals O, Dean J et al (2015) Distilling the knowledge in a neural network. arXiv preprint arXiv:1503.02531 2(7):1
Horng M-H, Yang C-W, Sun Y-N, Yang T-H (2020) Deepnerve: a new convolutional neural network for the localization and segmentation of the median nerve in ultrasound image sequences. Ultrasound Med Biol 46(9):2439–2452
Hosny A, Parmar C, Quackenbush J, Schwartz LH, Aerts HJWL (2018) Artificial intelligence in radiology. Nat Rev Cancer 18(8):500–510
Houssein EH, Emam MM, Ali AA, Suganthan PN (2021) Deep and machine learning techniques for medical imaging-based breast cancer: a comprehensive review. Expert Syst Appl 167:114161

Huang X, Shan J, Vaidya V (2017) Lung nodule detection in CT using 3D convolutional neural networks. In: 2017 IEEE 14th international symposium on biomedical imaging (ISBI 2017). IEEE, New York, pp 379–383
Huang W, Yang H, Liu X, Li C, Zhang I, Wang R, Zheng H, Wang S (2021) A coarse-to-fine deformable transformation framework for unsupervised multi-contrast MR image registration with dual consistency constraint. IEEE Trans Med Imaging 40(10):2589–2599
Hubara I, Courbariaux M, Soudry D, El-Yaniv R, Bengio Y (2016) Binarized neural networks. In: Proceedings of the 30th international conference on neural information processing systems, pp 4114–4122. Citeseer
Jagdev G, Singh S (2015) Implementation and applications of big data in health care industry. Int J Sci Tech Adv (IJSTA) 1(3):29–34
Kande NA, Dakhane R, Dukkipati A, Yalavarthy PK (2020) Siamesegan: a generative model for denoising of spectral domain optical coherence tomography images. IEEE Trans Med Imaging 40(1):180–192
Kang E, Min J, Ye JC (2017) A deep convolutional neural network using directional wavelets for low-dose X-ray CT reconstruction. Med Phys 44(10):e360–e375
Karamata B, Hassler K, Laubscher M, Lasser T (2005) Speckle statistics in optical coherence tomography. JOSA A 22(4):593–596
Karaoglu O, Hasan S, Bilge A, Uluer I (2022) Removal of speckle noises from ultra-sound images using five different deep learning networks. Eng Sci Technol Int J 29:101030
Khalid AM, Sharma S, Dubey AK (2020) Data gap analysis, indicator selection and index development: a case for developing economies. Soc Indic Res 148:893–960
Khorasani R, Erickson BJ, Patriarche J (2006) New opportunities in computer-aided diagnosis: change detection and characterization. J Am Coll Radiol 3(6):468–469
Kleppe A, Skrede O-J, De Raedt S, Liestøl K, Kerr DJ, Danielsen HE (2021) Designing deep learning studies in cancer diagnostics. Nat Rev Cancer 21(3):199–211
Kooi T, Litjens G, Van Ginneken B, Gubern-Merida A, Sanchez CI, Mann R, den Heeten A, Karssemeijer N (2017) Large scale deep learning for computer aided detection of mammographic lesions. Med Image Anal 35:303–312
Korolev S, Safiullin A, Belyaev M, Dodonova Y (2017) Residual and plain convolutional neural networks for 3D brain MRI classification. In: In 2017 IEEE 14th international symposium on biomedical imaging (ISBI 2017). IEEE, New York, pp 835–838
Kouw WM, Loog M (2019) A review of domain adaptation without target labels. IEEE Trans Pattern Anal Mach Intell 43(3):766–785
Kumar H, DeSouza SV, Petrov MS (2019) Automated pancreas segmentation from computed tomography and magnetic resonance images: a systematic review. Comput Methods Prog Biomed 178:319–328
Lee DH, Yoon SN (2021) Application of artificial intelligence-based technologies in the healthcare industry: opportunities and challenges. Int J Environ Res Public Health 18(1):271
Lehman CD, Wellman RD, Buist DSM, Kerlikowske K, Tosteson ANA, Miglioretti DL, Breast Cancer Surveillance Consortium et al (2015) Diagnostic accuracy of digital screening mammography with and without computer-aided detection. JAMA Intern Med 175(11):1828–1837
Li T, Sahu AK, Talwalkar A, Smith V (2020) Federated learning: challenges, methods, and future directions. IEEE Signal Process Mag 37(3):50–60
Liang T, Glossner J, Wang L, Shi S, Zhang X (2021) Pruning and quantization for deep neural network acceleration: a survey. Neurocomputing 461:370–403
Likas A, Vlassis N, Verbeek JJ (2003) The global k-means clustering algorithm. Pattern Recognition 36(2):451–461
Lim S-H, Kim YJ, Park Y-H, Kim D, Kim KG, Lee D-H (2022) Automated pancreas segmentation and volumetry using deep neural network on computed tomography. Sci Rep 12(1):4075
Liu X, Yoo C, Xing F, Hyejin O, El Fakhri G, Kang J-W, Woo J et al (2022) Deep unsupervised domain adaptation: a review of recent advances and perspectives. APSIPA Trans Signal Inf Processing 11(1):1

Lopez MM, Frederick JM, Ventura J (2021) Evaluation of MRI denoising methods using unsupervised learning. Front Artif Intell 4:642731

Lowe DG (2004) Distinctive image features from scale-invariant keypoints. Int J Comput Vis 60(2):91–110

Ma N, Zhang X, Zheng H-T, Sun J (2018) Shufflenet v2: practical guide-lines for efficient CNN architecture design. In: Proceedings of the European conference on computer vision (ECCV). Springer, Cham, pp 116–131

Mackiewicz A, Ratajczak W (1993) Principal components analysis (PCA). Comput Geosci 19(3):303–342

Mandal PK, Jindal K, Maroon JC, Chhikara R, Samkaria A, Joshi M, Roy S, Arora Y (2023) Brain imaging databases. ACS Chem Neurosci 14:1930

McDonald RJ, Schwartz KM, Eckel LJ, Diehn FE, Hunt CH, Bartholmai BJ, Erickson BJ, Kallmes DF (2015) The effects of changes in utilization and technological advancements of cross-sectional imaging on radiologist workload. Acad Radiol 22(9):1191–1198

McMahan B, Moore E, Ramage D, Hampson S, Aguera B, y Arcas. (2017) Communication-efficient learning of deep networks from decentralized data. In: Artificial intelligence and statistics, vol 54. PMLR, Cambridge, pp 1273–1282

Midas (2023). https://midas.iisc.ac.in/. Accessed 28 July 2023

Ministry of Health and Family Welfare (2019) TB Division of the Health Ministry signs MoU to explore Artificial Intelligence (AI) based solutions in combating TB. https://pib.gov. in/PressReleaseIframePage.aspx?PRID=1583584

Mirsadraee S, Oswal D, Alizadeh Y, Caulo A, van Beek EJR (2012) The 7th lung cancer TNM classification and staging system: review of the changes and implications. World J Radiol 4(4):128–134

Moeskops P, Viergever MA, Mendrik AM, De Vries LS, Benders MJNL, Isgum I (2016a) Automatic segmentation of MR brain images with a convolutional neural network. IEEE Trans Med Imaging 35(5):1252–1261

Moeskops P, Wolterink JM, van der Velden BHM, Gilhuijs KGA, Leiner T, Viergever MA, Isˇgum I (2016b) Deep learning for multi-task medical image segmentation in multiple modalities. In: International conference on medical image computing and computer-assisted intervention. Springer, pp 478–486

Mohapatra I, Giri P (2020) Artificial intelligence in healthcare and application in the fight against current pandemic-covid-19. Natl J Res Community Med 9(2):81–84

Mukherjee S, De Silva T, Grisso P, Henry Wiley DL, Tiarnan K, Thavikulwat AT, Chew E, Cukras C (2022) Retinal layer segmentation in optical coherence tomography (OCT) using a 3D deep-convolutional regression network for patients with age-related macular degeneration. Biomed Opt Express 13(6):3195–3210

Nath V, Yang D, Landman BA, Daguang X, Roth HR (2020) Diminishing uncertainty within the training pool: active learning for medical image segmentation. IEEE Trans Med Imaging 40(10):2534–2547

Nayak GK, Mopuri KR, Jain S, Chakraborty A (2021) Mining data impressions from deep models as substitute for the unavailable training data. IEEE Trans Pattern Anal Mach Intell 44:1

NITI Aayog (2020) Detection of diabetic retinopathy in primary care using AI. https://indiaai.gov.in/government/niti-aayog

Niu S, Liu M, Liu Y, Wang J, Song H (2021) Distant domain transfer learning for medical imaging. IEEE J Biomed Health Inform 25(10):3784–3793

Noble WS (2006) What is a support vector machine? Nat Biotechnol 24(12):1565–1567

Ouyang C, Biffi C, Chen C, Kart T, Qiu H, Rueckert D (2022) Self-supervised learning for few-shot medical image segmentation. IEEE Trans Med Imaging 41(7):1837–1848

Paluru N, Dayal A, Jenssen HB, Sakinis T, Cenkera-maddi LR, Prakash J, Yalavarthy PK (2021) Anam-net: anamorphic depth embedding-based lightweight CNN for segmentation of anomalies in Covid-19 chest CT images. IEEE Trans Neural Netw Learn Syst 32(3):932–946

Paluru N, Ravishankar H, Hegde S, Yalavarthy PK (2023) Self-distillation for improving the generalizability of retinal disease diagnosis using optical coherence tomography images. IEEE J Sel Top Quantum Electron 29(4: Biophotonics):1–12

Panch T, Pearson-Stuttard J, Greaves F, Atun R (2019) Artificial intelligence: opportunities and risks for public health. Lancet Digit Health 1(1):e13–e14

Patriarche JW, Erickson BJ (2007) Part 1. Automated change detection and characterization in serial MR studies of brain-tumor patients. J Digit Imaging 20(3):203–222

Paul Y, Hickok E, Sinha A, Tiwari U, Mohandas S, Ray S, Bidare PM (2018) Artificial intelligence in the healthcare industry in India. The Centre for Internet and Society, India, Bengaluru

Pinninti R, Rajappa S et al (2020) Artificial intelligence in health-care: how long to go? Cancer research. Stat Treatment 3(1):133

Porwal P, Pachade S, Kamble R, Kokare M, Deshmukh G, Sahasrabuddhe V, Meriaudeau F (2018) Indian diabetic retinopathy image dataset (IDRID): a database for diabetic retinopathy screening research. Data 3(3):25

Pradhan K, John P, Sandhu N (2021) Use of artificial intelligence in healthcare delivery in India. J Hospital Manag Health Policy 5:1

Prasoon A, Petersen K, Igel C, Lauze F, Dam E, Nielsen M (2013) Deep feature learning for knee cartilage segmentation using a triplanar convolutional neural network. In: International conference on medical image computing and computer-assisted intervention. Springer, Cham, pp 246–253

PWC (2020) AI: an opportunity amidst a crisis. https://www.pwc.in/consulting/technology/data-and-analytics/ai-an-opportunity-amidst-a-crisis.html

Raghu M, Zhang C, Kleinberg J, Bengio S (2019) Transfusion: under-standing transfer learning for medical imaging. Adv Neural Inf Proces Syst 32:1

Rastogi A, Yalavarthy PK (2021) Spinet: a deep neural network for Schatten p-norm regularized medical image reconstruction. Med Phys 48(5):2214–2229

Rathod SM, Patel NC, Prajapati BG, Prajapati JB, Singh S (2024) Artificial intelligence issues in healthcare. In: Concepts of artificial intelligence and its application in modern healthcare systems. CRC, Boca Raton, FL, pp 213–231

Rieke N, Hancox J, Li W, Milletari F, Roth HR, Albarqouni S, Bakas S, Galtier MN, Landman BA, Maier-Hein K et al (2020) The future of digital health with federated learning. NPJ Digit Med 3(1):1–7

Ronneberger O, Fischer P, Brox T (2015) U-net: convolutional networks for biomedical image segmentation. In: International conference on medical image computing and computer-assisted intervention. Springer, Cham, pp 234–241

Roth HR, Le L, Farag A, Shin H-C, Liu J, Turkbey EB, Summers RM (2015) Deeporgan: multi-level deep convolutional networks for automated pancreas segmentation. In: International conference on medical image computing and computer-assisted intervention. Springer, Cham, pp 556–564

Sandler M, Howard A, Zhu M, Zhmoginov A, Chen L-C (2018) Mobilenetv2: inverted residuals and linear bottlenecks. In: Proceedings of the IEEE conference on computer vision and pattern recognition. IEEE, New York, pp 4510–4520

Schmitt JM, Xiang SH, Yung KM (1999) Speckle in optical coherence tomography. J Biomed Opt 4(1):95–105

Sharma N, Aggarwal LM et al (2010) Automated medical image segmentation techniques. J Med Physics 35(1):3

Shaw J, Rudzicz F, Jamieson T, Goldfarb A (2019) Artificial intelligence and the implementation challenge. J Med Internet Res 21(7):e13659

Sivaswamy J, Thottupattu AJ, Mehta R, Sheelakumari R, Kesavadas C et al (2019) Construction of Indian human brain atlas. Neurol India 67(1):229

Song A, Lisheng X, Wang L, Wang B, Xiaofan Yang BX, Yang B, Greenwald SE (2022) Automatic coronary artery segmentation of CCTA images with an efficient feature-fusion-and-rectification 3D-unet. IEEE J Biomed Health Inform 26(8):4044–4055

Srinivasan Nagaraj G, Narasinga Rao K, Koteswararao et al (2010) The role of pattern recognition in computer-aided diagnosis and computer-aided detection in medical imaging: a clinical validation. Int J Comput Appl 8(5):18–22

Tajbakhsh N, Roth H, Terzopoulos D, Liang J (2021) Guest editorial annotation-efficient deep learning: the holy grail of medical imaging. IEEE Trans Med Imaging 40(10):2526–2533

Tan M, Le Q (2019) Efficientnet: rethinking model scaling for convolutional neural Networks. In: International conference on machine learning. PMLR, Cambridge, pp 6105–6114

Thiesse P, Ollivier L, Di Stefano-Louineau D, Negrier S, Savary J, Pignard K, Lasset C, Escudier B (1997) Response rate accuracy in oncology trials: reasons for interobserver variability. groupe franc¸ais d'immunothérapie of the fédération nationale des centres de lutte contre le cancer. J Clin Oncol 15(12):3507–3514

Tian F, Gao Y, Fang Z, Jia G (2021) Automatic coronary artery segmentation algorithm based on deep learning and digital image processing. Appl Intell 51:8881–8895

Tomar D, Lortkipanidze M, Vray G, Bozorgtabar B, Thiran J-P (2021) Self-attentive spatial adaptive normalization for cross-modality domain adaptation. IEEE Trans Med Imaging 40(10):2926–2938

Tsehay YK, Lay NS, Roth HR, Wang X, Kwak JT, Turkbey BI, Pinto PA, Wood BJ, Summers RM (2017) Convolutional neural network based deep-learning architecture for prostate cancer detection on multiparametric magnetic resonance images. In: Medical imaging 2017: computer-aided diagnosis, vol 10134. SPIE, Bellingham, pp 20–30

Wang L, Yoon K-J (2021) Knowledge distillation and student-teacher learning for visual intelligence: a review and new outlooks. IEEE Trans Pattern Anal Mach Intell 44:3048

Wang L, Chen R, Wang S, Zeng N, Huang X, Liu C (2019) Nested dilation network (NDN) for multi-task medical image segmentation. IEEE Access 7:44676–44685

Wildeboer RR, van Sloun RJG, Wijkstra H, Mischi M (2020) Artificial intelligence in multiparametric prostate cancer imaging with focus on deep-learning methods. Comput Methods Prog Biomed 189:105316

Wolterink JM, Leiner T, de Vos BD, van Hamersvelt RW, Viergever MA, Isˇgum I (2016) Automatic coronary artery calcium scoring in cardiac CT angiography using paired convolutional neural networks. Med Image Anal 34:123–136

Yang Q, Yan P, Zhang Y, Hengyong Y, Shi Y, Mou X, Kalra MK, Zhang Y, Sun L, Wang G (2018) Low-dose CT image denoising using a generative adversarial network with Wasserstein distance and perceptual loss. IEEE Trans Med Imaging 37(6):1348–1357

Yoon YH, Khan S, Huh J, Ye JC (2018) Efficient b-mode ultrasound image reconstruction from sub-sampled RF data using deep learning. IEEE Trans Med Imaging 38(2):325–336

Yousef R, Khan S, Gupta G, Siddiqui T, Albahlal BM, Alajlan S A-d, Haq MA (2023) U-net-based models towards optimal MR brain image segmentation. Diagnostics 13(9):1624

Zhao Y, Wang X, Che T, Bao G, Li S (2022) Multi-task deep learning for medical image computing and analysis: a review. Comput Biol Med 153:106496

Zhou T, Zhu S (2023) Uncertainty quantification and attention-aware fusion guided multi-modal MR brain tumor segmentation. Comput Biol Med 163:107142

Zhou K, Ziwei Liu Y, Qiao TX, Loy CC (2022) Domain generalization: a survey. IEEE Trans Pattern Anal Mach Intell 45:1

Zhuang Z, Tan M, Zhuang B, Liu J, Guo Y, Qingyao W, Huang J, Zhu J (2018) Discrimination-aware channel pruning for deep neural networks. Adv Neural Inf Proces Syst 31:1

Zhuang F, Qi Z, Duan K, Xi D, Zhu Y, Zhu H, Xiong H, He Q (2020) A comprehensive survey on transfer learning. Proc IEEE 109(1):43–76

Part IV

Industrial Biotechnology

Fermentation Technology: Driving Biotechnology to Work

9

Amulya K. Panda

Abstract

Fermentation helps in producing biologicals using either a living organism or a part of it and it is considered to be the core of bioprocess technology. Developments in fermentation technology have played a major role in the manufacturing of biologicals in India for the last 100 years. In India, the use of fermentation technology started after the second World War where industries started producing alcohol, vitamins, etc. using stirred tank reactors. The real boom of fermentation technology in India was seen after the establishment of antibiotic industries such as IDPL, HAL, Alembic, Sarabhai Chemicals, and many others. Simultaneously, fermentation research started in CSIR laboratories like CFTRI, Mysore; CDRI, Lucknow; NCL, Pune; and RRL, Jammu, and in a few other places. With the setting up of the Department of Biotechnology, India has started producing recombinant proteins and other biologicals using fermentation technology. However, in recent times more emphasis is given on Advance Therapy Medicinal Product (ATMP), which includes cell, gene therapy, and tissue engineering-based products. There are also demands for producing products using cellular agriculture and proteins using precision fermentation. Thus, there is a need to adopt fermentation technology to cater to the need for ATMP and production of many complex biologicals using multicellular organisms. To achieve this, a paradigm shift in fermentation technology is needed particularly in the area of generation of trained technical manpower, setting up of bioprocess engineering research centers, and more importantly manufacturing of equipment and control systems for the fermentation plant. This chapter describes the chronology of fermentation technologies and their use in the promotion of the biotechnology industry in

A. K. Panda (✉)
Panacea Biotec Limited, New Delhi, India
e-mail: amulya@nii.ac.in

G. John, A. K. Panda (eds.), *Biotechnology in India - Reworking A Strategy*,
https://doi.org/10.1007/978-981-97-0896-3_9

India. It also discusses a much-needed strategy to harness its huge potential for the benefit of society at large.

Keywords
Fermentation · Bioreactor · Recombinant products · Gene and cell therapy product

9.1 Introduction

Fermentation essentially involves a chemical transformation of molecules with the help of a living organism or a part of it. Fermentation has been used since 10,000 BC for making dairy and food products (Debret 2021; Taveira et al. 2021). It was Louis Pasteur who confirmed the involvement of a living microorganism in fermentation. Alcohol fermentation became more modern after that, and during World War II, the production of antibiotics through submerged fermentation changed the fermentation technology in the line of bulk chemical industry. In India, even though vaccines were being made by growing cells at the Central Research Institute, Kasauli, and Pasteur Institute of India, Coonoor, using fermenters, the real fermentation technology boomed with the production of antibiotics at Indian Drug and Pharmaceuticals Limited (IDPL), Hindustan Antibiotics Limited (HAL), Alembic Chemicals, Sarabhai Chemicals, and few others. Vaccines were produced using fermentation technology at the Serum Institute of India and then later at the Indian Immunologicals, Hyderabad. These were the prominent fermentation industries in the country during the 1960s. They profoundly contributed towards the development of fermentation technology while supplying antibiotics and vaccines to the country. The development of recombinant DNA technology and the setting of National Biotechnology Board by the Government of India in 1982 which later became the Department of Biotechnology (DBT) in 1986, promoted the growth of fermentation technology to cater to the needs of biotechnology-based products. With the continuous support of the Department of Biotechnology and Biotechnology Industry Research Assisted Council (BIRAC), fermentation technology has been used to produce novel therapeutics and biologicals such as therapeutic proteins, monoclonal antibodies, blood products, cells, and gene therapy-based products. It has also promoted the production of chemicals and secondary metabolites using different microorganisms. Thus, with the changing requirement of medicines from small molecules to antibiotics, monoclonal antibody, cell and gene therapy-based products, fermentation technology has advanced to meet the demand of vaccine and biopharma-based industries. The technology which was primarily used to provide sterile controlled environments to produce biologics has metamorphosed into different types to meet industry and end-user demand. The demonstration of India's capability to produce Covid-19 vaccine through fermentation showed that Indian industries have the capacity and talent to handle the challenges of novel biotechnology-based processes. India is the vaccine capital of the world due to the use of fermentation technology to produce quality biologicals at a very reasonable price. The Indian government along with the regulators has supported the fermentation technology sector and this is being used in healthcare, agriculture, environment sector, and in producing bioenergy. In a true sense, fermentation technology delivers the final product and has helped in the realization of the true potential of modern biotechnology. It is thus essential to know how fermentation technology evolved in the world and how India adopted the technology before suggesting the action for the future.

9.2 History of Fermentation Technology

Fermentation technology in terms of preserving and making milk products dates back to 10,000 BC with reports of alcoholic drinks by the Chinese around 7000 BC. Bread and pickles were made through fermentation around 2000 BC and around 200 BC, tea fermentation started (Debret 2021; Buchholz and Collins 2013). All these processes were due to the involvement of a microbe as shown by Louis Pasteur around 1856. From those times, fermentation industries started improving till the second World War where production of penicillin changed the fermentation technology from an art to a science-based process. The concept of sterility, oxygen transfer, and control of parameters with automation helped in better performance of fermentation-based processes. The concept of scale-up in high-volume bioreactors was optimized, which replaced the prevailing small culture flasks. Fermentation process and operation were further optimized to produce different biologics using living organisms. The high-volume fermenter with sterilization, aeration, and mixing facility became the norm for large-scale production of small biological molecules. Added advantages of fermentation technology became clear with the development of sterilizable sensors and control panels where different operations were fine-tuned to obtain the desired growth of living cells. This instrumentation and control system helped in improving the efficiency of fermentation technology and resulted in cost-effective production of antibiotics, small molecules, and biologicals (Stanburry et al. 2016).

9.2.1 Renaissance in Fermentation Technology

The inputs from submerged penicillin and other antibiotics fermentation process opened new possibilities of fermentation technology (Elander 2003). It led to the growth of other organisms such as actinomycetes, yeast, fungi, and bacteria to produce secondary metabolites such as steroids, enzymes, and other microbial products. The first renaissance of fermentation technology came with the extension of submerged technology to produce other small biologicals like citric acid, vitamins, organic acids, high fructose syrup, and steroids. China was a major producer of fermentation-based products in the last century (Chen et al. 2005). Different bioreactor configurations were also designed with aeration and agitation to grow cells (Arthur 1998). Attempts were also made to grow organisms aerobically or anaerobically to produce the desired product. Many industrial enzymes and enzyme-based biotransformations were achieved using fermentation process (Al-Maqtari and

Mahdi 2019). Shear sensitive cell-based fermentation such as plant cell fermentation was also developed using standard fermentation technology (Panda et al. 1989). Secondary metabolites from plants which took many years from natural plants were produced using plant cell fermentation processes (Yue et al. 2016; Wawrosch and Zotchev 2021). Modern fermentation technology was also further used to produce different types of food (Ranjana et al. 2020). The use of fermentation technology for the production of recombinant proteins and antibodies started in the last part of the twentieth century. Fermentation technology was mostly utilized to produce a huge number of recombinant proteins that led to the development of Biosimilars. This helped for the first time in producing human proteins outside of the human body. Recombinant insulin, human growth hormone, and monoclonal antibodies are the flagship products of the modern fermentation industry.

The 2nd renaissance of fermentation technology came with the development of high cell density and perfusion-based processes (Lee 1996). High cell density growth of *E. coli* for recombinant protein production helped in improving the volumetric productivity of many proteins (Joseph and Fassb 2005). This was extended to grow hybridoma and animal cells using perfusion technology for the production of monoclonal antibody and glycoproteins (Jyothilekshmi and Jayaprakash 2021). This led to the concept of cellular scale-up where increase in cell density became more common during the fermentation process. Developments in molecular biology and cell engineering leading to high performance cell lines made the fermentation process more advanced and user friendly. The concentrations of proteins and antibodies from a few 100 mg/L were enhanced to gm/L in the bioreactor (Bielsera 2018). The days of more than 10,000 L bioreactors, which were the gold standards in the early 1980s, became a low-volume high-density process. High cell density fed-batch fermentation and perfusion-based fermentation helped in improving the concentration of therapeutic recombinant proteins considerably (Subramaniam et al. 2018). This also helped in the large-scale production of viral vaccines using cell culture rather than using animal brain or eggs (Fang et al. 2022). The development of different fermentation processes such as fed-batch with re-cycling operation, perfusion-based fermentation, and multiphase fermentation involving gas, liquid, and solid improved the process performances. High cell density fermentation helped in reducing the volume of fermenters for large-scale production of biologicals.

The third renaissance came with the use of fermentation technology for gene, cell therapy, and tissue-engineered products (Ganeeva et al. 2022). Fermenters promoting the growth of multicellular organisms, organ culture, and cell therapy applications were designed. This resulted in the production of personalized medicine using different configurations of fermenters. Fermentation technology was used to produce stem cells, T and Dendritic cells (DC) for immunotherapy application, and CAR-T cell therapy for cancer immunotherapy (Gracia-Aponte et al. 2021). Stem cells and engineered stem cells were propagated in specialized bioreactors to meet the requirement (Cherian et al. 2020). The advanced therapy medicinal products (ATMP) which include products based on gene therapy, cell therapy, and tissue engineering required specialized fermentation design and operation to grow cells and tissue (Lim et al. 2022). Already, more than 20 ATMP products are FDA approved, and fermentation technology will play a major role in delivering the product to end users (Approved Cellular and Gene Therapy Products 2022). Starting from a simple vat with controlled environment to produce alcohol and solvents, fermentation technology changed over the last five decades in response to the growing demands of medical treatment. Similar trends were developed in growing plant cells, organ culture, along with embryo and hairy root culture for different applications. Fermenter design and operations were optimized to meet the changing requirements with advancements in modern biology. The use of control and monitoring systems provided an opportunity to communicate with the living system through fermentation.

From shaker flasks to vats to computer-controlled stainless steel vessels and to disposable bioreactors, the fermentation technology has progressed and advanced concomitant with the advancement in biology. In the last hundred years, the fermentation technology has changed in terms of geometry, control, and instrumentation considerably and this is depicted in Fig. 9.1. Even though the basic features of providing a controlled sterile environment inside the vessel remained the same, the design and operation have improved for growing cells of higher living organism, complex tissues, human organs, and even multicellular organisms. The single-use, perfusion-based fermenter to support adherent cells for viral vaccines and cell-based products is the hallmark of the latest advancement in fermentation technology.

9.2.2 Solid-State Fermentation Technology

In parallel with the development in submerged fermentation which started with modern penicillin manufacturing, solid-state fermentation process was also developed to produce biologicals (Pandey et al. 2000). Solid-state fermentation involves the utilization of solid matrix containing adequate moisture to promote the growth of microorganisms for the production of suitable biochemicals (Pandey et al. 2018; Steudler et al. 2019). The solid material can be a nutrient source or one that provides a support system for the growth of microorganisms or both. Solid-state fermentation is most suitable for the growth of filamentous organisms like ascomycetes, basidiomycetes, yeast, and various bacteria for the production of secondary metabolites, enzymes, and other biologics of diverse range. Traditionally, solid-state fermentation was used to produce fermented food and products of fungal origin which can be easily grown on the support system. Solid-state growth reduces the problems of aeration and agitation encountered in submerged fermentation and has been used for the production of many enzymes and secondary metabolites. It offers advantages in terms of low catabolite repression and substrate inhibition during fermentation, low energy consumption, and high volumetric outputs for metabolite production and most importantly less effluent generation. In the last fifty years, solid-state fermentation technology has advanced in terms of design, automation, introduction of moist phase for controlled growth of microorganisms, and online monitoring of microbial growth using respirometry (Yafetto 2022). This resulted in the use of

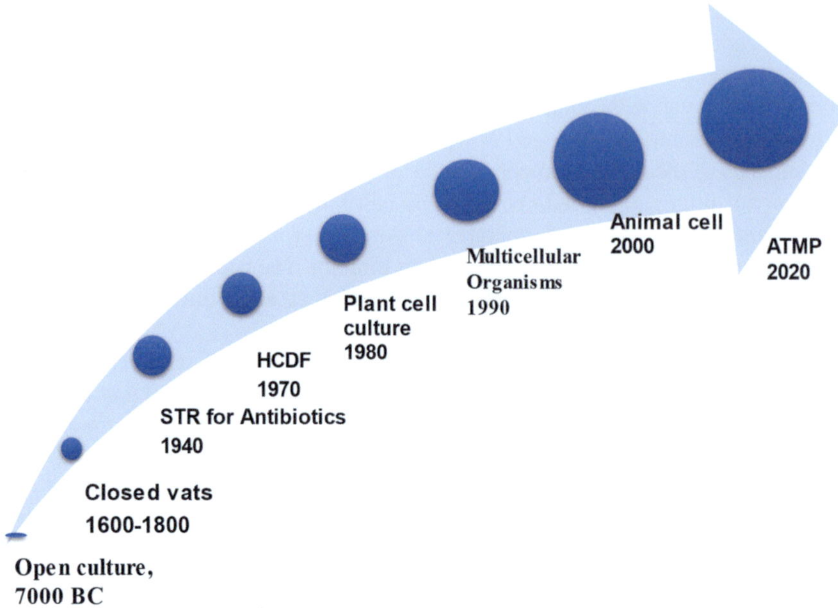

Fig. 9.1 Use of fermentation technology over the years for production of biologicals. The real changes in fermentation technology started in the early part of the twentieth century and during World War II. Large-scale requirement of antibiotics introduced the concept of aeration and agitation in fermentation. This further led to the development of high cell density fermentation (HCDF) and the production of plant metabolites using fermentation in the eighties. The growth of higher organism and animal cell using the perfusion system helped in large-scale manufacturing of complex metabolites, recombinant protein, and monoclonal antibodies. Currently, novel fermenter design is being used for producing cell and gene therapy and tissue-engineered products, which are collectively known as advanced therapy medicinal products (ATMP)

solid-state fermentation for the production of enzymes, secondary metabolites, organic acids, and also bioconversion of agriculture wastes to value added products (Kumar et al. 2021; Chilakamarry et al. 2022).

Different designs of bioreactors have been used for improving the volumetric productivity of the solid-state fermentation process (Pandey et al. 2018). Tray-type reactor, packed bed bioreactor, air pressure pulsation bioreactor, and moving bed reactor have been used for solid-state fermentation (Arora et al. 2018). Biocon India limited developed a PlaFractor bioreactor for fed-batch operation of solid-state fermentation process (Majumdar-Shaw and Suryanarayana 2003). Recently, new designs such as multi-stacked circular tray reactor and other reactor designs have been developed for solid-state fermentation (Manan and Webb 2020). These reactors helped in better operation and provided cell growth as achieved with submerged fermentation. With more and more higher organisms being used for the production of secondary metabolites and other biochemicals, solid-state fermentation will be very useful for the large-scale production of biologics using multicellular organisms.

9.2.3 Challenges with Fermentation Technology

With the application of fermentation technology, almost all types of biologicals such as proteins, nucleic acids, carbohydrates, and lipids are produced at industrial scale. Since most of the fermentation products are commodities, it is essential to analyze how fermentation technology competes with chemical manufacturing in terms of energy, water intensiveness, sustainability, and in being environment friendly (Berenjin 2019). The basic problem with the fermentation process is that it is water intensive as most of the biologics are around a maximum 10–25% of the reactor volume. The final products are most of the time thermolabile and need specific storage and formulation for different uses. New molecules and new technology evolving in biotechnology place demands on fermentation technology to evolve suitably (São Pedro et al. 2021). Accordingly, different reactors in terms of geometry and operation have been combined to develop novel fermentation processes. Different bioreactor geometry with operation has evolved in recent times for growing microorganisms and animal cells (Zhang et al. 2010). Airlift bioreactor, bubble column bioreactor, fixed bioreactor, tubular bioreactor, and hollow fiber-based bioreactor have evolved over time. Particularly for growth of anchorage-dependent mammalian cells, varieties of bioreactors which provide more surface area have been developed. Microcarrier-based fermentation, fermenters with matrix to support cell growth, and wave bioreactor have provided new opportunities for animal cell growth. The greatest challenges have been in designing bioreactors for tissue growth. Fermentation technology has been developed for growing cartilage, corneas, skin, and other tissues. In these cases, cells are grown on a scaffold which promotes cell and tissue growth in three dimensions inside the bioreactor. Different bioreactors have been designed to produce tissue-engineered products (Lim et al. 2022). The commercial availability of tissue-engineered products is an example of the application of modern fermentation technology in healthcare.

9.3 Fermentation-Based Industries in India

Fermentation-based industries have been one of the strengths of India for the last 100 years. These industries which came into existence by producing antibiotics and alcohol encouraged other pharmaceutical giants to venture into biological production. This led to the setting of large-scale antibiotic manufacturing plants by Indian Drugs and Pharmaceuticals Limited (IDPL), Hindustan Antibiotics Limited (HAL) at Pune, Alembic and Sarabhai Chemicals, and many more in different parts of the country and predominantly in the western part of India. Apart from antibiotics, fermentation technology-based industries were set up to produce vitamins, enzymes, organic acids, and alcohols in different parts of the country. The setting up of NBTB in 1982 energized the fermentation industry to divert into vaccines and therapeutics. This led to setting up of therapeutic and biologicals producing fermentation plants at Biocon Limited, Cadila Pharma, Intas Pharma, Lupin, Biological Evans, Dr. Reddy's Laboratory, Aurobindo Pharma, Wockhardt Ltd., US Vitamins, and Panacea

Biotec, New Delhi. This also promoted a series of new biotechnology companies exclusively for manufacturing biologics, such as Shantha Biotechnics, Bharath Biotech, Zydus, Cadila, Reliance Life Sciences, and many others. The DBT support in terms of generating manpower and infrastructure to do cutting-edge research also led to rapid developments in the modern biotechnology industry in the areas of agriculture, pharmaceuticals, and chemicals including biofertilizers and bioinsecticides. BIRAC support also helped in the establishment of many fermentation-based industries in many parts of the country. Anthem Biosciences in Bangalore, Aumgene Biosciences in Surat, IPCA Laboratories in Mumbai, Gennova and MJ Biopharm at Pune, Virchow Biotech in Hyderabad, and many others adopted modern fermentation technology to produce recombinant biologicals. The fermentation technology-based industry was further promoted by setting up research-based small fermentation laboratories in many academic research institutes. This helped in the optimization of novel technologies for the growth of fermentation-based industry. Currently, around 50 fermentation technology-based large-scale industries are operating in India and are mostly producing biologics for healthcare segments (Overview of Fermentation Industries 2022).

Apart from biopharma and organic chemical production through fermentation, India has a large base of fermentation industry for agriculture, veterinary, and aquaculture products. Large-scale manufacturing of bioinsecticides and biofertilizers is a major component of fermentation-based industry for agriculture. Fermentation Technology is currently being used for green agriculture by growing bacteria, fungi in fermenters for agricultural crop improvements. Veterinary and poultry vaccines are also being produced using fermentation technology almost in all State government laboratories to cater to the animal population. The fermentation technology-based industry thus has made inroads into agriculture and veterinary and aquaculture sectors. In the future, modern fermentation technology will certainly help in providing quality biologicals for the overall improvements of crops, animals, and fisheries.

The other areas where fermentation technology has made an impact are in controlling water and solid waste treatments where bio-based fermentation processes are used to reduce pollution loads in the water and environment (Narayanan and Narayan 2019). Both anaerobic and aerobic fermentation processes could be used to effectively solve the problems. The biggest challenge in the coming days is the production of bioenergy using renewable resources or biological waste using fermentation technology (Yadav et al. 2014). Major petrochemical giants in the country are in the process of setting up of 2 G ethanol plants and Bio-CNG plants where fermentation process is the heart of the system. A real challenge will be to optimize and design fermentation technology where process efficiency can be improved to produce bioenergy at reduced costs. With the development of both aerobic and anaerobic fermentation technology, novel processes can be developed to convert agricultural waste efficiently into energy.

9.3.1 Fermentation Industry with Inputs from Government Research Laboratories

With the involvement of the Department of Biotechnology, Government of India, and BIRAC, not only biotechnology but all collaterals of fermentation-based industry received a fillip in India (Ghose and Bisaria 2000; Nogardy 2018). This includes fermentation-based industry using chemoenzymatic methods and using cell culture for therapeutic proteins and different types of glycoproteins in the country. Being a core activity associated with biotechnology, it is expected to contribute more in future. Biotechnology is inching towards a $100 billion biomanufacturing hub with the objective of reaching around $300 billion by 2030 (Anon 2023). Fermentation technology will play a major role in achieving this grand objective. Application of fermentation technology for gene and cell therapy is at its infancy in the country in comparison to the western counterparts. However, Indian companies have started using novel fermenters for the growth of stem cells and T cells for therapeutic applications. This is important as it needs the involvement of both physicians and engineers to produce a desirable tissue or cell culture-based products for therapeutic use. Fermentation technology with high-density cultures is being used at appreciable scales in different industries around Bengaluru, Mumbai, Hyderabad, and Delhi to make cell therapy-based products. Currently these therapies with disposable fermenters are too costly and it is expected that with indigenous development of fermentation technology, the cost will be within the reach of many. This will be a major challenge in developing cell therapy using fermentation technology as it involves multiple stakeholders. It needs the involvement of physicians, engineers, biologists, and material scientists and government support. Multiple government departments including DBT and DST, Health and Pharmaceuticals along with material scientists and engineers will be required to develop novel fermentation technology-based products for future human health care. Similarly, agricultural and veterinary scientists will be working together to promote cellular agriculture, which is based on advanced fermentation protocols.

9.3.2 Contribution of Fermentation Technology to Indian Industry

From antibiotics to the next-generation protein therapeutics, the Indian fermentation industry has delivered using the existing technology. Almost all types of antibiotics which the country needs are being produced by fermentation. Penicillin, Cephalosporins, Streptomycin, Gentamycin, Tetracyclines, Amphotericin B, Erythromycins, Griseofulvin, Rifampicin, and many other antibiotics are produced through fermentation despite the recent Chinese intrusion. The initiation of production linked incentive scheme (PLI) by the Department of Pharmaceuticals of the Government of India for antibiotics is a very welcome step. It is expected that with the use of appropriate modern fermentation technology, Indian industries will be able to produce affordable antibiotics for public use. This will also help in producing other small molecules such as steroids, vitamins, antioxidants, and hormones using fermentation technology.

Fermentation technology has also been used by Indian industries to produce enzymes for use in food, poultry, and aquaculture industries. Anil Starch and Advanced Enzyme Technologies Ltd. and other industries have contributed towards the production of varieties of enzymes for various applications. Both submerged and solid-state fermentation technology have been developed and used by Indian industries to meet the demand of food, poultry, and other allied industries.

The Indian vaccine industry used fermentation technology to make the most of the bacterial-based and recombinant vaccines. Fermentation technology has contributed immensely to make India the vaccine capital of the world. Even polysaccharide-based vaccines against *Streptococcus pneumoniae* and other viral-based vaccines are being produced using fermentation technology. The recent success of two covid vaccines which were used during Covid-19 such as Covishield by Serum Institute of India and Covaxin by Bharath Biotech International was achieved through large-scale cell culture-based fermentation process. Other vaccine platforms such as subunit vaccine, mRNA-based vaccine, and plasmid-based DNA vaccine also need fermentation technology for mass production. Fermentation technology has thus contributed immensely towards vaccine manufacturing in India.

Recombinant insulin and hepatitis B vaccine production using fermentation process is the best example of the contribution of the Indian fermentation industry. This opens up more possibilities of recombinant fermentation products such as GCSF, interleukins, erythropoietin, and many other therapeutic proteins. The combination of fermentation technology with new biotechnology will make India the next biosimilar hub of the world.

Fermentation technology-based industries in India have contributed significantly towards the production of biologicals starting from antibiotics to high-value therapeutic proteins. With further developments in science and technology, even more biologics based on carbohydrates, nucleic acid, and lipids will be produced through fermentation for uses in healthcare and other industries. As these biomolecules are part of the living system, fermentation technology will be the only way to produce these biologicals for different uses. Polysaccharides are used as part of conjugate vaccine for pneumoniae and meningitis. As the science advances, different types of carbohydrates found in higher organisms like fungi, mushrooms, and higher plants will be used as therapeutic agents. In such a scenario, producing carbohydrates through fermentation will offer new challenges, and it will also be a reliable option. Similarly, the use of small lipids and nucleic acids is being explored for different medical interventions. Producing lipids through the fermentation process will be an exciting endeavor in future. Once proven, fermentation technology will be sought after to produce these biologics. In future, fermentation technology will be used for the cost-effective production of proteins, carbohydrates, lipids, and nucleic acid for different applications. It will be necessary to adopt technology for process development so that these novel biologics are produced for human use.

9.4 Challenges of Indian Fermentation Industry

The Indian fermentation industry is moving in the right direction but at a rather slow speed. We are the vaccine capital of the world whereas we are not adequately placed in producing low-value, high-volume biochemicals. We have not used our past strength of large volume fermentation process technology optimally. While cell-based therapeutics are produced, there is insufficient production of Vitamin C or B12. Insulin production is also not self-sufficient. We are in the very early stages of CAR-T cell therapy and other gene therapy and tissue-engineered products. Though biosimilars are produced they are not at globally competitive prices. We lack the technology for large-scale culture of nematodes, multicellular fungi, and other organisms using bioreactors. The Indian problem is thus both in scale and scope for using fermentation technology. The fermentation industry is at a crossroads, and there is a need to address both the issues: the first on producing high-volume low-cost biologicals and the second on low-volume and high-cost complex biologicals. This needs to be comprehensively addressed taking care of all stakeholders associated with the fermentation industry.

Currently, we are exploiting only a handful of living organisms for the production of suitable chemical commodities. Understanding the biology of higher organisms, marine organisms, and others opens up new possibilities of *in vitro* culturing of multicellular organisms in a bioreactor. The growth of multicellular organisms like sponges, coelenterates, nematodes, fungi, algae, and lichens opens up numerous opportunities. Producing protein through fermentation is another exciting area of research to supplement nutrition requirement. From the Indian point of view, it needs a different perspective and plan to exploit the full potential of fermentation technology. There is a clear need to invest in the development of large-scale fermentation processes where energy and water requirements are more critical for metabolite production. The design and operation of low-volume sophisticated fermenters to produce gene therapy, cell therapy, and tissue-engineered products, needs special attention. As cell therapy and personalized medicines are becoming a reality, substantial investment is called for both in terms of research and process development to make such products available at a reasonable price.

9.5 The Way Forward for Fermentation Technology-Based Industry in India

Having established a wonderful biotechnology research ecosystem by the Department of Biotechnology across the country, it is the right time to refocus the strategy with innovative ideas and sustainable policies to promote biological manufacturing to produce quality biologics at a competitive price. It is time to promote fermentation which has the potential to translate biological phenomena to tangible products. This will need redesigning and rethinking our strategy in the following three different areas. We need trained manpower and modern research centers in fermentation technology or bioprocess engineering with support systems in terms of equipment, facility, and ancillary services related to fermentation. Unless we focus on these areas, the true benefits of biotechnology through fermentation and product development will not be realized.

9.5.1 Generation of Trained Manpower

This is the most important requirement as the fermentation-based research is truly multidisciplinary. Initially, microbiologists and some chemical engineers were exposed to the art of fermentation technology. Currently as the product ranges from simple biologics to complex cell and tissue, it needs the involvement of physical scientists, engineers, and medical doctors to redesign fermentation technology to get the desired products. Similarly, if fermentation technology has to be a part of cellular agriculture, we need the involvement of experts from different discipline. The best solution will be to introduce the concept of fermentation in a variety of disciplines and allow the respective people to use it for manufacturing a particular product. This needs a more interdisciplinary approach in teaching particularly for professional students like engineers, agricultural scientists, and especially medical professionals, who will help in providing better treatment options to patients. Cell and molecular engineering should be the part of fermentation technology curriculum to produce novel products. Fermentation technology involving cellular engineering, upstream and downstream processes along with the formulation of complex biologicals should be the main curriculum of industrial biotechnology degree courses. It will be desirable to have detail fermentation technology courses in the B.Tech Industrial biotechnology syllabus with the aim to produce quality manpower for biological manufacturing industries. Similarly, a short course in the basics of fermentation technology needs to be taught to medical doctors or medical biotechnologists so that they can use cell and gene therapy and tissue engineering-based products for treatment. Students should be trained as technocrats than scientists or biologists with analytical ability for modern biology. We need to develop curricula with equal weightage given to modern biology and engineering. This will help in creating a much-needed skill pool that can handle interdisciplinary research to fully deploy fermentation technology.

9.5.2 Promoting Research in Fermentation Process Technology

The next requirement is a proper in-house research training in fermentation technology. Biotechnology is driven by the concept that "Product is the Process and Process is the Product." Thus, fermentation process development needs high-end research components to understand the complexities of living organisms. Previously, all the national research laboratories carrying out biological research had designated pilot plant facilities for fermentation process development. The facilities at erstwhile NCL Pune, CDRI Lucknow, RRL Jammu, and CFTRI Mysore are examples of how they contributed towards the development of fermentation technology in the country. We need to set up high-intensity fermentation pilot plants to develop future cell and organism-based products. Setting up of fermentation research laboratories at institutes of DBT, ICMR, and ICAR will help in evaluating the relevant process technology in a "regulatory environment" for quick translation of research leads into biologics. It will be appropriate to adopt fermentation technology-based research pertaining to organisms as in the European Union. Separate bioprocess-based fermentation research for bacteria, fungi, plant cells, animal cells, and higher organisms will help in producing biologicals from these living organisms. These centers with cross-disciplinary subject scientists will add value to the modern bioprocess development based on fermentation technology.

9.5.3 Focusing on Equipment and Instruments Used for Fermentation

Apart from different organisms and fermentation processes, it is also imperative to indigenously produce all the major equipment, instruments, and other ancillary parts associated with the fermentation process. It is essential that the Indian government and industry focus on manufacturing fermenters, probes, pumps, and control panels, which are essential part of the fermentation technology. There are manufacturers in India who have started supplying fermenters, instrumentation, and other accessories, but we need to go much beyond that. The Atmanirbhar Bharat program is a good platform to design and fabricate all the instruments used for bioprocess technology development. Manufacturing of other consumables such as tubes, filters, sterile connections, peristaltic pumps, and plastic disposables will also be helpful in running the fermentation plant smoothly. The importance of such instruments and accessories was badly felt due to poor supply chain during the Covid-19 pandemic. India has a vast number of engineering institutes, and it is high time that they are geared to produce different accessories for fermentation technology, both for small-scale and turnkey plants. Such an initiative by the Department of Biotechnology involving multiple stakeholders will be a critical support to the fermentation technology sector. It is time to support the industries who are designing vessels, heat exchangers, pumps, instrumentation and control systems, sensors, and other downstream operation equipment such as centrifuges and filtration system. This will make fermentation technology-based industry truly self-reliant in terms of manufacturing biologics at a competitive price. The way forward for promoting the Indian fermentation technology sector is to reorient our focus in these three main areas such as generating trained manpower in fermentation technology, promoting research in fermentation technology, and promoting indigenous equipment and instrumentation industries to provide plant and machinery associated with the fermentation industry.

9.6 Conclusion

Fermenters are being used extensively for bulk therapeutics, for personalized medicine, food industry, agricultural productivity, environmental sustainability, and energy. It will be worthwhile to develop appropriate fermentation technologies as new biology advances using different living organisms. Unlike other sectors such as automobiles, electronics, and computation, the requirement of fermentation technology is quite different. It is research intensive and shows a high variability due to the involvement of living organisms. It is thus essential to evaluate how fermentation technology will evolve for the production of bulk therapeutics, developing personalized medicines, food and organic acid, ensure environmental sustainability, and also for producing better biofuels and energy. Using fermentation technology for cellular agriculture for large-scale production of engineered foods and beverages will be beneficial for meeting the food demand. Solid waste disposal using fermentation process will not only reduce environmental problems but will lead to the production of energy and fuels. Algal bioreactors, fermentation of seaweed, and gas fermentation also offer novel opportunities to produce useful material from biological sources. Modern medicine is shifting from biologics to cell-based therapy. It requires appropriate fermenter design to grow the desired cell type. It is expected that numerous challenges associated with food, health care, environment, and biofuel along with wastewater treatments can be solved with the use of appropriate fermentation technology. Predicting the future of fermentation technology in the areas of healthcare, agriculture, industry, and the environment will be the key for its successful utilization.

References

Al-Maqtari QA, AL-Ansi W, Mahdi AA (2019) Microbial enzymes produced by fermentation and their applications in the food industry–a review. Int J Agric Innov Res 8(1):1

Anon (2023) India-inching towards a $ 100 Bn Bio-manufacturing hub. www.investindia.gov.in/sector/biotechnology

Approved Cellular and Gene Therapy Products (2022). https://www.fda.gov/vaccines-blood-biologics/cellular-gene-therapy-products/approved-cellular-and-gene-therapy-products

Arora S, Rani R, Ghosh S (2018) Bioreactors in solid state fermentation technology: design, applications and engineering aspects. J Biotechnol 269:16–34

Arthur H (1998) Shake flask to fermentor: what have we learned? Biotechnol Prog 14(1):3–7

Berenjin A (ed) (2019) Essentials in Fermentation Technology. Springer Nature, Switzerland

Bielsera J-M, Wolf M, Souquet J, Broly H, Morbidelli M (2018) Perfusion mammalian cell culture for recombinant protein manufacturing—a critical review. Biotechnol Adv 36:1328–1340

Buchholz K, Collins J (2013) The roots—a short history of industrial microbiology and biotechnology. Appl Microbiol Biotechnol 97:3747–3762. TPS://WWW.SFGATE.COM/LIFESTYLE/SLIDESHOW/HISTORY-OF-.PHP

Chen F, Jiang Y, Ouyang F (2005) Development of bioprocess engineering in China. Biotechnology 4(1):1–6

Cherian DS, Bhuvan T, Meagher L, Heng TSP (2020) Biological consideration for scaling up therapeutic cell manufacturing. Front Pharmacol 11:654

Chilakamarry CR, Mimi Sakinah AM, Zularisam AW, Sirohi R, Khilji IA, Ahmad N, Pandey A (2022) Advances in solid-state fermentation for bioconversion of agricultural wastes to value-added products: opportunities and challenges. Bioresour Technol 343:126065

Chelsea Debret (2021) History of fermentation around world. https://www.sfgate.com/lifestyle/slideshow/History-of-fermentation-around-the-world-228894.php

Elander RP (2003) Industrial production of β-lactam antibiotics. Appl Microbiol Biotechnol 61:385–392

Fang Z, Lyu J, Li J, Li C, Zhang Y, Guo Y, Zhang YWY, Chen K (2022) Application of bioreactor technology for cell culture-based viral vaccine production: present status and future prospects. Front Bioeng Biotechnol 10:921755

Ganeeva I, Zmievskaya E, Valiullina A, Kudriaeva A, Miftakhova R, Rybalov A, Bulatov E (2022) Recent advances in the development of bioreactors for manufacturing of adoptive cell immunotherapies. Bioengineering 9:808

Ghose TK, Bisaria VS (2000) Development of biotechnology in India. Adv Biochem Eng and Biotechnol 69:87–124

Gracia-Aponte OF, Herwig C, Kozma B (2021) Lymphocyte expansion in bioreactors: upgrading adoptive cell therapy. J Biol Eng 15:13

Joseph S, Fassb R (2005) Growing *E. coli* to high cell density—a historical perspective on method development. Biotechnol Adv 23:345–357

Jyothilekshmi I, Jayaprakash NS (2021) Trends in monoclonal antibody production using various bioreactor systems. Microbiol Biotechnol 31(3):349–357

Kumar V, Ahluwalia V, Saran S, Kumar J, Patel AK, Singhania RR (2021) Recent developments on solid-state fermentation for production of microbial secondary metabolites: challenges and solutions. Bioresour Technol 323:124566

Lee SY (1996) High cell density culture of *E. coli*. Trends Biotechnol 4:98–105

Lim D, Renteria ES et al (2022) Bioreactor design and validation for manufacturing strategies in tissue engineering. Biodesign Manufacturing 5(1):43–63

Majumdar-Shaw K, Suryanarayana S (2003) Commercialization of a novel fermentation concept. Adv Biochem Engg Biotechnol 85:29–42

Manan MA, Webb C (2020) Newly designed multi-stacked circular tray solid-state bioreactor: analysis of a distributed parameter gas balance during solid-state fermentation with influence of variable initial moisture content arrangements. Bioresour Bioprocess 7:16

Narayanan M, Narayan V (2019) Biological wastewater treatment and bioreactor design: a review. Sustain Environ Res 29:33

Nogardy B (2018) How Indian biotech is driving innovation. Nature 564:S53

Overview of Fermentation Industries (2022). https://www.biotechbug.in/2021/04/fermentation-industries.html?m=1

Panda AK, Mishra S, Bisaria VS, Bhojwani SS (1989) Plant cell reactor—a perspective. Enzym Microb Technol 11:386–397

Pandey A, Soccol CR, Mitchell D (2000) New developments in solid state fermentation: bioprocesses and products. Process Biochem 35:1153–1169

Pandey A, Larroche C, Soccol R (2018) Current developments in biotechnology and bioengineering: Current Advances in Solidstate Fermentation. Elsevier, Amsterdam

Ranjana S, Garg P, Kumar P, Bhatia SK, Kulshrestha S (2020) Microbial fermentation and its role in quality improvement of fermented foods. Fermentation 6:102

São Pedro MN, Silva TC, Patil R, Ottens M (2021) White paper on high-throughput process development for integrated continuous biomanufacturing. Biotechnol Bioeng 118(9):3275–3286

Stanburry PF, Whitakar A, Hail SJ (2016) Principle of Fermentation Technology. Elsevier Science & Technology, Amsterdam

Steudler S, Werner A, Cheng JJ (2019) Solid state fermentation: research and industrial applications. Advances in Biochemical Engineering and Biotechnology, vol 169. Springer, Cham

Subramaniam RV, Thirumal A, Chistoserdov R, Bajpai JB, Popovic M (2018) High-density cultivation in the production of microbial products. Chem Biochem Eng Q 32:451–464

Taveira I, Nogueira K, Oliveira D, Silva R (2021) Fermentation: Humanity's oldest biotechnological tool. Front Young Minds 9:568656. https://doi.org/10.3389/frym

Wawrosch C, Zotchev SB (2021) Production of bioactive plant secondary metabolites through in vitro technologies—status and outlook. Appl Microbiol Biotechnol 105:6649–6668

Yadav G, Kumar R, Ramakrishna S (2014) Fermentation techniques in bioenergy production. In: Kim S-K, Lee C-G (eds) Marine Bioenergy: Trends and Developments, 1st edn. CRC, London, pp 111–134

Yafetto L (2022) Application of solid-state fermentation by microbial biotechnology for bioprocessing of agro-industrial wastes from 1970 to 2020: a review and bibliometric analysis. Heliyon 8:e09173

Yue W, Lin Q-LMB, Rahman K, Qin C-JZTHL-P (2016) Medicinal plant cell suspension cultures: pharmaceutical applications and high-yielding strategies for the desired secondary metabolites. Crit Rev Biotechnol 36(2):215–232

Zhang S, Cao X, Chu J, Qian J, Zhuang Y (2010) Bioreactors and bioseparation. Adv Biochem Eng Biotechnol 122:105–150

Lignocellulosic Biomass Based Biorefinery Employing Fungus and its Enzymes: Indian Perspective and Global Trend

10

Sonu Saini, Astha Chhillar, Krishna Kant Sharma, and Ramesh Chander Kuhad

Abstract

Declining fossil resources, energy crisis, and environmental pollution are driving the shift from linear petroleum to circular bioeconomy. Lignocellulosic biomass, a globally available organic agricultural by-product is crucial for sustainable waste biomass management in a circular bioeconomy, second-generation biorefineries, and white biotechnology. However, the complex structures of lignocellulosic biopolymers hinder economic viability. Various physicochemical processes have been developed to fractionate and convert these substrates into

S. Saini
Laboratory of Enzymology and Gut Microbiology, Department of Microbiology, Maharshi Dayanand University, Rohtak, Haryana, India

Department of Renewable and Bio-Energy Engineering, College of Agricultural Engineering and Technology, Chaudhary Charan Singh Haryana Agricultural University, Hisar, Haryana, India

A. Chhillar
Department of Zoology, Maharshi Dayanand University, Rohtak, Haryana, India

K. K. Sharma (✉)
Laboratory of Enzymology and Gut Microbiology, Department of Microbiology, Maharshi Dayanand University, Rohtak, Haryana, India
e-mail: kksharma.microbiology@mdurohtak.ac.in

R. C. Kuhad (✉)
Academy of Microbiological Sciences, Department of Microbiology, Maharshi Dayanand University, Rohtak, Haryana, India

D P G, Institute of Technology and Management, Gurugram, India

School of Biological Sciences Research, Sharda University, Greater Noida, Uttar Pradesh, India
e-mail: Ramesh.kuhad@sharda.ac.in

G. John, A. K. Panda (eds.), *Biotechnology in India - Reworking A Strategy*, https://doi.org/10.1007/978-981-97-0896-3_10

valuable products. Lignocellulolytic fungi and their enzymes show potential for efficient and eco-friendly utilization of lignocellulosic biomass. Despite five decades of research focusing on lignocellulose bioprocessing, more emphasis is needed on adopting a biorefinery approach to make bioconversion processes economically viable. Exploring fungi from plant waste biomass-rich ecosystems or degraded plant materials can lead to more efficient plant cell wall degrading fungi and enzymes. This chapter highlights the contributions of fungi and their lignocellulolytic enzymes in developing plant biomass-based biorefineries globally, including India.

Keywords

Lignocellulosic biomass · Lignocellulolytic fungi · Lignocellulolytic enzymes · Biofuels · Biorefinery

10.1 Introduction: Linear Fossils Economy to Circular Bioeconomy

Industrial production of fuels, synthetic polymers, and their derivatives heavily rely on fossil resources and petroleum refineries. However, their extensive use over the past two centuries has led to the depletion of natural resources, which raises political, economic, and environmental concerns. The decline in fossil reserves not only threatens energy security but also environmental sustainability. Furthermore, around 90% of energy supply is contributed by fossil fuel and biomass that eventually cause greenhouse gas (GHG) emission, whereas other by-products go to waste (Cañete Vela et al. 2022). Current research focuses on renewable sources and synthetic polymers as alternatives to petroleum refineries, alongside developing technologies to minimize waste (Saini et al. 2022).

The Intergovernmental Panel on Climate Change (IPCC) report emphasizes the role of human activities, including burning fossil fuels and petroleum refineries, in causing climate change through greenhouse gas (GHG) emissions (Malla et al. 2022). India ranks fourth in terms of GHG emissions and energy consumption, necessitating sustainable energy sources to meet growing demands. It has been assumed that excessive consumption of non-renewable fossil fuels will deplete global oil reserves by the end of the twenty-first century (Deshmukh et al. 2021). Depletion of conventional energy resources has raised oil prices, prompting the need for novel and sustainable solutions (Hassan et al. 2022).

Scientists are actively exploring renewable feedstocks, such as lignocellulosic biomass, to produce fuels, chemicals, polymers, and bio-based materials following a biorefinery platform. Utilizing a biorefinery approach for lignocellulosic biomass offers both environmental feasibility and economic benefits (Shankar et al. 2022). The transition to a circular bioeconomy from a linear petroleum economy is crucial for sustainable industrialization, energy independence, and efficient management of greenhouse gas emissions. The circular bioeconomy promotes the use of renewable biomass, including underutilized and waste lignocellulosic materials; further, it also replaces non-renewable fossil-based products and reduces waste generation (Stegmann et al. 2020). The US Department of Energy has identified 15 target chemicals for biorefinery research and development, aiming for a zero-waste circular bioeconomy (Saini and Sharma 2021a). These lignocellulosic materials, such as agricultural and forest residues, municipal solid waste, and paper and pulp waste, have a significant potential in reducing the negative socio-economic and environmental impacts associated with the linear petroleum economy (Karimah et al. 2021). Importantly, the utilization of lignocellulosic materials does not require additional crop land, thereby avoiding adverse effects on food economies (Pereira et al. 2022).

Currently, various industries use plant biomass as raw materials for diverse applications, including cardboard, pulp and paper production, liquid fuels, biofertilizers, synthetic polymers, and animal feed (Babu et al. 2022). However, a substantial amount of biomass remains underutilized, and ends up as waste in farm land or landfills, both pre- and post-harvest, representing a missed economic opportunity (Chukwuma et al. 2021). It is crucial to develop systematic strategies to maximize the utilization of plant by-products, which can contribute in achieving holistic environmental security. In this context, fungal processes have been developed to fractionate the lignocellulosic biomass into their chemical constituents that can be further utilized to develop lignocellulosic biorefinery. Significant research has been conducted over the past decade on fungal enzymes and their applications in biorefineries, encompassing areas such as enzymatic degradation and product synthesis. However, there is limited literature on the detailed applications of wood-rotting fungi and their enzymatic machinery from the biorefinery perspective. Therefore, the purpose of this book chapter is to provide an overview of the valuable products developed using fungal cultures through biorefinery and white biotechnology strategies.

10.2 Lignocellulosic Biomass: The Future of Bioenergy and Bio-products

The utilization of renewable lignocellulosic materials as feedstock for biofuels, chemicals, and industrial products has gained significant interest due to its non-competition with food/feed crops and abundant availability (Saini and Sharma 2021a). India has a long history in lignocellulose-based ethanol production with significant pioneering efforts made by the Biochemical Engineering Research Centre at the Indian Institute of Technology, Delhi (IIT Delhi). As early as in 1980, this center established an innovative facility committed to an integrated bioprocess designed for the daily production of bioethanol of around 50 L. The basis of this novel process was to utilize the rice straw as the primary raw material. This advanced approach yielded 230 L/ton of rice straw and with an estimated cost of $0.54 per liter (Ghosh and Ghose 2003). The recalcitrant nature of lignocellulose and high processing costs pose major obstacles in the biorefinery platform. Therefore, a combination of pretreatment and enzymatic hydrolysis is required to overcome the recalcitrant architecture and efficiently utilize lignocellulosic components (Saini et al. 2020b).

Pretreatment methods generally account for more than 50–70% of the total operation cost and represent the most expensive step in lignocellulose processing (Ibarra-Gonzalez and Rong 2019). These methods can be classified into four types: mechanical/physical, chemical, physicochemical, and biological. Mechanical/physical methods involving heat and external force require costly equipment and consume substantial energy, making them economically unfeasible on an industrial scale (Chowdhury et al. 2018). Chemical and physicochemical methods utilize hazardous chemicals, posing risks to environment. To address these environmental concerns, lignocellulolytic fungi and their enzyme systems are being utilized for the fractionation of lignocellulosic biomass (Saini and Sharma 2021b).

In recent years, the use of biocatalysts or microorganisms in the entire lignocellulose processing route, known as white biotechnology, has emerged as an efficient strategy (Manzanares 2020). Within the context of biorefineries, circular bioeconomy, and white biotechnology, certain white-rot fungi have the potential to play a crucial role in achieving almost complete utilization of lignocellulosic components at each stage of the process. Wood-rotting fungi, in particular, produce a diverse array of active extracellular enzymes that can effectively break down polymeric units into their monomeric constituents. These monomers can then be converted into various valuable products by the same species of fungi or other fungal species following a biorefinery approach (Saini and Sharma 2021b).

10.3 Biorefinery: Definition, Types, and Classification

Initially, biorefinery was defined by the US Department of Energy (DOE) in 2004 and refers to a processing plant where biomass feedstocks are converted into a range of value-added products (Li and Takkellapati 2018). Technologies developed for biomass processing have enabled the production of various products such as biofuels (bioethanol, biobutanol, biodiesel, etc.), biogas, lubricants, food, microbial biomass, animal feed, xylitol, organic acids, furan, furfural, bio-chemicals, biofertilizers, synthetic polymers, single-cell proteins, antimicrobial compounds, and biomaterials such as fibers and pulp. This comprehensive utilization of biomass enhances its economic value. The biorefinery concept can contribute to economic improvement and rural development by fostering new business opportunities and generating employment, particularly in agriculture-based countries (Pawlak and Kołodziejczak 2020).

The biorefinery can be classified based on the type of feedstock used (Fig. 10.1). Following are the major types of biorefineries classified according to platforms:

- Starch biorefinery (Läufer 2019)
- Sugar biorefinery (Höfer 2015)
- Lignocellulosic biorefinery (Saini and Sharma 2021a)
- Lignin biorefinery (Cao et al. 2019)

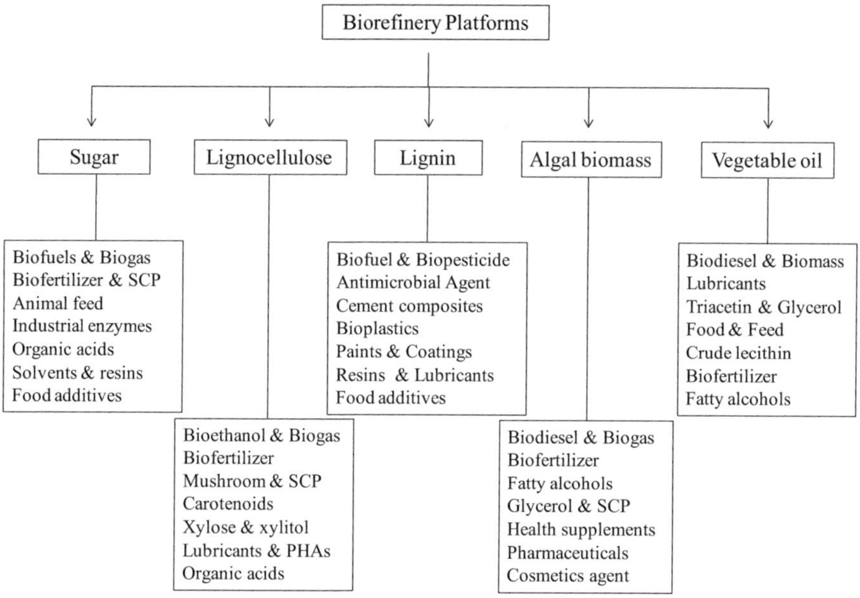

Fig. 10.1 Production of various industrial products from different types of biorefinery platforms

- Algal biorefinery (Chandra et al. 2019)
- Vegetable oil biorefinery (Kumar et al. 2016)

In a lignocellulosic biorefinery, the biomass is initially fractionated into its structural intermediates, including cellulose, hemicellulose, and lignin, which are then converted into value-added products and energy-rich compounds (Rajak et al. 2021). The overall lignocellulosic conversion process typically involves five major steps: (1) selection of appropriate raw materials, (2) assessment of pretreatment approaches, (3) conversion of plant cell wall polymers into monomeric constituents, (4) production of value-added products from the monomeric constituents, and (5) downstream processing. These steps can be planned and optimized based on the specific goals of the process. The selection of the conversion route, suitable bio-agents, and biocatalysts are crucial factors for achieving economically viable biorefineries (Pellis et al. 2018).

Currently, the majority of biorefinery industries worldwide employ thermochemical processes for lignocellulosic biomass fractionation (Table 10.1). Abellon Clean Energy, an Indian lignocellulose-based company, specializes in converting cotton stalks into briquettes using thermochemical processes. These cotton stalk briquettes serve as renewable fuel alternatives contributing to environmental sustainability and fuel crisis (https://abelloncleanenergy.com/). Comparing these thermochemical methods with biological processes especially fungal bioconversion strategies for carbon-to-energy conversion involves several crucial considerations like scale of operations, availability of resources, and environmental factors (Almomani et al.

Table 10.1 Some of the known commercial lignocellulosic biorefinery industries across the world

Industry and location	Feedstocks	Products	Website
Godavari Biorefineries Ltd, India	Sugarcane bagasse	Bioethanol, bio-based chemicals (ethyl acetate, bio-ethyl acetate, 1,3 butylene glycol, crotonaldehyde, acetaldehyde, acetic acid), sugar, rectified spirits, and other grades of alcohol and power	https://godavaribiorefineries.com/about-us
Assam Bio Refinery Private Limited, India	Wheat straw, corn stalks, reeds, bagasse, and residue from palm oil production	Bioethanol, acetic acid, furfural, and furfuryl alcohols	https://abrpl.co.in/
Beta Renewables, Italy	Rice straw and sugarcane bagasse	Bioethanol, fatty acids, ethylene glycol, aromatics, and terephthalic acid	http://betarenewables.st.e-one.it/it/proesa/che-cos-e-proesa
Kemijarvi Biorefinery, Finland	Wood	Bioethanol, microcrystalline cellulose, sugars, long fiber, pine oil, and turpentine	https://www.vataset.com/en/ekologinen-edellakavija/
Abellon Clean Energies, India	Agricultural residue, cattle dung, sugarcane press mud, municipal solid waste, and sewage treatment plant waste	Bioethanol, biogas, and solid fuel (carbon neutral solid biomass pellets)	https://abelloncleanenergy.com/our-solutions/biomass-based-heating-transport-solutions/

2023). Considering costs and investments, thermochemical methods involve capital for infrastructure, ongoing operating expenses, and maintenance. On the other hand, biological processes involve comparatively lower initial investments but require continuous subsidy for microorganism cultivation and nutrient supply (Sandesh and Ujwal 2021). Another crucial factor, i.e., return on investment (ROI) for thermochemical processes, relies on factors like fuel costs, energy input, and often requires years to yield returns. In contrast, biological methods owing to lower capital investment might offer quicker ROI. Lastly, with reference to carbon conversion efficiency, biological processes exhibit notable potential for high efficiency, approaching almost 100%, and can yield biofuels and biogas for direct energy production (Lim et al. 2022). However, biological processes involving fungi are at an intermediate stage towards designing a circular bioeconomy that prioritizes environmental sustainability. Due to their inherent ability to utilize a wide range of substrates and produce various industrially relevant compounds, wood-rotting fungi are highly suitable candidates for biorefineries, in near future (Bera et al. 2022).

10.4 Fungi: Introduction and Classification

Fungi, as part of the microbial community, play a crucial role in the recycling of waste materials and addressing major global environmental challenges. The Kingdom Fungi, also known as Eumycota, encompasses a diverse range of organisms with chitinous cell walls, ranging from single-celled yeasts to some of the largest microbes on Earth (Naranjo-Ortiz and Gabaldón 2019). Mycologists have classified fungi into eight phyla: Ascomycota, Basidiomycota, Zygomycota, Chytridiomycota, Microsporidia, Blastocladiomycota, Neocallimastigomycota, and Glomeromycota. Recent classification based on their life cycles categorizes fungi into three groups: unicellular, multicellular, and macro filamentous fungi. Furthermore, fungi can be defined as white-rot, brown-rot, or soft-rot fungi based on their mode of wood degradation (Saini and Sharma 2021a).

Wood-rotting fungi, in particular, are capable of completely fractionating lignocellulosic biomass into its structural monomers, making them valuable bio-agents in lignocellulosic biorefineries. These fungi participate in the bioconversion of plant cell walls through their activities, including the secretion of active extracellular enzymes that break down polymeric substrates (Saini and Sharma 2021a). The high stability of extracellular enzymes outside the fungal cell is another adaptive trait of fungi, which has been explored by scientists for the development of circular bioeconomy. More than 50% of globally utilized industrial enzymes are produced by fungi, thanks to their high protein stability and shelf life. Currently, fungal enzymes find applications in various industries, including pulp and paper, textiles, biofuels, biofertilizers, and leather, significantly reducing their harmful impact on the environment (Akhtar and Mannan 2020).

10.5 Role of Lignocellulolytic Fungi in Lignocellulosic Biorefinery

Wood-rotting fungi, belonging to the classes Ascomycetes, Basidiomycetes, and Zygomycetes, along with their enzymes, are valuable bio-agents in lignocellulosic biorefineries due to their ability to naturally produce and secrete various organic acids, which can be used as raw materials in chemical refineries (Janusz et al. 2017). The capability of wood-rotting fungi to fractionate lignocellulosic materials is attributed to their hyphal growth and highly efficient enzymatic systems. The active extracellular enzymatic machinery of wood-rotting fungi comprises oxidative and hydrolytic enzymes that catalyze a series of biochemical reactions to break down lignocellulosic materials (Veloz Villavicencio et al. 2020). Cellulolytic enzymes belonging to CAZymes are responsible for degrading carbohydrates of the plant cell wall, while oxidative enzymes primarily degrade lignin polymers (Hage and Rosso 2021).

Currently, there is significant research interest in the generation of biofuels and co-production of value-added products using wood-rotting fungi in lignocellulosic biorefineries, as it addresses the energy crisis and utilizes waste materials as the sole feedstock, thereby contributing to environmental management (Adewuyi 2022).

Lignocellulosic biomass is processed and delignified using ligninolytic fungi, particularly white-rot fungi, which are proficient producers of extracellular ligninases and cellulolytic enzymes that break down the polymeric material into structural monomeric units (Suryadi et al. 2022). White-rot fungi, known for their ability to preferentially degrade heteropolymeric lignin as the sole carbon and energy source, are widely utilized for the processing and delignification of lignocellulosic materials (Andlar et al. 2018). Based on their mode of lignin solubilization, white-rot fungi can be categorized into two groups: selective and nonselective lignin degraders (Saini and Sharma 2021a).

Selective lignin-degrading fungi primarily solubilize the heteropolymer lignin unit and have limited utilization of the cell wall polysaccharides. On the other hand, nonselective fungi delignifiers simultaneously solubilize all structural components of the plant cell wall (Hatakka and Hammel 2011). In recent years, selective lignin-degrading fungi have garnered more attention from researchers compared to nonselective delignifiers due to their potent ability to liberate lignin-free cellulosic biomass. The resulting readily available sugar fraction can be further converted by the enzymatic systems of the same white-rot fungi or other fungal species, such as brown-rot and soft-rot fungi, to produce liquid fuels and industrially important biological products (Fig. 10.2) (Shankar et al. 2022).

Unlike white-rot fungi, brown-rot fungi preferentially solubilize the structural polysaccharides of the plant cell wall, leaving behind the lignin residue (Arantes and Goodell 2014). The residual lignin fraction can serve as a raw feedstock for the production of various commercial products, including synthetic resins, lubricants, adhesives and natural binders, concrete admixtures, polyurethane foams, mosquito repellents, dispersants, biopesticides, and potentially polyhydroxyalkanoates (PHAs) (de Padilha et al. 2021).

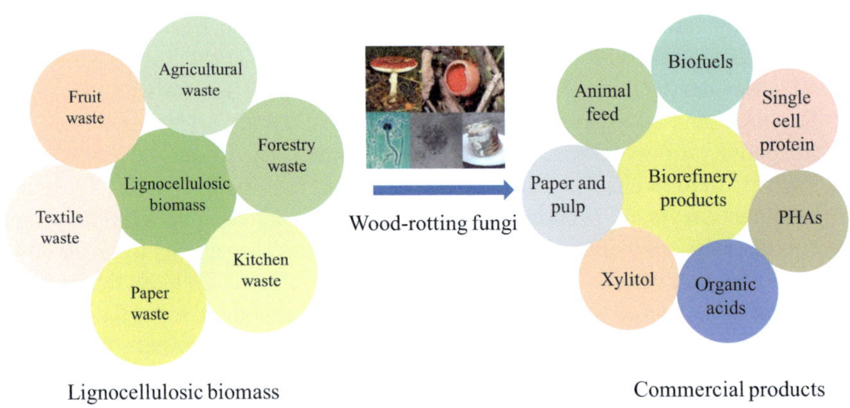

Fig. 10.2 Various sources of lignocellulosic biomass and its conversion into commercial products using wood-rotting fungi

Most brown-rot fungi belong to the major phylum Basidiomycota, while some species are categorized under the phylum Ascomycota (Saini and Sharma 2021a). Brown-rot fungi degrade structural cell wall carbohydrates through a non-enzymatic pathway, with low molecular weight compounds such as oxalic acid playing a crucial role. These fungi have been extensively studied for various micro biotechnological applications, including the development and improvement of animal feed, production of fermentation sugars and organic acids, liquid fuel production, and microbial enzyme production, following a biorefinery approach (Sharma et al. 2013; Saini and Sharma 2021a; Shankar et al. 2022; Saini et al. 2023).

Soft-rot fungi have a significant role in natural cycles and have been extensively studied for the industrial production of organic acids such as citric, gluconic, and itaconic acids, as well as various commercial enzymes (Porro and Branduardi 2017). They primarily solubilize plant cell wall carbohydrates but can also degrade structural lignin (Kuhad et al. 2013). Some species of soft-rot fungi are capable of lignin degradation through secondary cell wall disintegration, leading to the breakdown of the acid-insoluble klason lignin content in angiosperm wood (Saini and Sharma 2021b). These fungi are known to oxidize and mineralize the syringyl lignin unit through the production of potent extracellular peroxidases and oxidases. However, they are unable to mineralize the guaiacyl lignin unit (Janusz et al. 2017).

Unlike white-rot and brown-rot fungi, soft-rot fungi currently have limited applications in the field of lignocellulosic biorefinery. However, their ability to tolerate a wide range of temperature and moisture, along with their simultaneous solubilization of all structural plant cell wall polymers, may make them attractive for future fungal-integrated lignocellulosic biorefinery applications (Goodell et al. 2020). Additionally, modern genetic engineering approaches, such as CRISPR-Cas9 gene editing, have been used to improve the lignin degradation systems of soft-rot fungi. Genetic improvement has been achieved in over 40 species of Ascomycetes, with particular attention given to genera like *Aspergillus*, *Fusarium*, *Trichoderma*, and *Neurospora* (Liu et al. 2015; Matsu-Ura et al. 2015). These genetically improved fungal cell factories could be utilized in establishing fungal-integrated biorefineries for the production of multiple products.

In a recent study, the white-rot fungus *Ganoderma lucidum* MDU-7 demonstrated the co-production of edible fruiting bodies, industrially important enzymes, pharmaceutically important fungal polysaccharides, and liquid biofuel using lignocellulosic biomass as the sole feedstock through a biorefinery approach (Table 10.2) (Shankar et al. 2022). Similarly, the brown-rot fungus *Aspergillus* sp. and the soft-rot fungus *Neurospora intermedia* have been reported for the co-production of xylitol (an artificial sweetener), industrial enzymes, single-cell oil, bioethanol, biogas, and protein-rich fungal biomass suitable for food and feed applications using wheat straw biomass (Kamat et al. 2013; Nair et al. 2018).

Lignocellulolytic enzymes encompass a diverse group of active extracellular biocatalysts responsible for the fractionation of biomass into its components, including cellulose, hemicellulose, and lignin. These enzymes are classified into three main categories: ligninolytic/ligninases, cellulolytic/cellulases, and hemicellulolytic/hemicellulases (Kumar et al. 2015; Saini and Sharma 2021a).

Table 10.2 Role of various wood-rotting fungi in lignocellulosic biorefinery for co-production of biofuels and other industrially important products

Fungus	Lignocellulosic biomass	Products and applications	Reference(s)
G. lucidum MDU-7	Mixture of agricultural biomass including rice straw, wheat straw, rice bran, sorghum bagasse, wood chips, and cotton stalk	Industrial enzymes (laccases), pharmaceutical product (fungal polysaccharides), antimicrobial compound (fungal polysaccharides), and medicinal mushroom	Shankar et al. (2022)
A. terreus	Sugarcane bagasse residue	Artificial sugar (xylose), industrial enzymes (xylanases), liquid biofuels (single-cell oil and biodiesel), and xylitol	Kamat et al. (2013)
N. intermedia	Dilute acid pretreated wheat straw	Biogas and animal feed (edible fungal biomass)	Nair et al. (2018)
T. versicolor, *P. eryngii*, and *P. ostreatus*	Wheat straw	Industrial enzymes (laccase, manganese peroxidase, and lignin peroxidase) and biogas	Albornoz et al. (2018)
P. ostreatus	Sawdust, cottonseed hull, and corncob	Human food (oyster mushroom), industrial enzymes (laccase, manganese peroxidase, and lignin peroxidase, cellulases, xylanases)	Xiao et al. (2019)
Mucor circinelloides	Sugarcane bagasse	Liquid fuels (single-cell oil and ethyl esters), pharmaceutical product (carotenoids), nutraceutical compound (polyunsaturated fatty acids), and fungal biomass with lipolytic activity	Carvalho et al. (2019)
Hormonema sp.	Eucalyptus globulus wood	Biofuels platform (fermentable sugars), and industrial product (cellulosic pulp)	Fillat et al. (2017)
T. versicolor, *P. ostreatus*, *P. chrysosporium*, *Coriolopsis gallica*, *Pleurotus sajor-caju*, and *Lentinula edodes*	Corn stover	Human food (edible mushroom), industrial enzymes (ligninases and cellulases), artificial sugars (xylose and glucose)	Ding et al. (2019)

The group of extracellular oxidative fungal ligninases comprises several enzymes which are collectively responsible for lignin degradation (Table 10.3). Similarly, hemicellulolytic and cellulolytic enzymes belong to the group of glycoside hydrolases, which break down the glycosidic bonds between sugar residues or between

Table 10.3 Classification of fungal lignocellulolytic enzymes

Sr. No.	Enzyme group	Enzyme machinery	Major producer	Function
1.	Cellulolytic	Endo-(1,4)-β-D-glucanase (EC 3.2.1.4), Exo-(1,4)-β-D-glucanase (EC 3.2.1.91) β-glucosidase (EC 3.2.1.21)	White-rot and soft-rot fungi	Depolymerization of cellulose into glucose monomers
2.	Hemicellulolytic	Endo-1,4-β-xylanase (EC 3.2.1.8) Exoxylanase β-xylosidase (EC 3.2. 1.37) Xylanosomes α-glucuronidase Acetyl xylanesterase α-arabinofuranosidase Ferulic acid esterase	White-rot and soft-rot fungi	Depolymerization of hemicellulose into their monomeric constituents (pentoses and hexoses)
3.	Ligninolytic	Laccase (EC 1.10.3.2) Manganese peroxidases (MnPs) (EC 1.11.1.13) Lignin peroxidases (LiPs) (EC 1.11.1.14) Versatile peroxidases (VPs) (EC 1.11.1.16) Dye-decolorizing peroxidase (DyP) (EC:1.11.1.19) Glyoxal oxidases (EC 1.2.3.15) Alcohol oxidases (EC 1.1.3.13) Aryl-alcohol dehydrogenase (EC 1.1.1.90) Feruloyl esterase (EC 3.1.1.73) Quinone reductase (EC 1.6.5.5)	White-Rot Fungi and Brown-Rot Fungi	Depolymerization of lignin

sugar and non-sugar residues (Chutani and Sharma 2016). These enzymes play a crucial role in the complete fractionation of lignocellulosic carbohydrates into their monomeric fermentable sugars (hexoses and pentoses), thereby facilitating the production of biofuels (Saini et al. 2020a).

Wood-rotting fungi are potent producers of lignocelluloses and are integral to the lignocellulosic biorefinery platform (Saini and Sharma 2021a). Currently, lignocellulolytic enzymes are extensively used in the processing of lignocellulosic biomass for second-generation bioethanol production. The shift from second-generation bioethanol production to the principles of circular lignocellulosic biorefinery is likely to further explore their industrial importance in the near future (Table 10.4). However, a cost-benefit analysis is required for fungal-based bioconversion of lignocellulosic materials to yield energy and other products for

Table 10.4 Applications of lignocellulolytic enzymes for the production of various industrially important products following a biorefinery approach

Enzyme	Microbial source	Lignocellulosic biomass	Products	Reference(s)
Cellulases	*T. reesei*	Softwood bleached Kraft pulp	Co-production of nanocellulose and biofuels was achieved	Yarbrough John et al. (2017)
Laccases Cellulases Xylanases	Laccase from *G. lucidum* MDU-7. Cellulase and xylanases from various *Aspergillus* and *Trichoderma* sps	Rice straw	Co-production of bioethanol, biofertilizer, and biopesticides was achieved	Saini et al. (2023)
Cellulases	*T. reesei*	Ammonia fiber expansion pretreated corn stover	Co-production of bioethanol, vitamins, minerals, and protein-rich yeast biomass for human feed application was achieved	Lau et al. (2012)
Cellulases and hemicellulases	*Trichoderma longibrachiatum*	Hydrothermal pretreated wheat bran	Production of glucose, xylose, arabinose, galactose, amino acids, minerals (potassium, magnesium, and phosphate), glycerol, fatty acids, furans, HMF, formic acid, and levulinic acid was achieved	Reisinger et al. (2013)
Cellulases	–	Hydrothermal pretreated salt cedar	Co-production of monosaccharides (glucose, xylose, arabinose, and galactose), oligosaccharides, furans, HMF, formic acid, and acetic acid was achieved	Xiao et al. (2013)
Cellulases and xylanases	*Trichoderma longibrachiatum*	Organosolv pretreated wheat bran	Co-production of glucose, xylose, arabinose, galactose, glycerol, minerals, fatty acids, and amino acids was achieved	Reisinger et al. (2014)

commercial-scale implementation. To establish the optimal cost-benefit ratio, crucial factors like initial investment in fungal technology, operational costs, energy invention efficiency, and potential income from final products must be accounted for against the environmental benefits, reduced waste disposal costs, and sustainability advantages.

10.6 Conclusion

The circular bioeconomy and white biotechnology are interconnected concepts that offer a more secure and sustainable future for utilizing low-value bioresources. Biorefinery platform plays a crucial role in achieving the mandate of the economic viability and sustainable technology. The selection of cost-effective feedstocks, such as biodegradable lignocellulosic waste, is essential for ensuring the economic and sustainable viability of the biorefinery platform. Additionally, it is imperative to maximize the efficiency of biomass utilization, minimizing energy, and mass losses. Physicochemical and thermochemical conversion processes hold great potential in meeting the technical and economic objectives of a biorefinery framework. Among the microorganisms, wood-rotting fungi in particular, show potential for the bioconversion of a wide range of waste feedstocks into valuable commercial products. Their ability to synthesize highly active and stable extracellular enzymes involved in biomass conversion provides a significant opportunity for the scientific community to exploit them for the production of commercially valuable products. This book chapter provides an overview of the current applications of wood-rotting fungi in utilizing various lignocellulosic materials for the generation of biofuels/bioenergy and other bio-products employing a biorefinery approach. However, there are still numerous questions and challenges that need to be addressed for the development and execution of a successful biorefinery platform. The science of fungal-integrated lignocellulosic biorefineries requires careful exploration as it holds the potential to mitigate some of the challenges associated with environmental pollution.

Acknowledgments The authors acknowledge the infrastructural support from DST-FIST grant (Grant No. 1196SR/FST/LS-I/2017/4) & DBT-BUILDER grant (Grant No. BT/INF/22/SP43043/2021). S.S is thankful to MDU for University Research Scholarship (URS).

Conflict of Interest The authors declare that they have no conflict of interest with contents of this article.

References

Adewuyi A (2022) Underutilized lignocellulosic waste as sources of feedstock for biofuel production in developing countries. Front Energy Res 10:741570. https://doi.org/10.3389/fenrg.2022.741570

Akhtar N, Mannan MA (2020) Mycoremediation: expunging environmental pollutants. Biotechnol Rep 26:e00452. https://doi.org/10.1016/j.btre.2020.e00452

Albornoz S, Wyman V, Palma C, Carvajal A (2018) Understanding of the contribution of the fungal treatment conditions in a wheat straw biorefinery that produces enzymes and biogas. Biochem Eng J 140:140–147. https://doi.org/10.1016/j.bej.2018.09.011

Almomani F, Abdelbar A, Ghanimeh S (2023) A review of the recent advancement of bioconversion of carbon dioxide to added value products: a state of the art. Sustain For 15:10438. https://doi.org/10.3390/su151310438

Andlar M, Rezić T, Marđetko N et al (2018) Lignocellulose degradation: an overview of fungi and fungal enzymes involved in lignocellulose degradation. Eng Life Sci 18:768–778. https://doi.org/10.1002/elsc.201800039

Arantes V, Goodell B (2014) Current understanding of Brown-rot fungal biodegradation mechanisms: a review. In: Deterioration and protection of sustainable biomaterials. American Chemical Society, pp 3–21

Babu S, Singh Rathore S, Singh R et al (2022) Exploring agricultural waste biomass for energy, food and feed production and pollution mitigation: a review. Bioresour Technol 360:127566. https://doi.org/10.1016/j.biortech.2022.127566

Bera S, Mukherjee T, Das S et al (2022) Exploration of extremophiles for value-added products: a step towards sustainable bio-refinery. In: Extremophiles. CRC Press, Boca Raton

Cañete Vela I, Berdugo Vilches T, Berndes G et al (2022) Co-recycling of natural and synthetic carbon materials for a sustainable circular economy. J Clean Prod 365:132674. https://doi.org/10.1016/j.jclepro.2022.132674

Cao Y, Chen SS, Zhang S et al (2019) Advances in lignin valorization towards bio-based chemicals and fuels: lignin biorefinery. Bioresour Technol 291:121878. https://doi.org/10.1016/j.biortech.2019.121878

Carvalho AKF, Bento HBS, Reis CER, De Castro HF (2019) Sustainable enzymatic approaches in a fungal lipid biorefinery based in sugarcane bagasse hydrolysate as carbon source. Bioresour Technol 276:269–275. https://doi.org/10.1016/j.biortech.2018.12.118

Chandra R, Iqbal HMN, Vishal G et al (2019) Algal biorefinery: a sustainable approach to valorize algal-based biomass towards multiple product recovery. Bioresour Technol 278:346–359. https://doi.org/10.1016/j.biortech.2019.01.104

Chowdhury JI, Hu Y, Haltas I et al (2018) Reducing industrial energy demand in the UK: a review of energy efficiency technologies and energy saving potential in selected sectors. Renew Sust Energ Rev 94:1153–1178. https://doi.org/10.1016/j.rser.2018.06.040

Chukwuma OB, Rafatullah M, Tajarudin HA, Ismail N (2021) A review on bacterial contribution to lignocellulose breakdown into useful bio-products. Int J Environ Res Public Health 18:6001. https://doi.org/10.3390/ijerph18116001

Chutani P, Sharma KK (2016) Concomitant production of xylanases and cellulases from *Trichoderma longibrachiatum* MDU-6 selected for the deinking of paper waste. Bioprocess Biosyst Eng 39:747–758. https://doi.org/10.1007/s00449-016-1555-3

de Padilha CEA, da Nogueira CC, BRA A et al (2021) Production and application of lignin-based chemicals and materials in the cellulosic ethanol production: an overview on lignin closed-loop biorefinery approaches. Waste Biomass Valorization 12:6309–6337. https://doi.org/10.1007/s12649-021-01455-5

Deshmukh MKG, Sameeroddin M, Abdul D, Abdul Sattar M (2021) Renewable Energy in the 21st Century: a review. Mater Today Proc 80:1756. https://doi.org/10.1016/j.matpr.2021.05.501

Ding C, Wang X, Li M (2019) Evaluation of six white-rot fungal pretreatments on corn stover for the production of cellulolytic and ligninolytic enzymes, reducing sugars, and ethanol. Appl Microbiol Biotechnol 103:5641–5652. https://doi.org/10.1007/s00253-019-09884-y

Fillat Ú, Martín-Sampedro R, Ibarra D et al (2017) Potential of the new endophytic fungus Hormonema sp. CECT-13092 for improving processes in lignocellulosic biorefineries: biofuel production and cellulosic pulp manufacture. J Chem Technol Biotechnol 92:997–1005. https://doi.org/10.1002/jctb.5075

Ghosh P, Ghose TK (2003) Bioethanol in India: recent past and emerging future. Adv Biochem Eng Biotechnol 85:1–27. https://doi.org/10.1007/3-540-36466-8_1

Goodell B, Winandy JE, Morrell JJ (2020) Fungal degradation of wood: emerging data, new insights and changing perceptions. Coatings 10:1210. https://doi.org/10.3390/coatings10121210

Hage H, Rosso M-N (2021) Evolution of fungal carbohydrate-active enzyme portfolios and adaptation to plant cell-wall polymers. J Fungi 7:185. https://doi.org/10.3390/jof7030185

Hassan ST, Batool B, Sadiq M, Zhu B (2022) How do green energy investment, economic policy uncertainty, and natural resources affect greenhouse gas emissions? A Markov-switching equilibrium approach. Environ Impact Assess Rev 97:106887. https://doi.org/10.1016/j.eiar.2022.106887

Hatakka A, Hammel KE (2011) Fungal biodegradation of lignocelluloses. In: Industrial Applications. Springer, pp 319–340

Höfer R (2015) Sugar-and starch-based biorefineries. In: Industrial Biorefineries and White Biotechnology, pp 157–235

Ibarra-Gonzalez P, Rong B-G (2019) A review of the current state of biofuels production from lignocellulosic biomass using thermochemical conversion routes. Chin J Chem Eng 27:1523–1535. https://doi.org/10.1016/j.cjche.2018.09.018

Janusz G, Pawlik A, Sulej J et al (2017) Lignin degradation: microorganisms, enzymes involved, genomes analysis and evolution. FEMS Microbiol Rev 41:941–962. https://doi.org/10.1093/femsre/fux049

Kamat S, Khot M, Zinjarde S et al (2013) Coupled production of single cell oil as biodiesel feedstock, xylitol and xylanase from sugarcane bagasse in a biorefinery concept using fungi from the tropical mangrove wetlands. Bioresour Technol 135:246–253. https://doi.org/10.1016/j.biortech.2012.11.059

Karimah A, Ridho MR, Munawar SS et al (2021) A comprehensive review on natural fibers: technological and socio-economical aspects. Polymers 13:4280. https://doi.org/10.3390/polym13244280

Kuhad RC, Kuhar S, Sharma KK, Shrivastava B (2013) Microorganisms and enzymes involved in lignin degradation Vis-à-Vis production of nutritionally rich animal feed: an overview. In: Kuhad RC, Singh A (eds) Biotechnology for environmental management and resource recovery. Springer, New Delhi, pp 3–44

Kumar A, Sharma KK, Kumar P, Ramchiary N (2015) Laccase isozymes from *Ganoderma lucidum* MDU-7: isolation, characterization, catalytic properties and differential role during oxidative stress. J Mol Catal B Enzym 113:68–75. https://doi.org/10.1016/j.molcatb.2015.01.010

Kumar R, Sundari IS, Sen S et al (2016) Chapter 19- animal fat- and vegetable oil-based platform chemical biorefinery. In: Kaur Brar S, Jyoti Sarma S, Pakshirajan K (eds) Platform Chemical Biorefinery. Elsevier, Amsterdam, pp 361–377

Lau MW, Bals BD, Chundawat SPS et al (2012) An integrated paradigm for cellulosic biorefineries: utilization of lignocellulosic biomass as self-sufficient feedstocks for fuel, food precursors and saccharolytic enzyme production. Energy Environ Sci 5:7100–7110. https://doi.org/10.1039/C2EE03596K

Läufer A (2019) Starch biorefinery enzymes. Adv Biochem Eng Biotechnol 166:137–152. https://doi.org/10.1007/10_2016_60

Li T, Takkellapati S (2018) The current and emerging sources of technical lignins and their applications. Biofuel Bioprod Biorefin 0:1–32. https://doi.org/10.1002/bbb.1913

Lim KL, Wong WY, James Rubinsin N et al (2022) Techno-economic analysis of an integrated bio-refinery for the production of biofuels and value-added chemicals from oil palm empty fruit bunches. PRO 10:1965. https://doi.org/10.3390/pr10101965

Liu R, Chen L, Jiang Y et al (2015) Efficient genome editing in filamentous fungus *Trichoderma reesei* using the CRISPR/Cas 9 system. Cell Discov 1:15007. https://doi.org/10.1038/celldisc.2015.7

Malla FA, Mushtaq A, Bandh SA et al (2022) Understanding climate change: scientific opinion and public perspective. In: Bandh SA (ed) Climate change: the social and scientific construct. Springer International Publishing, Cham, pp 1–20

Manzanares P (2020) The role of biorefinering research in the development of a modern bioeconomy. Acta Innov 37:47. https://doi.org/10.32933/ActaInnovations.37.4

Matsu-Ura T, Baek M, Kwon J, Hong C (2015) Efficient gene editing in *Neurospora crassa* with CRISPR technology. Fungal Biol Biotechnol 2:4. https://doi.org/10.1186/s40694-015-0015-1
Nair RB, Kabir MM, Lennartsson PR et al (2018) Integrated process for ethanol, biogas, and edible filamentous fungi-based animal feed production from dilute phosphoric acid-pretreated wheat straw. Appl Biochem Biotechnol 184:48–62. https://doi.org/10.1007/s12010-017-2525-1
Naranjo-Ortiz MA, Gabaldón T (2019) Fungal evolution: diversity, taxonomy and phylogeny of the fungi. Biol Rev 94:2101–2137. https://doi.org/10.1111/brv.12550
Pawlak K, Kołodziejczak M (2020) The role of agriculture in ensuring food security in developing countries: considerations in the context of the problem of sustainable food production. Sustain For 12:5488. https://doi.org/10.3390/su12135488
Pellis A, Cantone S, Ebert C, Gardossi L (2018) Evolving biocatalysis to meet bioeconomy challenges and opportunities. New Biotechnol 40:154–169. https://doi.org/10.1016/j.nbt.2017.07.005
Pereira AG, Fraga-Corral M, Garcia-Oliveira P et al (2022) Single-cell proteins obtained by circular economy intended as a feed ingredient in aquaculture. Food Secur 11:2831. https://doi.org/10.3390/foods11182831
Porro D, Branduardi P (2017) Production of organic acids by yeasts and filamentous fungi. In: Sibirny AA (ed) Biotechnology of yeasts and filamentous fungi. Springer International Publishing, Cham, pp 205–223
Rajak RC, Saha P, Singhvi M et al (2021) An eco-friendly biomass pretreatment strategy utilizing reusable enzyme mimicking nanoparticles for lignin depolymerization and biofuel production. Green Chem 23:5584–5599. https://doi.org/10.1039/D1GC01456K
Reisinger M, Tirpanalan Ö, Prückler M et al (2013) Wheat bran biorefinery—a detailed investigation on hydrothermal and enzymatic treatment. Bioresour Technol 144:179–185. https://doi.org/10.1016/j.biortech.2013.06.088
Reisinger M, Tirpanalan Ö, Huber F et al (2014) Investigations on a wheat bran biorefinery involving organosolv fractionation and enzymatic treatment. Bioresour Technol 170:53–61. https://doi.org/10.1016/j.biortech.2014.07.068
Saini S, Sharma KK (2021a) Fungal lignocellulolytic enzymes and lignocellulose: a critical review on their contribution to multiproduct biorefinery and global biofuel research. Int J Biol Macromol 193:2304–2319. https://doi.org/10.1016/j.ijbiomac.2021.11.063
Saini S, Sharma KK (2021b) Ligninolytic fungi from the Indian subcontinent and their contribution to enzyme biotechnology. In: Satyanarayana T, Deshmukh SK, Deshpande MV (eds) Progress in mycology: biology and biotechnological applications. Springer, Singapore, pp 139–184
Saini S, Chandel AK, Sharma KK (2020a) Past practices and current trends in the recovery and purification of first generation ethanol: a learning curve for lignocellulosic ethanol. J Clean Prod 268:122357. https://doi.org/10.1016/j.jclepro.2020.122357
Saini S, Chutani P, Kumar P, Sharma KK (2020b) Development of an eco-friendly deinking process for the production of bioethanol using diverse hazardous paper wastes. Renew Energy 146:2362–2373. https://doi.org/10.1016/j.renene.2019.08.087
Saini S, Kumar A, Singhal B et al (2022) Fungal oxidoreductases and CAZymes effectively degrade lignocellulosic component of switchgrass for bioethanol production. Fuel 328:125341. https://doi.org/10.1016/j.fuel.2022.125341
Saini S, Kuhad RC, Sharma KK (2023) Valorization of rice straw biomass for co-production of bioethanol, biopesticide and biofertilizer following an eco-friendly biorefinery process. Process Saf Environ Prot 173:823–836. https://doi.org/10.1016/j.psep.2023.03.044
Sandesh K, Ujwal P (2021) Trends and perspectives of liquid biofuel—process and industrial viability. Energy Convers Manag X 10:100075. https://doi.org/10.1016/j.ecmx.2020.100075
Shankar A, Saini S, Sharma KK (2022) Fungal-integrated second-generation lignocellulosic biorefinery: utilization of agricultural biomass for co-production of lignocellulolytic enzymes, mushroom, fungal polysaccharides, and bioethanol. Biomass Convers Biorefinery 14:1117. https://doi.org/10.1007/s13399-022-02969-1

Sharma K, Shrivastava B, Sastry V et al (2013) Middle-redox potential laccase from Ganoderma sp.: its application in improvement of feed for monogastric animals. Sci Rep 3:1299. https://doi.org/10.1038/srep01299

Stegmann P, Londo M, Junginger M (2020) The circular bioeconomy: its elements and role in European bioeconomy clusters. Resour Conserv Recycl X 6:100029. https://doi.org/10.1016/j.rcrx.2019.100029

Suryadi H, Judono JJ, Putri MR et al (2022) Biodelignification of lignocellulose using ligninolytic enzymes from white-rot fungi. Heliyon 8:e08865. https://doi.org/10.1016/j.heliyon.2022.e08865

Veloz Villavicencio E, Mali T, Mattila HK, Lundell T (2020) Enzyme activity profiles produced on wood and straw by four fungi of different decay strategies. Microorganisms 8:73. https://doi.org/10.3390/microorganisms8010073

Xiao L-P, Shi Z-J, Xu F, Sun R-C (2013) Hydrothermal treatment and enzymatic hydrolysis of *Tamarix ramosissima*: evaluation of the process as a conversion method in a biorefinery concept. Bioresour Technol 135:73–81. https://doi.org/10.1016/j.biortech.2012.08.143

Xiao Q, Yu H, Zhang J et al (2019) The potential of cottonseed hull as biorefinery substrate after biopretreatment by *Pleurotus ostreatus* and the mechanism analysis based on comparative proteomics. Ind Crop Prod 130:151–161. https://doi.org/10.1016/j.indcrop.2018.12.057

Yarbrough John M, Zhang R, Mittal A et al (2017) Multifunctional cellulolytic enzymes outperform processive fungal cellulases for coproduction of nanocellulose and biofuels. ACS Nano 11:3101–3109. https://doi.org/10.1021/acsnano.7b00086

11 Innovations in the Pharma and Biopharma Industries: Role of Academia–Industry Interactions

Pallu Reddanna, Jameel Shaik, and Jagadeesh Gandla

Abstract

Global investments in biotech have increased remarkably, with a combined value of $2.0 trillion by the end of 2021. The pharma and biopharma industries in India have been global players in the area of generics and biosimilars, but with no significant entry into original innovations, which are the major cash cows. Innovative discoveries toward novel product development have always been associated with high risk, long gestation, and huge expenditure. These hurdles have been overcome globally through fruitful academia-industry interactions with innovative discovery research and risk mitigation by academic institutions, with the industry stepping in for the later part of drug development. The Indian government has set up several universities and institutes as well as launched several initiatives to promote industry-academia collaboration. Even though these efforts have led to the creation of several breakthrough items, they are yet to have a global impact. This overview seeks to present some of the success stories in driving innovative product development, highlight the issues and challenges, and provide strategies to move forward.

P. Reddanna (✉)
School of Life Sciences, University of Hyderabad, Hyderabad, India

Brane Enterprises, Hyderabad, India
e-mail: prsl@uohyd.ac.in

J. Shaik
Biomedical Engineering, HomeoWealth, Passion Life Skills Academy, Kadapa, India; http://homeowealth.com; https://jameelshaik.com

J. Gandla
The Federation of Asian Biotech Associations, IIIT Campus, Hyderabad, India

G. John, A. K. Panda (eds.), *Biotechnology in India - Reworking A Strategy*, https://doi.org/10.1007/978-981-97-0896-3_11

11.1 The Pharma and Biopharma Industries

India has made significant progress in various spheres of the economy, healthcare, education, space, and technology, among others, in the last 76 years of India's independence (Financialexpress 2022). The average Indian life expectancy has increased from 32 years in 1947 to 70.6 years in 2024 (United Nations - World Population Prospects - India Life Expectancy 1950–2024. www.macrotrends.net. Retrieved 2024-06-22.) This is mainly due to the control of communicable diseases and infant, child, and maternal mortalities as a result of improved treatment availability, affordable medicines, and evolving technology. The Indian pharmaceutical and biopharma businesses deserve credit for developing life-saving treatments and making them accessible not only nationally but also globally. The R&D spent in the biotech sector crossed US $ 1 billion and almost tripled in a year from US $ 320 million in 2020 to US $ 1.02 billion in 2021 (PIB 2022), though India still spends less than 1% of its GDP on R&D.

The growth of India from 2014 to 2022 in the pharma and biopharma sectors is presented in Table 11.1.

11.1.1 Pharma Industry

Today, India is the largest manufacturer of generic drugs in the world, meeting more than 50% of the global demand for various vaccines, 40% of the generic demand in the USA, and 25% of all medicines in the UK. India has nearly 3000 drug companies and 10,500 manufacturing units. About 80% of the antiretroviral drugs used globally against AIDS are supplied from India, the "pharmacy of the world," courtesy of low-cost but high-quality drugs (IBEF 2022).

The Indian pharmaceutical industry, valued at US $ 42 billion in 2021, is estimated to reach ~US $ 65 billion by 2024 and US $ 120–130 billion by 2030. This has become a flourishing industry, growing at a compound annual growth rate (CAGR) of 9.43% over the past 9 years. It is currently ranked third in pharmaceutical production by volume. However, in the international pharmaceutical trade (expected to reach US $1–1.3 trillion by 2030), India's current share is just 2.5%! (The Hindu 2022).

The Indian pharma sector can reach a share of 6–7% by 2030 by focusing on innovative product development, expanding the existing export corridors, and developing new ones (The Hindu 2022). It is high time to identify new avenues and

Table 11.1 Impressive growth statistics of Indian pharma and biopharma sectors from 2014 to 2022

Year	Bioeconomy (in US $ billion) (growth ~8 times)	Biotech start-ups (growth ~100 times)	Investment in bioeconomy (in INR million) (growth ~400 times)
2014	10	52	100
2022	80	5300+	42,000

possibilities to support Indian pharma's positioning as a global leader, thereby boosting the Indian economy.

11.1.2 Biopharma Industry

With a solid foundation in the pharma sector based on India's expertise in academic chemistry, the big pharmaceutical companies entered the biopharma sector with the production of biosimilars (vaccines and therapeutic molecules) and emerged as global players in the biopharma industry. Around 300+ companies produce hormones, insulin, blood products, and vaccines, the sales of which registered INR 33.067 billion in 2020–2021. During the COVID-19 pandemic, it registered a remarkable growth rate of 13% (Biospectrumindia 2022). India administered nearly 4 million doses of COVID-19 vaccines per day, for a total of 1.45 billion doses in 2021. India will soon enter the big league of the world biotech ecosystem (PIB 2022). The Indian COVID economy had a share of 18.1% (the second largest contributor to the total bioeconomy of US $80.1 billion) in 2021. The bioeconomy is projected to reach US $ 277.8 billion by the year 2030 with government support, as seen during the COVID-19 crisis (Table 11.2), or a realistic projection (Table 11.3) of US $ 210 billion (India Bioeconomy Report 2022).

The Indian biotech industry's focus has been on vaccines and therapeutic molecules, and there is great scope to enter specialized areas like cell and gene therapies, a fast-growing segment of the biopharma industry.

In general, the focus of pharma and biopharma has so far been on generics, biosimilars, and me-too products, with little or no emphasis on innovations. Table 11.4 presents the strengths and weaknesses of the pharma and biopharma industries (UKEssays 2022).

Table 11.2 Optimistic projection: Bioeconomy 2025–2030 with support from the government (adapted from India Bioeconomy Report 2022, BIRAC) (India Bioeconomy Report 2022)

Segment/year	2020	2021	2022	2023	2024	2025	2026	2027	2028	2029	2030
Biopharma	38	39.4	45.4	51.5	57.3	63.9	70.9	79.2	88.6	101.1	112.2
Bioagri	11.1	10.5	12.5	14.3	16.9	21.1	28.6	32.8	36.3	40.3	45.3
Bioindustrial	5.1	10.3	14.5	18.5	23.1	28.9	32.6	36.1	41.1	46.1	53.1
CRO/bio-IT/research	10.5	5.4	10.6	14.5	19.5	26.6	30.2	35.7	41.8	48.8	57.2
COVID economy/others	5.5	14.5	10	10	10	10	10	10	10	10	10
Total bioeconomy	70.2	80.1	93	108.8	126.8	150.5	172.3	193.8	217.8	246.3	277.8

Table 11.3 Realistic projection: Bioeconomy 2020–2030 (adapted from India Bioeconomy Report 2022, BIRAC) (India Bioeconomy Report 2022)

Segment/year	2020	2021	2022	2023	2024	2025	2026	2027	2028	2029	2030
Biopharma	38	39.4	45.4	50.5	55.9	61.3	66.7	72.2	77.6	83.2	89.8
Bioagri	11.1	10.5	12.5	14.3	16.4	18.9	21.8	25.1	28.6	32.8	36.3
Bioindustrial	5.1	10.3	13.5	15.5	18.1	21.1	24.1	26.9	29.6	32.6	36.1
CRO/bio-IT/research	10.5	5.4	10.6	13.5	16.5	19.5	23.1	26.6	30.2	33.8	37.8
COVID economy/others	5.5	14.5	10	10	10	10	10	10	10	10	10
Total bioeconomy	70.2	80.1	92	103.8	116.9	130.8	145.7	160.8	176	192.4	210

Table 11.4 Strengths and weaknesses of the pharma and biopharma industries in India

Strengths	Weaknesses
India has skilled scientists and technicians at affordable costs, leading to low costs for innovation, manufacturing, and compliance	Low investments in innovative R&D are a major weakness of the Indian pharmaceutical industry
Indian pharma and biopharma focus is more on generics, biosimilars, and me-too products which have a huge demand globally for being highly affordable in price	Many companies lack the ability to compete with MNCs for new drug discovery, research, and commercialization
India has a well-developed chemistry, R&D, and manufacturing infrastructure	The diffused nature of the industry means that only a small number of companies are large enough for sustained exports
The domestic market is healthy and growing, allowing for economies of scale. The country also has a strong marketing and distribution network	There are limited linkages between industry and academia in India
India is a desirable destination for offshore data management for clinical trials due to its rich biodiversity and strength in chemistry	The domestic market is small due to low healthcare spending
The domestic regulatory environment has been conducive to the growth of the pharmaceutical industry	Competency in API/formulation, IP creation, facility design, and global regulatory affairs is limited to a few players

11.2 Academic Setup in the Country

Education and economic development go hand in hand in any country. The seventh 5-year plan emphasized the development of human resources as a key strategy. Thus, skilled and specialized workers can handle complex machines and techniques, while unskilled workers lack that capacity. This skilled human capital helps increase the productivity of the physical capital, leading to an increase in productivity and creating economic growth (Toppr 2022).

While efforts are going on to expand the number of educational institutions and develop skilled human resources, these institutions have been systematically compartmentalized into basic and applied or professional institutions, which are under the administrative control of different agencies and governed by different ministries.

11.2.1 Basic Institutions

A central body, the University Grants Commission (UGC), established under the UGC Act of 1956, has been governing higher education in the country with periodic reforms. As of 22nd June 2024, there are 487 state, 56 central, 482 state private, and 136 deemed-to-be universities, totaling 1161 (UGC 2023). In addition there are seven IISERs (the Indian Institute of Scientific Education and Research), which are governed directly by the central government.

11.2.2 Professional Institutions

11.2.2.1 Engineering: IITs, NITs, IIITs, Polytechnics

The All-India Council for Technical Education (AICTE), established in November 1945 first as an advisory body and later in 1987 by an Act of Parliament, is responsible for proper planning and coordinated development of technical education and management of the Indian education system.

The Indian Institutes of Technology (IITs) and the Indian Institute of Science (IISc) are world-class educational institutions. Next are the National Institutes of Technology (NITs). Indian Institutes of Information Technology (IIITs) are new entrants. Polytechnics and Industrial Training Institutes (ITIs) offer diploma and certificate courses in engineering, respectively.

Pharmacy education and research are also regulated by the AICTE. Pharmacy universities and colleges offer diploma, graduate, and post-graduate courses, imparting students the latest techniques in pharmaceutical science. The National Institutes of Pharmaceutical Education and Research (NIPERs), a group of Institutes of National Importance, are autonomous bodies under the aegis of the Department of Pharmaceuticals, Ministry of Chemicals and Fertilizers.

11.2.2.2 Others

The Medical Council of India (MCI), set up by the Indian Medical Council Act, 1956, amended in 1993, prescribes minimum standards of medical education essential for granting recognized medical qualifications by universities or medical institutions in India. At the top are the All-India Institutes of Medical Science (AIIMS). Besides New Delhi, they are in 23 other places, out of which 20 are functional, as of May 11 2024. Quality research publications by the faculty involved here are evident from journals covered by the Science Citation Index that emerge every year (AIIMS 2022).

11.2.3 National Laboratories

To promote research and development, address societal problems, and offer innovative technologies and products, the government of India has set up national laboratories in different areas across the country. Based on the discipline, the ministries concerned have set up organizations to establish and monitor their activities. These are listed in Table 11.5. Ayurveda, Yoga & Naturopathy, Unani, Siddha, and Homeopathy (AYUSH) come under the Ministry of AYUSH (Homeowealth 2022).

The CSIR, known for its cutting-edge R&D knowledge base in diverse areas of Science and Technology (S&T), with 37 national laboratories, 39 outreach centers, 1 innovation complex, and 3 units with a pan-Indian presence, oversees a wide spectrum of branches of S&T (CSIR 2022a). These laboratories generate well-trained manpower and do cutting-edge research leading to publications, patents, and innovative technologies. It filed 229 Indian and 202 foreign patents during 2021–2022. CSIR's total unique patent portfolio is 1132, with 140 in commercial use and 2587 granted abroad. In 2021, it published 5769 research papers in science citation

Table 11.5 List of major government organizations and the ministries concerned

S. no.	Organization	Concerned ministry
1.	Council of Scientific and Industrial Research (CSIR)	Ministry of Science and Technology
2.	Department of Science and Technology (DST)	Ministry of Science and Technology
3.	Department of Biotechnology (DBT)	Ministry of Science nd Technology
4.	Indian Council of Medical Research (ICMR)	Ministry of Health and Family Welfare (MHFW)
5.	Directorate of Health Research (DHR)	Ministry of Health and Family Welfare (MHFW)
6.	Indian Council of Agricultural Research (ICAR)	Ministry of Agriculture
7.	Department of Agricultural Research and Education (DARE)	Ministry of Agriculture
8.	Defence Research and Development Organisation (DRDO)	Ministry of Defence

indexed (SCI) journals with an average impact factor of 5.401 per paper (CSIR 2022b). In addition, it is playing a major role in Human Resources Development (HRD) in S&T through the award of fellowships at various levels.

The DST, established in May 1971 to promote new areas of S&T, is also playing a pivotal role in the organization, coordination, and promotion of S&T activities in India. It has major responsibilities in formulating S&T policies, promoting new and emerging areas of S&T, promoting R&D, coordinating and integrating cross-sectoral S&T areas, providing funding and grants-in-aid to research institutions, and promoting the application and dissemination of S&T to the public and disadvantaged groups (DST 2022). "Innovation in Science Pursuit for Inspired Research (INSPIRE)" is an innovative programme of the Department of Science & Technology to attract talent to the study of science at an early age and thus build the human resource pool for strengthening and expanding the Science & Technology system and R & D base.

The DBT, established in 1986 as a separate department dedicated to Biotechnology (BT), has several cutting-edge research institutes listed in Table 11.6 (DBTINDIA 2022). These promote cutting-edge research and innovations in different areas of biotechnology leading to the generation of quality manpower, publications, and patents (Table 11.6).

DBT also plays a major role in HRD through research fellowships at the doctoral and postdoctoral levels. To imbue the students with the right skills and make them industry-ready, it has set up Biotechnology Consortium India Limited (BCIL), which provides internships in industries. In addition to promoting state-of-the-art research in different disciplines of biotechnology, DBT is the major funding agency supporting innovative research in academia and industry. The Biotechnology Industry Research Assistance Council (BIRAC), a non-profit public sector enterprise of DBT, provides an industry-academia interface, implementing its mandate through a wide range of initiatives through targeted funding, technology transfer, intellectual property (IP) management, and handholding schemes that help bring innovation excellence to biotech firms and make them globally competitive. It has

Table 11.6 Major research institutes of the Department of Biotechnology

S.no.	Institute
1	National Institute of Immunology
2	National Centre for Cell Science
3	National Institute of Plant Genome Research
4	National Brain Research Centre
5	Centre for DNA Fingerprinting and Diagnostics
6	Institute of Bioresources and Sustainable Development
7	Institute of Life Sciences
8	Translational Health Science and Technology Institute
9	Institute for Stem Cell Science and Regenerative Medicine
10	National Agri-Food Biotechnology Institute
11	National Institute of Biomedical Genomics
12	National Institute of Animal Biotechnology
13	Rajiv Gandhi Centre for Biotechnology
14	Regional Centre for Biotechnology
15	Center of Innovative and Applied Bioprocessing
16	International Centre for Genetic Engineering and Biotechnology

been promoting innovations starting from the proof of concept all the way to pilot plant setup, clinical trials, etc., through various programs such as E-YUVA, SITARE, BIG, SBIRI, BIPP, PACE, SIIP, SPARSH, etc.

The ICMR, established in 1949, is the apex body for formulating, coordinating, and promoting biomedical research in medicine. It is a premier autonomous organization of the Ministry of Health and Family Welfare. It has established around 31 national institutes across India to address the growing demand for scientific advances in biomedical research and find practical solutions to India's health problems. It too generates well-trained manpower to work in different disciplines of biomedical research and creates innovative solutions to promote the health and well-being of the population (ICMR 2022). It is the major funding agency for projects in biomedical areas and generates HRD through fellowships. ICMR is the nodal agency for the control and management of infectious diseases. It is credited for the effective handling of the unprecedented challenges of the recent Covid-19 pandemic.

Very recently, on 9th August 2023, the Indian Parliament passed the Anusandhan National Research Foundation Bill, 2023, that paves the way for a National Research Foundation (NRF), modeled on the lines of the US National Science Foundation, to ramp up research funding across universities, colleges, institutes, and laboratories. The Act aims to pump in 500 billion rupees (US $6 billion) over the next 5 years, 70% of which is expected to be sourced from non-government sources, industry, and philanthropists.

Thus, these agencies have become the major funding bodies for research at academic institutions and promote HRD via scholarships and awards. While the mandate of the national laboratories is to come up with innovative solutions to address the problems in society, they have also become glorified universities churning out PhDs. The focus of these institutions, therefore, should be on developing innovative technologies and address the problems in the society rather than simply guiding students for PhD and publishing papers. This requires a policy change to be brought by the Government to make the National Laboratories as centers of innovation (Table 11.7).

The focus of Indian educational institutions has been on teaching, research, and thus generating human resources, with little attention on innovation and IPR generation. This is reflected in the poor ranking of India in the global innovation index, as it ranked at 40. Even the manpower generated by these institutions largely lacks the skills required in the industry. As a result, their contribution to the industry in innovative drug discovery and development is very minimal (Dutta et al. 2022). There is a compelling need to address the issues and make academic institutions globally competitive.

Table 11.7 Strengths and weaknesses of the academic institutes in India

Strengths	Weaknesses
Cutting-edge infrastructure, at least in IITs, IISc, IISERs, central universities, and national laboratories	Scientists, engineers, and clinicians working in silos
Faculty/scientists with global recognition	• Little to no training on industry ready skill sets • Lack of research translation capacities and business insight
Can think outside of the box looking for novel solutions	Lack of interactions with the industry
Reasonably well funded by the Govt. of India and international agencies	Lack of access to big funding which is needed to do big projects, e.g., to perform clinical trials
Access to highly motivated talent pool	Focus on publications with little emphasis on innovation and IPR

11.3 Academia-Industry Interactions

As an instance of successful academia-industry linkages, Elsevier, in its comment, says:

> *University researchers are good at finding difficult problems and having the freedom to pursue different solutions; companies are good at taking discoveries and developing them* (Elsevier 2022).

11.3.1 Global Scenario

Globally, in the USA, Europe, and other advanced countries, the interactions between academia and industry have become the driving force behind innovations in the industry. The reasons for this may be due to very high allocation of budgets for promoting research, innovation, and entrepreneurship. Worldwide, innovative high-risk discoveries, especially in pharma, biopharma, and related fields, have come mostly from academic institutes. These are later taken up by big pharma companies for further development. The resulting innovative biomedical scientific discoveries are also the result of active collaborations among basic and clinical researchers working under one roof in academic institutions (Reddanna 2020a; Hughes et al. 2011).

11.3.2 Indian Scenario

11.3.2.1 Initiatives from the Academic Institutions

Innovative biomedical discoveries emanate globally from academic institutions where basic scientists and clinicians work together. In India, there is a systematic isolation of basic science institutions from medical schools and hospital setups. As a result, they do not have a collaborative and enabling environment for working with clinical researchers (Reddanna 2020a; Hughes et al. 2011). Several biomedical scientists do not have any idea of the clinical setting. Similarly, the clinicians working in various hospitals and medical institutions do not have access to the proper instrumentation or expertise to understand the mechanisms behind the pathogenesis of diseases. Also, there are very few collaborations between basic scientists and clinician researchers to translate some of their innovations. Realizing this problem, initiatives are being taken to establish medical schools and hospitals in some of the IITs and IISc. This needs expansion to other institutions as well.

The new education policy (NEP) of 2020 envisages the end of this fragmentation of universities and colleges. It suggests plans to consolidate and restructure existing institutions and establish new world-class model institutes. This is a much-desired initiative of the government, i.e., the phasing out of single-stream Higher Education Institutions (HEIs) such as engineering colleges and medical colleges. It also envisages greater importance in promoting innovation, entrepreneurship, and academia-industry interactions. Hopefully, these initiatives envisaged in the NEP will soon be implemented and bridge the gap between basic scientists, clinicians, and industry. Similarly, in many universities, both government and private, faculty are encouraged to conduct conferences, workshops, etc. in which industry experts are invited to deliver talks about their experiences and knowledge. Universities have started including experts from the industry as members of the Board of Studies, Executive Councils, and other policy bodies. The faculty outreach programs by academic institutions also help in the dissemination of knowledge by giving faculty and students from universities the opportunity to showcase their expertise, infrastructure, and innovations to the industry. Also, the proposed new "professors of practice" is a right step in the bidirectional mobility of personnel from industry to academia and this will promote bilateral interactions and innovative product development (Ahmed et al. 2022).

Traditionally, academic institutions in India have focused on excellence in teaching and research. Scientific research in academic institutions can change the world only when these findings can be transformed into innovations and IP. Realizing this, the focus has also been on innovation, IP generation, and discoveries in the last 3 to 4 decades. Today, India is one of the top 10 countries for filing patents, with an overall ranking of 6.22. It filed 61,573 patents in 2021, which was 8.5% more than in 2020 (Zeebiz 2022). As of 2021, the total number of patents in force was 110,879 (Wipo 2022). Most of these patents have originated from national laboratories, IITs, IISc, and central universities, though state and private universities are much larger in numbers.

In order to transform the innovations protected through IPRs into products, the Government of India established the National Research and Development Corporation (NRDC) in 1953, which forged strong links with the scientific and industrial communities in India and abroad and made formal arrangements with them for the commercialization of the know-how developed. It has licensed more than 4800 technologies to entrepreneurs in small, medium, and large-scale industries, not only in India but also abroad.

Biotechnology Consortium of India Limited (BCIL), incorporated in 1990 with the government in an advisory role and with development banks as the major funding agencies is also facilitating accelerated commercialization of biotech products, and processes (Biotech 2022). It has so far transferred 67 technologies in the areas of biology and biotechnology to various industries. The Council of Scientific and Industrial Research (CSIR), through its 37 laboratories in 25 subject disciplines, has come up with 3200+ active patents, developed 900+ technologies, and had 128+ success stories (Techindiacsir 2022).

ICMR has also accorded significant priority to the development of products and processes by seeking partnerships within and outside India through its Health Technology Accelerated Commercialization (HTAC) Program. The HTAC program, in collaboration with FICCI, is focusing on the affordability, availability, and accessibility of new health products from the technologies developed by the ICMR labs.

DRDO, on the other hand, is making efforts to engage with the industry through various models, including Development cum Production Partner (DcPP), Transfer of Technology (ToT), and the Technology Development Fund (scheme for R&D funding), to encourage industry and start-ups to work in collaboration with academia and research institutions for innovation, research, and development, and to promote self-reliance in defense technology.

11.3.2.2 Initiatives from the Industry

The Indian industry's indifference to academia is exactly the opposite of the frequent interactions and vibrant engagement exhibited by the industry in the developed world. Even the industry leaders serving on the governing boards of HEIs tend to play more of a decorative role than an executive or proactive one (Educationworld 2022).

The Indian industry is losing heavily in terms of downtime and training costs for failing to engage with academia. The lack of an "industry-ready" skilled workforce has led industries to pay a heavy price. For example, Infosys spends INR 2,00,000 on each student it recruits to make him or her industry-ready (Educationworld 2022). A few industries, due to the huge demand, are planning to form consortia and establish their own institutions to instill quality, 'industry-ready or specific' education and an example is the Biocon Academy (Biocon Academy 2022).

Though Government of India (GoI) has made efforts to promote innovation, IPR generation, and the transfer of the technologies developed in the academy, they have not made much impact in coming up with innovative products from the industry for making a global impact. This is mainly due to the risk-averse attitude of the industry. Talking about innovations in the Indian Pharma industry, Mr. Satish Reddy, Chairman of Dr. Reddy's Laboratories and the President of the Indian Pharmaceutical Alliance (IPA), says, "It would require a strong research push from the government through allocations in the budget for infrastructure by way of research-oriented innovation clusters and innovation hubs, incentives to industry to invest in research, and also backed by regulatory reforms" (The Hindu Business Line 2022).

11.3.2.3 Role of the Government

To accelerate innovations and transfer of technologies to the industry, GoI came up with several initiatives to provide grants to academia and industry. Some of these include the New Millennium Indian Technology Leadership Initiative (NMITLI), a program of CSIR; the Pharmaceutical Promotion and Development Scheme (PPDS), a program promoted by DST; the Small Business Innovation Research Initiative (SBIRI) scheme promoted by DBT; etc.

The Technology Development Board (TDB), established in 1995, has been promoting the development and commercialization of indigenous technology and the adaptation of imported technology for much wider application. It played a key role in the promotion and consolidation of the biotech industry in the country, in addition to other pharma and other industries in life sciences and healthcare.

The "Make in India" initiative is helping India make sweeping changes in the industrial sector. Besides providing fellowships, government agencies have come up with schemes to promote innovations not only from the faculty but also from research scholars and students at all levels. Several of these schemes have been earmarked for academia-industry collaborative proposals. These initiatives have resulted in many start-ups across the country in the pharma and biopharma sectors. Some of the schemes launched by the GOI are listed in Table 11.8.

Table 11.8 Some of the academia–industry partnership programs and schemes undertaken by the Government of India (India Science and Technology 2022)

Title	Ministry/Department
AICTE-INAE- Distinguished Visiting Professorship Scheme	Ministry of Education (MoE), Govt. of India
Biotech Urjit Cluster Scheme	Department of Biotechnology (DBT), Govt. of India
BIRAC-Wellcome Trust Partnership	Department of Biotechnology (DBT), Govt. of India
Contract Research Scheme (CRS)	Department of Biotechnology (DBT), Govt. of India
India Innovation Growth Programme (IIGP) 2.0	Department of Science and Technology (DST), Govt. of India
Industry Relevant R&D (IRRD)	Department of Science and Technology (DST), Govt. of India
Innovate in India (i3) under National Biopharma Mission	Department of Biotechnology (DBT), Govt. of India
Patent Acquisition and Collaborative Research and Technology Development (PACE)	Department of Scientific and Industrial Research
Prime Minister's Fellowship Scheme for Doctoral Research	Science and Engineering Research Board (SERB)
Scheme for Funding Industry Relevant R&D	Science and Engineering Research Board (SERB)
Scientific Infrastructure Access for Harnessing Academia University Research Joint Collaboration (SAHAJ)	Department of Biotechnology (DBT), Govt. of India
SEED Support Scheme for STEP/TBIs	Department of Science and Technology (DST), Govt. of India
TDB-ASSOCHAM Partnership	Department of Science and Technology (DST), Govt. of India
TDB-BPI-France Partnership	Department of Science and Technology (DST), Govt. of India
TDB-ICCO Partnership	Department of Science and Technology (DST), Govt. of India
TDB-PHDCCI Partnership	Department of Science and Technology (DST), Govt. of India
TDB-TIFAC Partnership	Department of Science and Technology (DST), Govt. of India
TDB-WWF Partnership	Department of Science and Technology (DST), Govt. of India
TDB-ASSOCHAM Partnership	Technology Development Board (TDB)
TDB-BIRAC Partnership	Technology Development Board (TDB)

(continued)

Table 11.8 (continued)

Title	Ministry/Department
TDB-BIRAC Partnership	Department of Science and Technology (DST), Govt. of India
TDB-Bpifrance Partnership	Technology Development Board (TDB)
TDB-CEFIPRA-Bpifranse	Department of Science and Technology (DST), Govt. of India
TDB-ICCO Partnership	Technology Development Board (TDB)
TDB-PHDCCI Partnership	Technology Development Board (TDB)
TDB-TIFAC Partnership	Technology Development Board (TDB)
Technology Acquisition and Development Fund (TADF)	Department of Industrial Policy and Promotion(DIPP)
Technology Acquisition Fund Programme (TAFP)	Department of Heavy Industries (DHI)
Technology Development Fund (TDF)	Defence Research and Development Organisation (DRDO), Govt. of India
UAY (Ucchatar Avishkar Yojana)	Ministry of Education (MoE), Govt. of India
Venture Capital Funds	Department of Science and Technology (DST), Govt. of India

11.3.2.4 Role of Start-Ups

Start-ups play a crucial role in bridging the gap between academia and industry. By working closely with universities and research institutions, they help to commercialize new technologies and ideas that have been developed in academia. This will help bring cutting-edge innovations to the market, leading to a positive impact on society and the economy. Additionally, by providing opportunities for researchers to work in industry, start-ups facilitate the transfer of knowledge and expertise from academia to the broader business community.

With the changed vision of the Union government, efforts are now being made to promote innovation and entrepreneurship through the establishment of incubation centers and the promotion of the start-up ecosystem in the country. As of 30th November 2022, there were 84,012 start-ups (58% of them located in five states) (Smefutures 2022), 700+ incubators, 108 unicorns, and more. Around 10,000+ start-ups are established in various universities and national laboratories, and most of these have been set up by students or faculty members. Some of the organizations involved in setting up incubation centers include the National Science and Technology Entrepreneurship Development Board (NSTEDB) set up by the DST, which has established 326 Technology Business Incubation (TBI) centers across the country; the Technology Incubation and Development of Entrepreneurs (TIDE) set up by the Ministry of Electronics and Information Technology, which has 51 incubation centers in various educational institutions; and the Atal Innovation Mission, which has set up 69 Atal Incubation Centers across the country. Similarly, BIRAC has helped in the creation of 3500+ start-ups, entrepreneurs/companies, supported 750+ products and technologies, generated 32,474 employments, through the establishment of 75 bio-incubation and pre-incubation centers. BIRAC also helped raise INR 4000 crores in follow-on funding for more than 125 BIRAC-supported start-ups (India Bioeconomy Report 2022).

Some of the well-established incubation centers and research parks are located at IITs. Other leading bio-incubators include the IKP Knowledge Park and SBTIC located in the Genome Valley in Hyderabad, the Centre for Cellular and Molecular Platforms (C-CAMP) in Bangalore, the NCL Venture Center in Pune, the KIIT Incubation Center in Bhubaneswar, the Atal Incubation Center (CCMB) in Hyderabad, the ASPIRE BioNEST incubation center at the University of Hyderabad, the Bangalore Bio-innovation Center, the TICEL BioPark in Chennai, the PSG Science and Technology Entrepreneurial Park in Coimbatore, etc.

11.3.2.5 Role of the Ecosystem and Its Enablers

Academy–industry interactions happen through an ecosystem that enables collaborations between various stakeholders (Fig. 11.1). These include:

1. Academic institutions with state-of-the-art infrastructure and expert faculty involved in generating skilled manpower.
2. Liberal funding for the faculty to focus on cutting-edge research, leading to innovation and IPR generation.
3. Establishment of Technology Transfer Offices which serve as a bridge between academia and industry and empower them to explore the transfer of technologies around the globe.
4. An enabling environment for the faculty to start their ventures and promote a strong start-up ecosystem.
5. Financial support for promoting innovative research and entrepreneurship from the government and private funding agencies.
6. Change in the mindset of the big pharmaceutical and biotech companies to take risks for the development of innovative products.
7. Providing a platform for the academy and industry to come together and promote interactions through the organization of symposia and conferences by public and private agencies.
8. Providing a platform for the academy and industry to promote collaborations between them and avail a global platform for interactions by public or private organizations. In this connection, organizations such as the Federation of Asian Biotech Associations (FABA) and the Association of Biotech Led Enterprises (ABLE) have been organizing BioAsia and Bangalore Bio, respectively, which have provided global bio-business platforms.

In addition to the above, the following steps will foster successful collaborations between academia and industry:

- **A shared commitment to innovation and progress:** Both academia and industry need to be committed to working together to provide innovative solutions for problems in society.
- **Clear communication and collaboration:** Effective communication and collaboration are essential for successful academia–industry partnerships. These include regular meetings, open discussions, and the sharing of information and resources.

- **Flexible and adaptable organizational structures:** For academia–industry collaborations to be successful, organizations need to be flexible and adaptable to respond to changing needs and opportunities.
- **Push and Pull Models to be Adopted:** Push and pull models have to be created where "push" from the owners of the IP and "pull" from the pharma/biotech companies will help in effective translation of discoveries into innovative products.
- **A supportive policy and regulatory environment:** There is a need to create a favorable policy and regulatory environment that promotes academia and industry collaboration.
- **Bidirectional mobility of the faculty/scientists:** The movement of faculty to different industries and vice versa would bring in new and diverse expertise, skills and ideas to the private sector and academia, respectively.

Overall, the key to successful academia–industry collaboration is a willingness to work together and a commitment to address the problems in society with innovative solutions. By fostering a supportive ecosystem and enabling effective partnerships, as shown in Fig. 11.1, organizations may help bridge the gap between academia and industry and drive innovative product development.

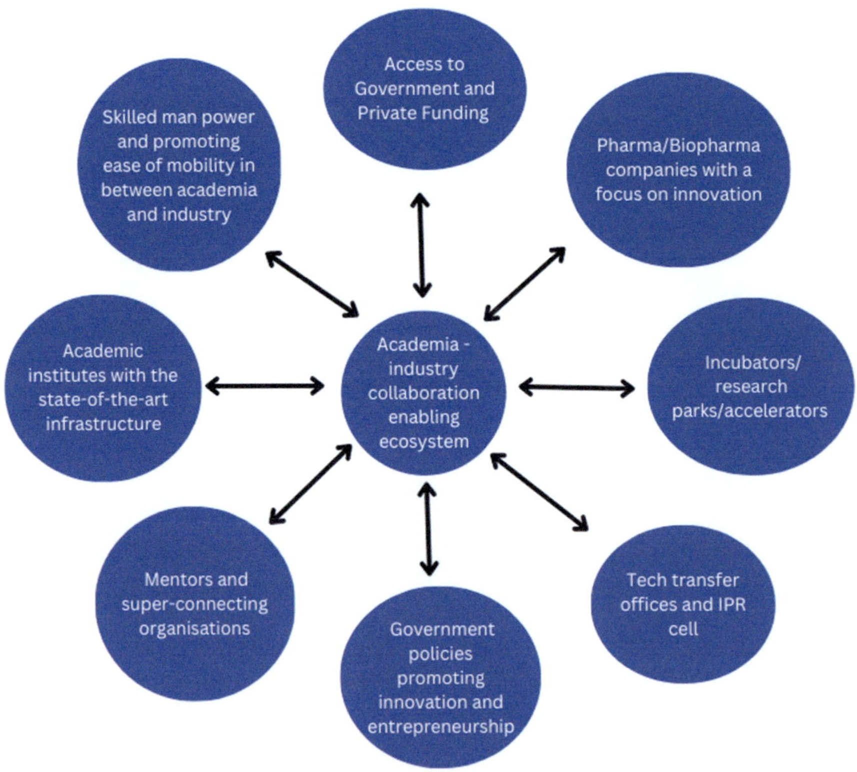

Fig. 11.1 Key enablers promoting innovations in pharma and biopharma sectors

11.3.2.6 Novel Models to Promote Innovative Product Development

Three-way partnerships between industry, academia, and government have been aptly termed the triple helix'. Strategic business-university research collaborations provide several benefits to participants (Elsevier 2022). According to an article published in Nature (Padmanaban 2016), there are many reasons why many of India's innovations never make it to commercial scale. For example, (1) socio-economic reasons in academia; (2) an unclear commercial pathway for academics; and (3) a lack of regulatory support, especially when clinical trials come into the picture (Rangarajan 2022), (4) Lack of sustained funding from the government from the early stage all the way to the commercial stage; and (5) Risk-averse attitude of the industry in taking up innovative discoveries from academic institutions.

So far, the Indian industry has taken the path of least resistance by just copying what has been done in the West. Indian pharma and biopharma companies must quickly graduate from reverse engineering to research leading to innovative product development.

The state-of-the-art infrastructure, strong IT base, and large pool of scientific manpower have to be exploited for the design and development of innovative drugs. There is a need to refocus pharmaceutical research away from high-risk early discovery activities by involving academia early on and focusing on later stages of drug development. This shift in the focus of the industry from early to late-stage drug discovery will enable the translation of publicly funded research, leading to cost-effective development of drugs. This method, if adopted, will overcome the challenges in innovations like increasing costs, decreasing productivity, and attrition of projects.

To make these novel partnership models successful, sustainable funding strategies are needed to create specialized core facilities within academic institutions and drive them toward innovative research. In addition, there is a need to create an ideal environment for science-based enterprises focusing on innovation and creativity. On the same lines, nurturing young graduates with innovative ideas through liberal funds will drive innovation and entrepreneurship activities. These efforts create job opportunities for young graduates coming out of universities and institutes and direct them toward innovative translational research and nation-building (Reddanna 2020b).

11.3.2.7 Academy Industry Interactions: Success Stories

The sustained efforts of the Government of India through several focused schemes in the last two to three decades have resulted in promoting many collaborations between the academy and the industry. Industry has also started reaching out to academic institutions to address some of the problems they encountered in their efforts to develop products. Academic institutions and national laboratories also started approaching industries through Tech Transfer offices to commercialize the discoveries. In this connection, agencies like NRDC, BCIL, and other agencies also played a key role. These efforts in the last two to three decades have resulted in the development of several innovative products. Some of these are presented in Table 11.9.

Table 11.9 List of innovative products/technologies from academia and industry collaborations

	Industry	Academic institute	Product
1	Bharat Serums and Vaccines	National Chemical Laboratory	Recombinant cell line technologies (Economic Times 2023)
2	Serum Institute of India Pvt. ltd. (SIIPL)	Translational Health Science and Technology Institute	Development of a bivalent dengue-chikungunya vaccine, novel cell and gene therapy to cure HIV patients and other innovative technologies (THSTI 2023)
3	AstraZeneca	SASTRA University	Clinical research and Development (The Hindu Business Line 2023)
4	Biological E	DBT and BIRAC	CORBEVAXTM (PIB 2023a)
5	Bharat Biotech	CSIR	Developing monoclonal antibodies for COVID (Times of India 2023a)
6	Bharat Biotech	ICGEB	Malaria vaccine (Bharat Biotech 2023a)
7	M/s Sapigen Biologix Private Limited	Technology Development Board	Intranasal COVID-19 vaccine and RTS, S malaria vaccine (TDB 2023)
8	Ahammune Biosciences Pvt	NCL, IGIB of ICMR	Vitiligo drug development (Times of India 2023b)
9	Reliance Life Sciences	IIT-K	Gene therapy for hereditary eye diseases (BioSpectrum India 2023)

10	Wipro GE Healthcare	IIT Madras	Fellowship partnership to provide increased opportunities for research scholars to build and further improve the innovation ecosystem (Healthcare Executive 2023)
11	Asian Institute of Gastroenterology (AIG)	IIT Hyderabad	Master of technology—medical device Innovation (BME 2023)
12	Many industries	BETIC at IIT Bombay	BETIC labs developed several medical devices and these have been licensed to local industry partners (BETIC 2023)
13	HCL	IIT Kharagpur	Several medical devices (HCL 2023)
14	RICOVR Inc., a US-based oral diagnostics medical device company	IIT Madras	Point-of-care medical device development (Education Times 2023)
15	Apollo Hospitals	IIT Kanpur	Medical technologies (Indian Express 2023)
16	Bharat Biotech Int. ltd., Hyderabad, India	Indian Council of Medical Research, National Institute of Virology	COVAXIN (Bharat Biotech 2023b)
17	Cadila Healthcare ltd., Ahmedabad, India	Department of Biotechnology, India	ZyCoV-D (PIB 2023b)

(continued)

Table 11.9 (continued)

	Industry	Academic institute	Product
18	Bharat Biotech	Several scientific organizations such as DBT, AIIMS, etc.	ROTAVAC (Bharat Biotech 2023c)
19	Shanta Biotech	CCMB	Hepatitis B vaccine (Shanvac-B) (Chakma et al. 2011)
20	Johari Digital Healthcare ltd. (JDHL)	IIT Jodhpur Technology Park	Johari Centre of Excellence (CoE) (Healthcare Radius 2023)
21	Dr. Reddy's Laboratories	University of Hyderabad	Anti-psoriasis drug discovery (Herald 2023a)
21	VINS Bioproducts ltd.	University of Hyderabad and CCMB	Vincovin-19, first antidote against COVID-19 (Herald 2023b)
22	Natco Pharma	University of Hyderabad	Anti-cancer drug agents (UOHYD 2023)
23	KnowDis	IIT-Delhi	AI-driven antibody design (Anon 2023)
24	Sun Pharma	CSIR-IICT	Sun Pharma gets exclusive global license for CSIR-IICT's patents related to certain compounds with potential therapeutic activity across multiple indications (Sun Pharma 2023)
25	Piramal	DBT	Drug discovery initiative (The Hindu 2023)
26	Gennova	University of Delhi South Campus (UDSC)	Therapeutic neutralizing antibodies for COVID (Gennova 2023)
27	Sun Pharma	ICGEB	Collaboration to develop novel dengue vaccine (Anon n.d.)

28	Indian Immunologicals	Bharati Vidyapeeth Deemed University Medical College and Hospital, Pune, India, and Apollo DRDO Hospital, Hyderabad, Telangana, India	Hepatitis-A vaccine development (Susarla et al. 2021)
29	Zydus Cadila	CSIR-IMTECH	Natural streptokinase (IMTech 2023a)
30	McDowell and Company	CSIR-IMTECH	Energy efficient alcohol technology (IMTech 2023b)
31	Laurus Labs	IIT Bombay (Immuno ACT)	CAR-T cell therapy (https://economictimes.indiatimes.com/industry/healthcare/biotech/healthcare/laurus-labs-acquires-stake-in-iit-bombay-incubated-car-t-cell-therapy-company/articleshow/87798427.cms?from=mdr)

As indicated above, there are several innovative products from successful academy–industry collaborations, most of them from the IITs and National Laboratories. However, only a handful of them, like Shanvac-B, Rotavac, Covaxin, and Vincov-19, have made a global impact. Several innovations such as diagnostics, vaccines, ventilators, and other biomedical devices that came out of academy–industry interactions during the COVID-19 pandemic in record time exemplify the potential of such collaborations. With the lessons learned during the COVID-19 pandemic, both the academia and industry need to make sustained efforts to come up with innovative products of global impact. Also, there is a need for the Pharma and Biopharma industries to redesign their business models to embrace innovative technologies in order to sustain and be competitive at the global level (Reddanna 2023).

While discussing success stories, we must point out the failure stories that are noteworthy. Scientists from Indian Institute of Science developed DNA vaccine for rabies and filed an international patent, way back in 2002. While inaugurating the "International workshop on harmonizing testing procedures for rabies vaccine" Prof. G. Padmanaban, former Director of IISc, said that IISc would be the first in the world to bring out a DNA vaccine for rabies and stressed the need for India to be a global vaccine player (http://test.pharmabiz.com/news/international-workshop-on-harmonizing-testing-procedures-for-rabies-vaccine-11707). Dr. Manju Sharma, the then Secretary, Department of Biotechnology announced that "In six months' time, we will experiment with the vaccine on monkeys" (http://test.pharmabiz.com/news/dna-rabies-vaccine-to-be-ready-for-human-trials-by-next-month-12756). If it had materialized, it would have been the first DNA vaccine in the world, rolled out of academy industry interactions from India.

11.4 Conclusions and Way Forward

The pharma and biopharma industries have demonstrated their potential by coming up with innovative processes to develop generics and biosimilars and make them available at affordable prices globally. However, they are yet to make an impact in coming up with innovative product development. Academic institutions, at least elite ones such as IITs, Central Universities, IISc, IISERs, and National Laboratories, with state-of-the infrastructure and expert faculty, have been working on cutting-edge research, as demonstrated by publications in the leading scientific journals and patents filed. However, they are yet to make a significant impact on collaborations with industries for developing innovative technologies and products. Hence, there is a need to bring in several radical changes to harness the true potential of academic institutions in driving innovations and the industry through strong collaboration between them for making an impact in the global scenario. The following steps are pertinent.

11.5 Driving Innovations in Academic Institutions

- India needs to take radical steps to bring academic institutions involved in biomedical research closer to teaching and research hospitals and medical schools.
- All central universities, IITs, IISERs, and national laboratories should have teaching and research hospitals attached to them, with basic and clinical researchers in equal numbers. Similarly, all the AIIMS should be provided with state-of-the-art infrastructure and they should also recruit basic scientists and clinicians in equal numbers.
- The National laboratories should refocus their strategies toward innovative technologies to address society related problems rather than being "glorified universities."
- The faculty entrepreneurship scheme, introduced in India a decade ago, should be effectively implemented to start ventures by the faculty. Incubation centers or translation accelerators should form key units of premier academic institutions to accommodate faculty entrepreneurs and drive innovative discoveries.
- Establishment of innovation parks within the academic institutions and encouraging the big pharmaceutical and biotech companies to set up their R&D units within the academic institutions should be a priority.
- Revamping the teaching programs in science subjects with more practical orientation, hands-on training, and problem-solving approach.
- Introduction of innovative Programs that would result in creation of a start-up based on research done during the PhD, on the lines of BITS Pilani PhD program.
- Facilitate bidirectional mobility of academy and industry personnel to provide an understanding of the requirements in the industry and the constraints under which the academic institutions work.

11.6 Innovative Product Development by Industries

- The mindset of the industry should change and they should prepare to take risks in developing innovative products by leveraging innovative discoveries coming out of academic institutions and start-ups.
- The push and pull models, push by academia, and pull by industry, are to be adopted to promote academia-industry linkages.
- Indian industry should exploit the state-of-the-art infrastructure, highly qualified faculty, and human resources available at academic institutions for innovative product development.
- Academia and industry, supported with policy initiatives from the government, should take these radical steps to place India on the global map of innovations in R&D (Reddanna 2020a).

With the New Education Policy rolled out along with the establishment of National Research Foundation and several innovative schemes announced by the Government of India it is hoped that there will be closer interactions between the academic institutions and industries to translate innovative discoveries into novel product development and thus address the problems in the society.

References

Ahmed F, Fattani MT, Ali SR, Enam RN (2022) Strengthening the bridge between academic and the industry through the academia-industry collaboration plan design model. Front Psychol 13:875940. https://doi.org/10.3389/fpsyg.2022.875940

AIIMS (2022) Research. https://www.aiims.edu/en/research.html. Accessed 7 Dec 2022

Anon (2023) Generative AI meets drug discovery: KnowDis and IIT-Delhi collaboration for AI-driven antibody design. https://www.business-standard.com/content/press-releases-ani/generative-ai-meets-drug-discovery-knowdis-and-iit-delhi-collaboration-for-ai-driven-antibody-design-123022301017_1.html. Accessed 7 Sept 2023

Anon (n.d.) Sun pharma & ICGEB announce new exclusive collaboration to develop novel dengue vaccine for India & global markets

BETIC (2023) Impact|BETIC|India. https://www.betic.org/impact. Accessed 7 Sept 2023

Bharatbiotech (2023a) ICGEB's malaria vaccine ready for clinical trials. https://www.bharatbiotech.com/pdf/NewsUpdates/2004/ICGEB%27s-Malaria-Vaccine-Ready-for-Clinical-Trials-Financial-Express-Mar-03-2004.pdf. Accessed 7 Sept 2023

Bharatbiotech (2023b) COVAXIN—India's first indigenous covid-19 vaccine|Bharat Biotech. https://www.bharatbiotech.com/covaxin.html. Accessed 7 Sept 2023

Bharatbiotech (2023c) ROTAVAC|First rota virus vaccine manufacturer|Bharat Biotech. https://www.bharatbiotech.com/rotavac.html. Accessed 7 Sept 2023

Biocon Academy (2022). https://www.bioconacademy.com/. Accessed 26 Dec 2022

Biospectrumindia (2022) Indian biopharma industry hits Rs 33k crore with 13% growth in 2020–21. https://www.biospectrumindia.com/features/18/21144/indian-biopharma-industry-hits-rs-33k-crore-with-13-growth-in-2020-21.html. Accessed 28 Dec 2022

Biospectrumindia (2023) IIT-K and reliance life sciences to revolutionise gene therapy for hereditary eye diseases. https://www.biospectrumindia.com/news/96/22774/iit-k-and-reliance-life-sciences-to-revolutionise-gene-therapy-for-hereditary-eye-diseases.html. Accessed 7 Sept 2023

Biotech (2022) Technologies transferred|Page-1|Biotech Consortium India Limited (BCIL), pioneering biotechnology commercialization since 1990. https://www.biotech.co.in/en/activities/technology-transfer/technologies-transferred. Accessed 26 Dec 2022

BME (2023) BME @ IITH. https://bme.iith.ac.in/assets/pages/admtechmdi.html. Accessed 7 Sept 2023

Chakma J, Masum H, Perampaladas K, Heys J, Singer PA (2011) Indian vaccine innovation: the case of shantha biotechnics. Glob Health 7:9. https://doi.org/10.1186/1744-8603-7-9

CSIR (2022a) Council of Scientific & Industrial Research|CSIR|GoI. https://www.csir.res.in/. Accessed 4 Dec 2022

CSIR (2022b) About CSIR|Council of Scientific & Industrial Research|CSIR|GoI. https://www.csir.res.in/about-us/about-csir. Accessed 8 Dec 2022

DBTINDIA (2022) Introduction|Department of Biotechnology. https://dbtindia.gov.in/about-us/introduction. Accessed 4 Dec 2022

DST (2022) Introduction|Department of Science & Technology. https://dst.gov.in/introduction. Accessed 4 Dec 2022

Dutta S, Lanvin B, Rivera León L, Wunsch-Vincent S (2022) In partnership with academic network members global innovation index 2022 what is the future of innovation-driven growth? 15th edn. https://doi.org/10.34667/tind.46596

Economictimes (2023) Bharat serums and vaccines ltd: BSV forges industry-academia collaboration with national chemical laboratory, ET health world. https://health.economictimes.indiatimes.com/news/industry/bsv-forges-industry-academia-collaboration-with-national-chemical-laboratory/92182633. Accessed 7 Sept 2023

Educationtimes (2023) IIT Madras to research on point-of-care medical device development—EducationTimes.com. https://www.educationtimes.com/article/campus-beat-college-life/72913496/iit-madras-to-research-on-point-of-care-medical-device-development. Accessed 7 Sept 2023

Educationworld (2022) Why is Indian industry divorced from academia?—EducationWorld. https://www.educationworld.in/why-is-indian-industry-divorced-from-academia/. Accessed 4 Dec 2022

Elsevier (2022) University-industry collaboration: a closer look for research leaders. https://www.elsevier.com/research-intelligence/university-industry-collaboration. Accessed 4 Dec 2022

Financialexpress (2022) India at 75: Important healthcare achievements of the country since gaining independence|The Financial Express. https://www.financialexpress.com/healthcare/news-healthcare/india-at-75-important-healthcare-achievement-of-the-country-since-gaining-independence/2621945/. Accessed 13 Dec 2022

Gennova (2023) Antibody—Gennova Bio. https://gennova.bio/antibody/. Accessed 7 Sept 2023

HCL (2023) HCL inks pact with IIT Kharagpur, MIT Harvard to assess emerging markets for medical devices. http://test.pharmabiz.com/news/hcl-inks-pact-with-iit-kharagpur-mit-harvard-to-assess-emerging-markets-for-medical-devices-48218. Accessed 7 Sept 2023

Healthcareexecutive (2023) Wipro GE healthcare join hands with IIT madras to foster industry-academia collaboration—healthcare executive. https://www.healthcareexecutive.in/blog/industry-academia-collaboration. Accessed 7 Sept 2023

Healthcareradius (2023) Johari digital healthcare inaugurates the 'Johari Center of Excellence' in IIT Jodhpur technology park—healthcare radius. https://www.healthcareradius.in/features/management/johari-digital-healthcare-inaugurates-the-johari-center-of-excellence-in-iit-jodhpur-technology-park. Accessed 7 Sept 2023

Herald (2023a) "Lipoxygenase inhibitors to treat psoriasis and other skin disorders"—patent granted to University of Hyderabad and Dr. Reddy's Institute of Life Sciences. https://herald.uohyd.ac.in/lipoxygenase-inhibitors-to-treat-psoriasis-and-other-skin-disorders-patent-granted-to-university-of-hyderabad-and-dr-reddys-institute-of-life-sciences/. Accessed 7 Sept 2023

Herald (2023b) First antidote against COVID-19 developed by UoH, CCMB and Vins Cov 19—successful completion of phase 2 clinical trials across multiple centres in India. https://herald.uohyd.ac.in/uoh-ccmb-and-vins-cov-19-an-antidote-and-a-cure-against-covid-19-successfully-completes-phase-2-clinical-trials/. Accessed 7 Sept 2023

Homeowealth (2022) HomeoWealth—Wealth of Homeopathy in your hands. http://homeowealth.com/. Accessed 21 Dec 2022

Hughes JP, Rees SS, Kalindjian SB, Philpott KL (2011) Principles of early drug discovery. Br J Pharmacol 162(6):1239–1249. https://doi.org/10.1111/J.1476-5381.2010.01127.X

IBEF (2022) Pharmaceutical companies in India, Indian Pharma Industry- IBEF. https://www.ibef.org/industry/pharmaceutical-india. Accessed 3 Dec 2022

ICMR (2022) Indian council of medical research|Government of India. https://main.icmr.nic.in/. Accessed 4 Dec 2022

IMTech (2023a) Welcome to CSIR-IMTech—EN. https://www.imtech.res.in/achievements/technologies-commercialized/natural-streptokinase. Accessed 7 Sept 2023

IMTech (2023b) Welcome to CSIR-IMTech—EN. https://www.imtech.res.in/achievements/technologies-commercialized/energy-efficient-alcohol-technology. Accessed 7 Sept 2023

India Bioeconomy Report (2022). https://birac.nic.in/webcontent/1658318307_India_Bioeconomy_Report_2022.pdf. Accessed 26 Dec 2022

Indianexpress (2023) IIT Kanpur, Apollo sign MoU for research collaboration in medical technology|Education News—The Indian Express. https://indianexpress.com/article/education/iit-kanpur-apollo-join-hands-for-research-collaboration-in-medical-technology-8359309/. Accessed 7 Sept 2023

Indiascienceandtechnology (2022) Academia industry partnerships|India science, technology & innovation—ISTI Portal. https://www.indiascienceandtechnology.gov.in/programme-schemes/academia-industry-associations. Accessed 26 Dec 2022

Padmanaban G (2016) Biotech research in health sector: what has academia done? Nature India 2016:26. https://doi.org/10.1038/nindia.2016.28; Published online February

PIB (2022). https://pib.gov.in/PressReleaseIframePage.aspx?PRID=1880666. Accessed 6 Dec 2022

PIB (2023a) Department of Biotechnology Mission COVID Suraksha Supported Biological E Limited Novel COVID-19 vaccine candidate—CORBEVAXTM receives DCGI approval for Emergency Use Authorization. https://pib.gov.in/Pressreleaseshare.aspx?PRID=1786152. Accessed 7 Sept 2023

PIB (2023b) Press Information Bureau. https://www.pib.gov.in/PressReleasePage.aspx?PRID=1747669. Accessed 7 Sept 2023

Rangarajan PN (2022) A strategy for the development of a combination SARS-CoV-2 vaccine containing a DNA vaccine (ZyCoV-D) and an inactivated virus vaccine (Covaxin); https://currentscience.ac.in/. https://currentscience.ac.in/Volumes/121/11/1402.pdf. Accessed 25 Dec 2022

Reddanna P (2020a) Indian pharma and biotech industry-will they graduate pharma bioworld. Pharm Bio World 18(12):16–19

Reddanna P (2020b) Pharmaceutical research in India—new models of drug discovery pharma BioWorld, September, 2020. Pharma Bio World; Published online September

Reddanna P (2023) Reimagining business models: future of pharma and biopharma sectors. Pharma BioWorld 21(12):29–32

Smefutures (2022) Govt recognises over 84,012 startups; 58% are in 5 states. https://smefutures.com/govt-recognises-over-84012-startups-58-are-in-5-states/. Accessed 28 Dec 2022

Sunpharma (2023) Sun pharma announces licensing agreement with CSIR-IICT. www.sunpharma.com. Accessed 7 Sept 2023

Susarla SK, Palkar S et al (2021) Safety and immunogenicity of inactivated hepatitis-a vaccine developed by human biologicals institute in two age groups of healthy subjects: a phase I open label study. Vaccine 39(15):2088–2093. https://doi.org/10.1016/J.VACCINE.2021.03.012

TDB (2023) TDB supports M/s Sapigen Biologix Private Limited to develop and commercialize two indigenous vaccines- Intranasal Vaccine for COVID & Malaria Vaccine—Technology Development Board | Department of Science & Technology. http://tdb.gov.in/tdb-supports-ms-sapigen-biologix-private-limited-develop-commercialize-two-indigenous-vaccines-intranasal-vaccine-covid-malaria-vaccine/. Accessed 7 Sept 2023

Techindiacsir (2022) CSIR India technology portal. https://techindiacsir.anusandhan.net/online/Control.do?_main=488t3s. Accessed 26 Dec 2022

The Hindu (2022) The giant leap of India's pharmaceutical industry—the hindu. https://www.thehindu.com/brandhub/the-giant-leap-of-indias-pharmaceutical-industry/article65670866.ece. Accessed 5 Dec 2022

The Hindu (2023) Piramal, DBT begin second stage of drug discovery initiative|Science and Technology News|The Hindu. https://www.thehindu.com/business/companies/piramal-dbt-begin-second-stage-of-drug-discovery-initiative/article2230236.ece. Accessed 7 Sept 2023

Thehindubusinessline (2022) Pharma industry needs govt's research-push to scale innovation challenges: Satish Reddy—The Hindu BusinessLine. https://www.thehindubusinessline.com/companies/pharma-industry-needs-govts-research-push-to-scale-innovation-challenges-satish-reddy/article66282123.ece. Accessed 26 Dec 2022

Thehindubusinessline (2023) AstraZeneca India strengthens partnership with SASTRA University—The Hindu BusinessLine. https://www.thehindubusinessline.com/companies/astrazeneca-india-strengthens-partnership-with-sastra-university/article67123541.ece. Accessed 7 Sept 2023

THSTI (2023) THSTI—translational health science and technology institute. https://thsti.res.in/en/NewsDetails/230/translational-health-science-institute-and-serum-institute-of-india-sign-a-letter-of-intent-for-collaborative-efforts-in-vaccine-development-and-cell-and-gene-therapy. Accessed 7 Sept 2023

Timesofindia (2023a) Bharat Biotech leads CSIR approved project to develop human monoclonal antibodies|India Business News—Times of India. https://timesofindia.indiatimes.com/business/india-business/covid-19-bharat-biotech-leads-csir-approved-project-to-develop-human-monoclonal-antibodies/articleshow/75624804.cms. Accessed 7 Sept 2023

Timesofindia (2023b) NCL, startup sign ink pact for vitiligo drug development|Pune news—times of India. https://timesofindia.indiatimes.com/city/pune/ncl-startup-sign-ink-pact-for-vitiligo-drug-development/articleshow/57503501.cms. Accessed 7 Sept 2023

Toppr (2022) Education in India: importance, growth and future prospects. https://www.toppr.com/guides/economics/human-capital-formation-in-india/education-in-india/. Accessed 7 Dec 2022

UGC (2023) UNIVERSITY GRANTS COMMISSION. Total no. of universities in the country as on 25.01.2023. https://www.ugc.gov.in/oldpdf/consolidated%20list%20of%20All%20universities.pdf. Accessed 7 Dec 2022

UKEssays (2022) SWOT analysis Indian pharmaceutical industry. https://www.ukessays.com/essays/economics/swot-analysis-of-indian-pharmaceutical-industry-economics-essay.php. Accessed 15 Dec 2022

UOHYD (2023) Collaboration with natco pharma limited. https://herald.uohyd.ac.in/collaboration-with-natco-pharma-limited/. Accessed 7 Sept 2023

Wipo (2022) Statistical country profiles. https://www.wipo.int/ipstats/en/statistics/country_profile/profile.jsp?code=IN. Accessed 27 Dec 2022

Zeebiz (2022) Innovation: China files almost half of the world's patents, India lags far behind—Check details|Zee Business. https://www.zeebiz.com/zeebiz-exclusive/news-innovation-china-files-almost-half-of-the-worlds-patents-india-lagging-far-behind-check-details-209741. Accessed 29 Dec 2022

Catalysing and Accelerating a New Phase in Indian Biopharma

12

Girish Sahni, Abhay H. Pande, and Anil K. Angrish

Abstract

The Indian pharmaceutical industry became 'the pharmacy of the world' primarily because of its capabilities to produce quality generic drugs cost-effectively through 'reverse engineering' skills. A similar success story lies behind biogenerics produced through the biotech route. These remarkable successes have laid a solid foundation for a cash-rich and confident pharmaceutical industry to successfully cater to the ever-growing need for affordable medicines across the globe. The recent pandemic, and the daunting possibility of more such pandemics in the future, open a great opportunity for both vaccines and biopharmaceuticals. All stakeholders need to create a vibrant ecosystem in India, where the development of not only biogeneric medicines but their improved versions (biobetters) as well as new molecular entities are energized to foster a Pharma Revolution 2.0 for creating public goods in the health sector for a world increasingly beset by life-threatening situations and well-being challenges. To take advantage of this opportunity, India has a lot to learn from the leading economies in biopharmaceuticals. There is a need to rapidly evolve from smart processes (for biosimilar) to smart(er) new molecules. The pre-existing policy framework needs to be tweaked to encour-

G. Sahni (✉)
Panjab University, CSIR-Institute of Himalayan Biotechnology (IHBT), Palampur, and CSIR-Institute of Microbial Technology IMTECH), Chandigarh, India

A. H. Pande
Department of Biotechnology, National Institute of Pharmaceutical Education and Research (NIPER), SAS Nagar, Mohali, Punjab, India
e-mail: apande@niper.ac.in

A. K. Angrish
Department of Pharmaceutical Management, National Institute of Pharmaceutical Education and Research (NIPER), SAS Nagar, Mohali, Punjab, India
e-mail: anil@niper.ac.in

G. John, A. K. Panda (eds.), *Biotechnology in India - Reworking A Strategy*,
https://doi.org/10.1007/978-981-97-0896-3_12

age innovation (e.g., a fast-track regulatory pathway for structurally incremental but therapeutically beneficial biobetters) on the one hand, and to bring risky capital, so as to evolve a skilful and well-thought-out road map, on the other hand, for this transition. A close connection is required among organizations operating in upstream and downstream functions of biopharmaceuticals. To make it work, analysis of past successes—as well as gaps in the value chain—that need improvement are discussed to propose a viable action plan.

12.1 Spectacular Progress of the Indian Pharmaceutical Industry in the Last Four Decades

India is known as the 'Pharmacy of the World', and this is amply justified by the facts such as the Indian pharmaceutical industry's substantial contribution to the global generics market, with a volume share of over 20%, and more than 62% towards meeting the global demand for vaccines (Pharmaceuticals 2023). The fact that the Indian pharmaceutical industry currently ranks globally as third in terms of production volume (Pharmaceuticals 2023) and fourteenth in terms of dollar value attests to the economic benefit it offers to the economically underprivileged groups both nationally and internationally. The domestic pharmaceutical market, which is now estimated to be worth US $50 billion, is expected to expand to US $65 billion by 2024 and surpass US $130 billion by 2030, with an annual growth rate of 18.7% reported in FY21 (Assets 2021; Investindia 2023). Approximately 40% of the generic pharmaceutical demands in the USA, about 25% of all medicines in the UK, and over 50% of the generic drug needs in Africa have been met by the Indian pharmaceuticals and drugs, which make up nearly 5.92% of global exports (Assets 2021). Both the depth and breadth of the Indian pharmaceutical industry can be understood from the fact that there are more than 60,000 generic brands across 60 therapeutic categories (Investindia 2023; James 2020; Newcommittee 2013). Similarly, almost 60% of the global vaccine demand (in volume terms) is met by the Indian pharmaceutical industry and 70% of the World Health Organization's (WHO) vaccines, as per the essential Immunization Schedule, are sourced from India (Investindia 2023). The major share of India's pharmaceutical exports (i.e., 73.31%) is of formulations and biologicals while bulk drugs and drug intermediates form other major components (Assets 2021). The top five pharmaceutical export destinations are the USA, UK, South Africa, Russia, and Nigeria, and it is worthwhile to note that about 55% of pharmaceutical exports from India cater to 'highly regulated' markets. According to CARE Ratings data, the contribution of the domestic market to exports has been shifting over the years, with the ratio being 52:48 in FY17, 47:53 in FY21, and an expected ratio of 46:53 for FY23 (Livemint 2021; Economic Times 2024). In addition, India's pharmaceutical exports have significantly increased, rising from Rs. 90,415 Cr in 2013–2014 to Rs. 1,83,422 Cr in 2021–2022 (PIB 2022a; Business 2022a). From April 2000 to September 2021, the Indian pharmaceutical industry was successful in luring a total of about US $18.55 billion in Foreign Direct Investment (FDI), showing India's allure as a desirable FDI destination. An additional benefit is that, India had the most US-FDA-compliant pharmaceutical factories outside of the USA, with a total of 741 (Pharmaceuticals 2023). Notably, manufacturing expenses in India are about 33% cheaper than those in western countries.

In the recent past, the strength of the Indian pharmaceutical industry became quite visible when the whole world was finding it difficult to deal with COVID-19 and the industry came up with an effective vaccine at the required vast scales. AstraZeneca and Oxford University collaborated with the Serum Institute of India Private Limited (SII) for the supply of COVID-19 vaccine—'Covishield', as SII was already the largest vaccine manufacturer by the number of doses produced and sold globally (>1.5 billion doses) (Seruminstitute 2021). Based on the strength of organizations like SII, the Indian government could support 98 countries through 'Vaccine Maitri' initiative (Business 2022b). By supplying more than 23.5 Cr (over 235 million) doses, the Government of India (GoI) could lower the gap of 'vaccine inequity' as many countries did not have access to COVID-19 vaccine. The Prime Minister of India stated that the initiative is in line with the Indian ethos of caring for the entire planet. In a way, India's vaccination efforts had made the global fight against COVID-19 stronger. In January 2021, Jair Bolsonaro, President of Brazil, expressed his gratitude for the supply of vaccines by using the analogy of *God Hanuman* carrying a mountain with vaccines from India to Brazil (Indiatoday 2021). Because of India's strength in vaccine manufacturing, India could use the COVID-19 vaccine which was the world's most in-demand commodity, as a currency for international diplomacy (Investindia 2021).

12.2 Bold Policy Changes, Vibrant Academic and Strong Public Sector Industry Ambience are Behind the Growth of the Indian Pharmaceutical Industry

The Indian pharmaceutical industry has achieved the distinction of becoming 'the pharmacy of the world' due to a combination of various factors. Four major interventions largely contributed to the growth of the Indian pharmaceutical industry; enactment of the Indian Patents Act, implementation of the Drug (Prices Control) Order, the Foreign Exchange Regulation Act, along with setting up of Central Public Sector Undertakings (CPSUs). The Enactment of the Indian Patents Act in 1970 helped the Indian pharmaceutical industry to go in for 'reverse engineering' as 'process patent' was permitted in 1972 with the passage of the Indian Patents Act (Greene 2007). An appropriate example to show the impact is of erstwhile Ranbaxy Laboratories Limited (now, part of Sun Pharma) which came up with 'Calmpose'—a copy of Roche's patented Valium tranquilizer (Ghoshal et al. 2002). To block the entry of Roche's product (Valium) in India, the company reduced Calmpose's price by 25% and issued credit notes for a similar discount on existing products. Ranbaxy also made history in 1978 when it became the first Indian pharmaceutical business to develop a 'novel process' for making the antibiotic 'doxycycline', which was later patented in 1990. In 1985, Ranbaxy found a new way to manufacture 'ranitidine', which was the generic version of Glaxo's Zantac (Ghoshal et al. 2002). The Drug (Prices Control) Order of 1979 capped the prices for drugs that were sold in the domestic market and through this order, the government severely restricted 'profit' and 'growth opportunities' for pharmaceutical companies. Ranbaxy capitalized on chances in this instance as well by emphasizing more on exports, attaining an excellent average annual growth rate of 34% during the course of a decade-long period from 1986 to 1996. The same was the case for many other pharmaceutical companies from India (Ghoshal et al. 2002). Foreign Exchange Regulation Act was introduced in 1974 and it required multinational corporations (MNCs) to dilute their equity holding as MNCs were not allowed to hold more than 40% equity (Mondal and Pingali 2017). As a result, the market share of MNCs declined during 1970 and 1979. The establishment of Central Public Sector Undertaking (CPSU) units around this time gave the domestic pharmaceutical industry a boost. Hindustan Antibiotics Limited (HAL) came up in 1954 with the assistance from WHO and UNICEF. Indian Drugs and Pharmaceuticals Limited (IDPL) came into existence in 1961. Rajasthan Drugs & Pharmaceuticals (RDPL) was incorporated in 1978 with a 51% equity infusion from the Government of India and a 49% equity contribution from the Government of Rajasthan (through RIICO Limited). Karnataka Antibiotics & Pharmaceuticals Limited (KAPL) was set up in March 1981 as a joint sector undertaking promoted by Hindustan Antibiotics Limited (HAL) and Karnataka State Industrial and Infrastructure Development Corporation (KSIIDC). Bengal Chemicals & Pharmaceuticals Limited (BCPL), founded in 1892 as Bengal Chemicals & Pharmaceutical Works by Acharya Prafulla Chandra Ray, became sick and was nationalized in December 1980 (Rakshit 2016). These interventions played a very important role in the supply-side dynamics of the Indian pharmaceutical industry and by 2019 the Indian pharmaceutical industry contributed almost 2% to India's Gross Domestic Product (GDP) (RBI 2021).

From a demand-side dynamics point of view, the USA and European economies are dependent heavily on generic drugs. Due to high healthcare expenditures, including pharmaceutical products, these economies are under heavy pressure to reduce their expenditure on healthcare (Greene 2007). Generic products come to their rescue for this. This can be observed from the facts that for 3 years period, i.e., 2018–19, 2019–20 and 2020–21, estimated savings from generic drugs approved by the US-FDA, based on sales during 12-months following approvals, were US $17.8 billion, US $24.8 billion, US $10.7 billion, and US $16.6 billion, respectively (Conrad et al. 2022).

Thus, the learning from the journey of the pharmaceutical industry can be replicated to a large extent in the biopharmaceutical industry as well, as the latter is part of the larger set of Bioeconomy with industrial and business roots in the present pharmaceutical industry.

12.3 Early Rise of Bioeconomy in India and Opportunities for Further Growth

Bioeconomy in India has grown from US $10 billion to US $80 billion, i.e., 8 times in the last 8 years, and is projected to become one of the Top-10 nations in the overall global ecosystem of biotechnology (PIB 2022b). During the same period, the number of investors has risen by nearly nine times as also the available 'funding' for them has increased by seven times. The number of Biotech Incubators was just 6 in the year 2014, whereas in the year 2022 the number was 75, supported by Biotechnology Industry Research Assistance Council (BIRAC) in addition to one incubator set up by DBT-National Institute of Animal Biotechnology, Hyderabad (GoI, Ministry of Science & Technology, Department of Biotechnology, 2022) The number of biotech products has also increased from just 10 to more than 700 during this period. Out of the total 70,000-plus start-ups, more than 5000 start-ups are associated with biotech only; in a way, every 14th start-up is in the biotechnology sector. A notable fact is that 1128 biotech start-ups were incorporated in 2021 alone, further emphasizing the industry's rapid expansion and relevance (PIB 2022b). The Prime Minister of India, while addressing Biotech Startup Expo 2022, hailed the contribution of the India's Biotechnology Industry Research Assistance Council (BIRAC), and listed five underlying factors behind the growth of biotech in India, namely 'diverse population and diverse climatic zones', 'talented human capital pool', 'existing track record of success in biotech', 'improvement in ease of doing business in India', and 'increase in demand for bio-products' (Indianexpress 2022). These factors are relevant for the development of the biopharmaceutical industry too.

12.4 Booming Global Growth Potential of the Biopharmaceuticals Market

Biopharmaceuticals are pharmaceuticals produced using biotechnology routes rather than pure chemical processes. As per the latest definition, biopharmaceuticals are defined as *complex molecules derived from biological sources, with the purpose to diagnose, prevent, treat, or cure diseases or conditions of human beings (*Rasmussen et al. 2021*).* A variety of terms/synonyms are often used interchangeably to describe biopharmaceuticals: biopharma, biologics, biotherapeutics, biologicals, biological product, biological drug, biological medicine, biological medicinal product, etc. Biopharmaceuticals differ from pharmaceuticals in many ways in that they are more complex and bigger in size, are considered less hazardous by the regulatory authorities, have complex and sensitive manufacturing processes, and their toxicity issues are also substantially different from pharmaceuticals. Protein biopharmaceuticals can be roughly divided into three categories.

- Biosimilars are similar but not identical to the 'innovator' molecules and differ in their production process and other properties. They are also called 'follow-on biologics', 'biogenerics', 'similar biologics', 'similar biotherapeutic products', 'generic biologics', etc.
- Biobetters are an 'upgraded' version of the 'innovator' molecules and possess additional properties. They are also called 'me-betters', 'bio-superiors', etc. Biobetters represent one of the fastest-growing segments in the global biopharmaceutical market.
- Innovator molecules are New Molecular Entities (NME).

The share of biopharmaceuticals within approved pharmaceuticals is steadily increasing. The market for biopharmaceuticals was estimated to be worth about US

Table 12.1 Product type of biopharmaceuticals and percentage share of global revenue

	Year-wise percentage share of global revenue	
Product type	Year 2019	Year 2026
Biologics	67%	65–70%
Bio-similars	4%	8%–10%
Vaccines	13%	13–15%
Other product types	16%	10%

Source: Based on the graphs from 'Make in India' to fuel 4x growth in biosimilars and vaccines by 2026 (2021, September 15) (Kearney 2022)

$267 billion in 2019 by Kearney, a well-known international management consulting firm (formerly known as A.T. Kearney until January 2020). According to their predictions, the market is anticipated to earn $528 billion in sales by 2026 (Kearney 2022). Similarly, the share of biosimilars is likely to be double of what it was in 2019 due to the high growth rate expected in this category as per forecast for the year 2026 (Kearney 2022) (Table 12.1).

12.5 Pharma Revolution 2.0: the Leap from Pharmaceutical to Biopharmaceutical

The nature of the pharma industry is changing given the fact that eight out of ten top molecules currently in the world are 'biological'. The Indian Biopharma Industry touched sales of Rs. 33,067 Cr for the year 2020–2021, up from Rs. 29,176 Cr in the previous year, thereby attaining a growth of 13%. Additionally, the Indian Biopharma industry now has over 300 companies, which reflects its growing importance and prominence. When it comes to the export value of biosimilars, India has a market share of approximately 1.5% worldwide. Furthermore, the Indian Biopharma industry makes up about 8% of the whole market for biopharmaceuticals on a global scale. This increased proportion is primarily because of the significant contribution made by the Indian Biopharma industry, where vaccines alone account for almost 40% of the market in terms of volume (Biospectrumindia 2022). But this achievement needs to be seen in the background of 'concentration' in terms of market share and geographic presence, and 'significant presence of MNCs'. The biopharmaceutical business in India is dominated by selected companies as the top 20 biopharma companies contribute to 90% of the total biopharma industry. Moreover, only 10 companies contribute to 73% of the total biopharma industry. The sector is mostly fuelled by the top three players; SII, Biocon, and Sanofi Healthcare India, who collectively account for 45% of the market. This shows that the majority of biopharmaceutical companies are comparatively smaller in terms of revenue. In terms of geographic concentration, 9 out of the top 20 biopharma companies have their base in Mumbai, followed by 4 in Hyderabad, 3 in Ahmedabad, and 2 each in Bengaluru and Pune. Another notable point is that 6 out of the top 20 biopharmaceutical companies are MNCs of foreign origin and 4 out of these 6 MNCs are affiliates/associate companies of their parent companies and are listed on Indian stock exchanges indicating that they raised capital in the domestic market only. This shows that there is no dearth of capital for companies raising funds with appropriate business models. However, this does not hold true for early-stage companies as evident from concentration in terms of market share as well as support from organizations such as BIRAC.

India-based biopharmaceutical businesses' success is the consequence of many years of persistent work and deliberate expansion. Dr. Cyrus Poonawalla launched SII in 1966, and Kiran Mazumdar-Shaw formed Biocon Limited in 1978, both of which serve as examples of this journey of development and success. For instance, Biocon Limited started its operations in 1978 with the manufacture of enzymes. As of 2024, it will be a corporation with a broad portfolio that includes research services, new biologics and biosimilars, as well as generic medications and formulations. When Glargine (Semglee, a brand of Viatris), a product of Biocon Biologics, a subsidiary of Biocon Limited, received interchangeability approval by the US Food and Drug Administration (US-FDA) in July 2021, it marked a critical turning point for the company (Biocon Group Fact Sheet 2024). Additionally, in FY22, Biocon Biologics acquired the global biosimilars business of Viatris by paying US $3.335 billion (Biocon Annu Rep 2022). These tactical decisions and successes show how the company has evolved from an 'Enzyme-based company' to a company with a diverse portfolio of bulk drugs and formulations, novel biologics and biosimilars as well as research services.

Exciting opportunities are currently available for the Indian Biopharma industry. Biologics worth over US $70 billion are going to lose exclusivity over the next 5 years [IQVIA, 2021], thus offering a huge opportunity for the biosimilars sector of the Indian Biopharma industry. The 'China Plus One' factor also enters the picture as the USA and European Union are decoupling the business from China (Lebedenko 2022). This also provides an advantage for India, because biopharmaceutical companies across the world have realized that they need to diversify their supply chain post-COVID and for this purpose, these entities have started sourcing from India in order to reduce their dependence on China (The Hindu 2024).

Most recently, in the first week of June 2024 an alliance – Biopharmaceutical Alliance has been formed among South Korea, India, the United States, Japan, and the European Union (EU). Initially, the U.S. and South Korea had agreed to form the alliance but it was expanded to include, the EU, Japan, and India. The Alliance came into existence as a response to the drug supply shortages experienced during Covid-19. Inaugural meeting of this Alliance was held in San Diego during the Bio International Convention 2024 and had representatives of government and bio as well as pharmaceutical companies from participating nations. The purpose behind this Alliance is to come together and put joint efforts into building a resilient supply chain in the biopharmaceutical sector as the production of essential raw materials and ingredients was concentrated in few countries. Consensus was arrived at among participants on coordinating the respective countries' bio-policies, regulations, and R&D support measures (The Hindu 2024).

The Indian Biopharma industry will be able to avail these opportunities if appropriate mechanisms are in place. Four major factors which can place the Indian Biopharma industry on sound footing and be enablers for the growth of biopharmaceuticals are 'existing institutions in biotechnology in general and biopharmaceutical in particular', 'biotechnology parks/incubators', 'PSUs in biopharmaceutical space', and 'Production-linked Incentive (PLI) scheme for biopharmaceuticals'.

Existing 'Infrastructure' in India and 'Policy Framework' to Support Biopharmaceutical Industry in India—India has extensive infrastructure (both intellectual and logistical) in place to support the biopharmaceutical industry. In India, the Department of Biotechnology (DBT) has important institutes in the biotechnology domain, such as the National Institute of Immunology—New Delhi, Institute of Life Sciences—Bhubaneswar, Rajiv Gandhi Centre for Biotechnology—Thiruvananthapuram, Regional Centre for Biotechnology—Faridabad, etc. Similarly, the Council of Scientific and Industrial Research (CSIR) also has prominent institutes such as CSIR-Central Drug Research Institute (Lucknow), CSIR-Institute of Microbial Technology (Chandigarh), CSIR-Centre for Cellular Molecular Biology (Hyderabad), to name a few (CSIR 2022). These institutions have made notable discoveries such as enfenamic acid, an anti-inflammatory agent by IICT in 1980, risorine, a combination preparation for tuberculosis by IIIM in 2009 (Differding 2017), among others. In addition to this, a number of drugs or standard extracts with identified active compounds such as 'gum guggul' derived from *Commiphora mukul* had led to the identification of guggulsterones (farnesoid X receptor antagonists with lipid-lowering properties), due to knowledge of ayurvedic remedies. Similarly, 'brahmi' prepared from *Bacopa monnieri* gave triterpenoid glycoside modulators of the serotonin system as memory enhancers (Differding 2017). Several biotechnology Parks/Incubators such as Biotech Park—Lucknow, Guwahati Business Park—Assam, Biotechnology Incubation Centre—Hyderabad (Committee 2020), etc. have come up. Similarly, certain PSUs, such as Bharat Immunologicals and Biologicals Corporation Limited and BIRAC, have also contributed to the biopharmaceutical sector through their unique efforts. The role of these leading national units has been commendable, although more is needed in terms of invention as well as innovation.

Production-Linked Incentive Scheme for Promotion of Biopharmaceuticals— On 20.03.2020, the Government of India approved a scheme—'Production Linked Incentive Scheme' for the promotion of domestic manufacturing of critical Key Starting Materials (KSMs), Drug Intermediates (DIs), and Active Pharmaceutical Ingredients (API) in India (Pharmaceuticals 2022). Out of four segments, Target Segment-I and Target Segment-II cover 'Key Fermentation based KSMs/DIs' and

Table 12.2 Segment-wise details of investment till December 2021

Sr. No.	Target segment	Total applicants approved	Total committed investment (Rs. in Cr)	Actual investment up to December 2021 (Rs. In Cr)
1.	Key fermentation based KSMs/Drug Intermediates	3	2114.17	22.91
2.	Key fermentation based niche KSMs/Drug Intermediates	6	357.27	35.31

'Key Fermentation Based Niche KSMs/DIs', respectively. Further, for fermentation-based products, an incentive for FY 2023–2024 to FY 2026–2027 would be 20%, an incentive for 2027–28 would be 15%, and the incentive would be 5% for 2028–2029 (Pharmaceuticals 2022). Segment-wise details of investment made are given in Table 12.2. As can be seen, the number of applicants approved were few and the actual investment was just about 10% of committed investment.

12.6 Positive Macro-Economic Developments for Innovations and Biopharmaceuticals

In recent years, certain notable macro-economic developments have taken place that has proved positive for innovations and the biopharmaceutical sector in India. A supportive intellectual property rights (IPR) regime is one such development. IPR regime change in the 1970s supported the pharmaceutical sector in India, and those changes became a game changer for the Indian pharmaceutical industry. In the same way, recent developments in the Indian IP system are facilitators, and through these changes, the IPR regime has been strengthened. This is reflected in the increase in the number of examiners and controllers dealing with IP systems which has improved the number of patents granted over a period. The number of examiners has become 3.38 times by October 31, 2020, of those in March 2015, and the number of controllers has increased by 2.75 times by October 31, 2020, in comparison to the number of controllers in March 2015 (Table 12.3) (DPIIT 2021). With this, the number of patents granted in 2021–2022 was 5 times of what it used to be in 2014–2015. Table 12.3 and Fig. 12.1 provide the details regarding an increase in the manpower dealing with the IP system in India and patents granted *vis-à-vis* patents filed, respectively (Ahaskar 2022; PIB 2022c; DPIIT 2021).

Augmentation of manpower in the IP system is reflected in the filing of patents, the number of patents filed as well as granted. During January and March 2022, the number of domestic patent filings from India surpassed the number of international patent filings at the Indian patent office. For the second quarter of 2023, Indian applicants filed 19,796 patents in contrast to 9090 applications filed by non-Indian applicants (Intellectual Property India Annual Report 2022-23). This emphasizes the development of India's domestic patent ecosystem and draws attention to the expanding representation and involvement of Indian innovators.

More recently, certain procedural developments in the IPR regime in India like e-filing and processing, incentivizing e-filings, and implementation of electronic modes of communication are likely to help the sector. An example is the Scheme for Facilitating Start-ups Intellectual Property Protection (SIPP). Under the scheme,

Table 12.3 Increase in manpower dealing with IP system in India (DPIIT 2021)

		March 31, 2015	October 31, 2020
Patents	Examiners	183	619
	Controllers	89	245
Trademarks	Examiners & senior examiners	65	159
	Registrars	13	15

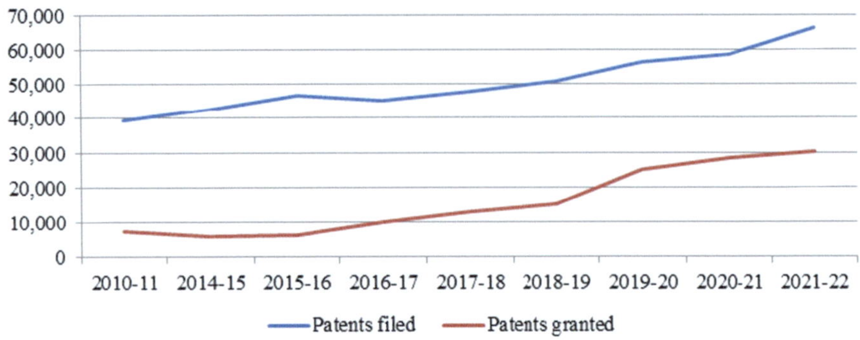

Fig. 12.1 Patents filed and patents granted over a period (Ahaskar 2022; PIB 2022c)

there is a provision for reimbursement of fee that is paid to facilitators for filing the patents and trademarks and about 5000 patent applications and 12,000 Trademark applications have been filed. Further, 510 patent facilitators have been empanelled by Indian IP Office. For Micro, Small, and Medium Enterprises (MSMEs), filing and processing fee for patent applications has been reduced by 80% in comparison to large entities. IPR reforms with reference to Institutional Mechanism have resulted in 'reduction of backlogs', digitization of processes', and 'manpower augmentation'. Regarding 'Compliance and Forms', there is a reduction in procedural complications, and a reduction in the number of forms, e.g., the number of Trademark forms has gone down from 74 to 8. The impact is visible in quantitative terms as well, e.g., from FY 2016–2017 to FY 2019–2020, the number of patents examined has increased from 28,967 to 80,080 which translate into a 176% increase (Ipindia 2019).

There has been a considerable shift in the importance given to innovation since India attained independence in 1947. Important policy efforts that have been put in place over time can be used to observe this development. A significant turning point in India's independent science policy occurred with the formulation of the Scientific Policy Resolution in 1958. The Technology Policy Statement was subsequently released in 1985 and emphasized the value of technical independence. It was in 2003 that Science and Technology Policy recognized the inter-linkages between Science and Technology. In 2013, Science, Technology and Innovation (STI) Policy was launched and 2010–2020 was declared as the 'decade of innovation'. This shift is evident from the emphasis on innovation that has become the focus across the globe. Global Innovation Index (GII) started capturing information about innovations across various nations and started ranking countries on the basis of innovation. This ranking reflected an emphasis on various contributors to innovation in countries covered in the Index. It is understood that innovation did not work in isolation; rather, it is an outcome of conscious and concerted efforts which get channelized. Table 12.4 shows the GII ranking of the top countries along with India.

It is apparent that the USA, Japan, and European countries are well positioned to take advantage of innovation-intensive domains. The biopharmaceutical domain is not an exception, rather it is highly innovation dependent in nature. India is also catching up given the fact that India's rank in GII has moved up from 81 to 40 from

Table 12.4 Countries and their rank in Global Innovation Index

	Year			
Country	2023	2022	2021	2020
USA	3	02	03	03
China	12	11	12	12
Japan	13	13	13	16
Europe				
Switzerland	1	01	01	01
Sweden	2	3	2	2
UK	4	4	4	4
Netherlands	7	5	6	6
Germany	8	8	10	9
Finland	6	9	7	8
Denmark	9	10	9	6
France	11	12	11	12
India	40	40	46	48

Source: Compiled from the GII of respective years

Table 12.5 Improvement in India's Rank in Global Innovation Index (GII)

Year	2015	2016	2017	2018	2019	2020	2021	2022
India's Rank in GII								40
							46	
						48		
					52			
				57				
			60					
		66						
	81							

the year 2015 to 2022 (Table 12.5). The 15th Edition of this index comprises 81 indicators, and these indicators are broadly covered under the headings—*Institutions* (covering Political Environment, Regulatory Environment, and Business Environment), *Human Capital and Research* (covering Education, Tertiary Education, and Research & Development), *Infrastructure* (covering Information and Communication Technologies, General Infrastructure, and Ecological Sustainability), *Market Sophistication* (covering Credit, Investment, and Trade, diversification, and market scale), *Business Sophistication* (covering Knowledge Workers, Innovation Linkages, and Knowledge Absorption), *Knowledge and Technology Outputs* (covering Knowledge Creation, Knowledge Impact, and Knowledge Diffusion), and *Creative Outputs* (covering Intangible Assets, Creative Goods and Services, and Online Creativity). Key indicators and sub-components reflect the overall improvement that is positive for innovation-intensive areas such as biopharmaceuticals.

However, India is lagging in biopharmaceuticals if we compare it with other developed economies. Life Sciences Institutes are seen as major game changers in the field of biopharmaceuticals. These institutes provide research facilities, germination and nurturing of ideas, space for incubation, trained manpower for the industry,

Table 12.6 Life Sciences Institutes among the World's Top 100 (Times Higher Education 2024)

Country/Region	Number of Life Sciences Institutes among Top 100
	Year 2023
USA	34
Europe	42
China (including Hong Kong)	8
Japan	2
India	NIL
Singapore	2
Canada	3
Australia	7
South Korea	2

Table 12.7 Number of drugs and country share of Pipeline Phase-I to regulatory submission based on company headquarters (2006–2021)

Year	Country/Region					
	USA	Europe	Japan	China	South Korea	Others
2006	43	31	11	2	2	11
2011	44	30	11	3	3	9
2016	46	28	11	4	3	8
2021	44	25	6	12	4	9

Source: IQVIA Pipeline Intelligence (2021, December)

opportunities for collaboration of academia and industry, etc. The growth of the Chinese biopharmaceutical industry is also attributed to the increase in the number of their Life Sciences Institutes which figure among the best institutes globally. India lags in this space as Life Sciences Institutes from India are nowhere among the top-ranking global institutions (Table 12.6) (Times Higher Education 2024).

A key indicator in the case of biopharmaceuticals success is the 'number of drugs' and 'country share' in terms of Pipeline Phase-I to regulatory submission based on company headquarters. From 2006 to 2021, the number of drugs and country share of Pipeline Phase-I to regulatory submission based on company headquarters has remained nearly static for the USA (Table 12.7). For Europe, the number has gone down from 31 to 25 during the same period. Interestingly, for almost a decade (2006–2016), the number was static at 11 for Japan, and by 2021, it went down to 6, showing a significant decline. In contrast to this, the number has gone up from 2 to 12, i.e., six times in a period of 15 years for China. Even in the case of South Korea, it has doubled in a period of 15 years (Table 12.7).

12.7 Macro Scenario on 'Patents Granted' and Time Taken for 'Granting Patents': A Comparison

There is definitely an improvement in the IPR system over recent periods, but the Indian system is lagging if we compare it with the IPR system of other countries which are doing much better in the biopharmaceutical sector. This is reflected in the number of patents granted as well as in the time taken for granting patents in

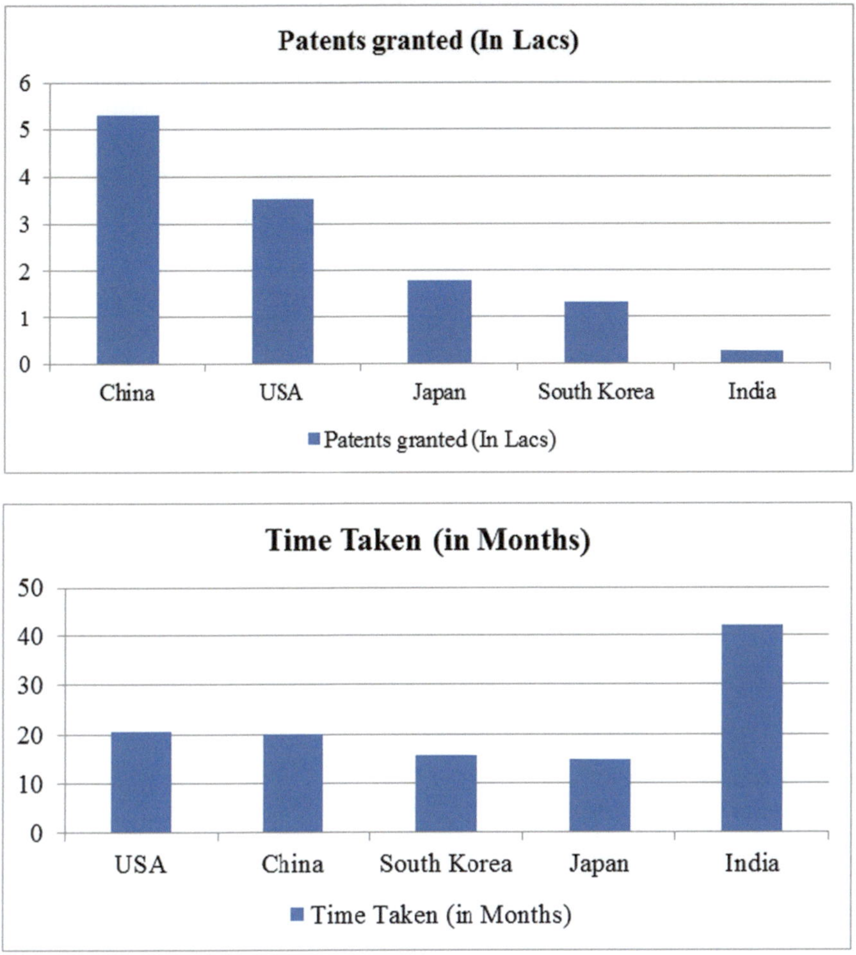

Fig. 12.2 Patents granted across selected countries in 2020–2021 and the average time taken for a final call on granting patents in 2020. (Source: Compiled from Economic Survey 2021–2022 (Indiabudget 2021))

India (Fig. 12.2). As can be seen, patents granted in India are meagre in comparison to countries like China and the USA. Even smaller countries like Japan and South Korea have done far better in terms of granting patents than India (Indiabudget 2021).

As can be seen (Fig. 12.2) the time taken for patent granting in India was on the higher side but improvement is also visible. The time taken for patent examination used to be 72 months in December 2016. In 2017, the time taken used to be 64 months that came down to 52 months in 2019 and 42 months in 2020. Most recent numbers reported in March 2022 show the reduction in time taken for granting patents to 5–23 months for different technological areas. There are exceptional cases such as the fastest granted patent is the one in which the patent was granted in 41 days after filing a request for expedited examination (DPIIT 2021). These are positive developments. Further improvement is desirable especially when this parameter is compared to other countries which are dominant in the biopharmaceutical space. Regulatory strength needs to be enhanced further. For example, China has increased the number of reviewers at the Center for Drug Evaluation (CDE) from 150 (in 2015) to more than 700 in 2018, and the number was higher by August 2022 (Jeffrey et al. 2022; Han et al. 2021).

12.8 Expectations for Pharma Revolution 2.0

The share of human vaccines in terms of overall exports of biopharmaceuticals from India in 2020–2021 stood at Rs. 6492 Cr as against exports of Rs. 5723 Cr in 2019–2020, thereby showing an increase of 13.42%. Human vaccines' export was Rs. 4453 Cr in 2018–2019. At the same time, Indian companies also imported vaccines worth Rs. 2628 Cr in 2020–2021 registering a decline of 6.25% if compared with 2019–2020 when India imported vaccines worth Rs. 2804 Cr. The export share of insulin injections is showing a positive trend as India exported insulin injections worth Rs. 372 Cr in 2020–2021 in comparison to insulin injection exports of Rs. 247 Cr in 2019–2020, registering a growth of 50.47%. In contrast to this, India imported insulin injections worth Rs. 1538 Cr in 2020–2021 as compared to Rs. 1328 Cr insulin injections imported in 2019–2020, thereby registering a growth of 15.84%. Exports of insulin and its salt increased to Rs. 229 Cr in 2020–2021 from Rs. 200 Cr in 2019–2020, thereby showing a growth of 14.62%. This reflects the need for 'import substitution' and 'domestic manufacturing' of vaccines and biopharmaceuticals in India.

In addition to this, attention is required on the following areas with respect to healthcare and medical biotechnology:

- Search and production of new molecular entities (Mckinsey 2020)
- Developing antibodies against diseases relevant to India
- New leads and improved biopharmaceuticals
- Second and third-generation biopharmaceuticals
- Antibody-drug conjugates (ADCs)
- Research on stem cell therapy
- Research on nano-biotechnology
- Developing cost-effective new drugs (Mckinsey 2020) against TB, cardiovascular diseases, chronic obstructive pulmonary diseases, and cancer
- Developing low-cost technologies for diagnosis and cheaper as well as effective biopharmaceuticals (Mckinsey 2020)
- Development of biosimilars for the treatment of diabetes, cancer, and arthritis
- Focus on Biologics License Application (BLA) in contrast to Abbreviated New Drug Applications (ANDAs)

12.9 Road Map for Pharma Revolution 2.0

12.9.1 A 'Crying' Need to Broaden PLI Scheme to 'Research and Production Linked Incentive' (RPLI) for Biopharmaceuticals

Industry representatives demanded the same while expressing their views during BioAsia 2022 in February 2022 (Das 2022). There is a strong need to prioritize biopharmaceuticals like biosimilars, cell and gene therapies, and other cutting-edge treatments because the PLI scheme is currently skewed towards generic pharmaceutical drugs. For biopharmaceuticals, a Biologics License Application (BLA) is required to be filed for seeking permission to launch them in the USA market. To quote Kiran Mazumdar Shaw—'*One BLA application is equivalent to almost 20 Abbreviated New Drug Applications (ANDAs) in terms of investments made*'. However, throughout time, there have been fewer incentives for making expenditures on research and development (R&D). For R&D investment, the weighted deduction of 200% that used to be given was reduced to 150% in 2017 and further, it was lowered to 100% in 2020 (Assets 2021; Babu 2021). With this background, the incentive to invest in R&D is considerably diminished, especially when research requires intensive hits and trials. The Parliamentary Standing Committee on Commerce (report submitted in February 2021) has sought the restoration of weighted tax deduction to 200% (Indiaenvironmentportal 2021). In its absence, the PLI scheme needs to be linked to research too. This will yield significant benefits as the 'slant' will shift towards short to medium-term business-oriented research in R&D laboratories.

12.9.2 R&D Expenditure Needs to be Increased in Biopharmaceuticals

While emphasizing the need to increase R&D expenses, the Parliamentary Committee in its 158th report entitled 'Attracting Investment in post-COVID Economy: Challenges and Opportunities for India' and submitted on February 2021 observed that R&D expenses by industries in India are about 8.4% of their total sales value. It is much below the global level of 11–12%. As highlighted above 65–70% market share of overall biopharmaceuticals of US $528 billion by 2026 is expected to come from biopharmaceuticals. In order to take a bigger pie of total revenue from biopharmaceuticals, India has to take the lead in biopharmaceuticals. However, this is a capital-intensive activity that involves high-value innovations. Interventions in the 'upstream' biopharmaceutical segment must be made in order to place the Indian Biopharma industry among the leading global players.

12.9.3 Risk Capital Is Required for R&D-Intensive Biopharmaceutical Projects

In China, during 2018–2020, US $12 billion came in the form of venture capital financing for biopharmaceuticals projects. In contrast to this, one major initiative in India is led by BIRAC, which has set up a 'Fund of Funds' called Biotechnology Innovation Fund—Accelerating Entrepreneurs (AcE) and made an investment of Rs. 733 Cr in 65 biotech start-ups, Small and Medium Enterprises (SMEs) through 10 AcE Daughter Funds (BIRAC n.d.). AcE daughter funds are private funds registered with the Securities and Exchange Board of India (SEBI) which invest equity capital in start-ups, thereby providing the risk capital to undertake innovation, research, and product development. While China has experienced significant venture capital investment in this area, Indian strategy through programmes like the Biotechnology Innovation Fund—AcE aspires to provide vital help to biotech start-ups and SMEs in getting the funds they require for their R&D pursuits. This needs to be further strengthened.

12.9.4 Public Sector Undertakings in the Biopharmaceutical Domain Need to be Strengthened

In April 2021, DBT decided to allocate funds to Indian Immunologicals Limited (Hyderabad) and Bharat Immunologicals and Biologicals Corporation Limited (Bulandshahr) along with Maharashtra state-owned Haffkine Biopharmaceuticals Corporation Limited (Mumbai) to increase the production of indigenously developed vaccine—Covaxin (Aninews 2021; Leo 2021). It was opted under 'Mission COVID Suraksha—the Indian COVID-19 Vaccine Development Mission', that was the need of the hour. Covaxin has shown to neutralize the variants—B.1.1.7 (Alpha) first isolated in the UK, P.1-B.1.1.28 (Gamma) & P.2-B.1.1.28 (Zeta) first isolated in Brazil, B.1.617 (Kappa) first isolated in India, B.1.351 & B.1.617.2 (Beta & Delta) first isolated in RSA and India. Further, efficacy data showed 65.2% protection against the SARS-CoV-2, B.1.617.2 Delta variant (Bharatbiotech n.d.). It was reported in July 2022 (*Times of India dated July 17, 2022*) that India achieved the 200 crore milestone in a period of 269 days. Further, COWIN dashboard data provided bifurcation—159.44 crore Covishield doses, 33.48 crore Covaxin doses, and Corbevax, 6.4 crore doses.

Support for PSUs is required to augment facilities for other biopharmaceutical products too. PSU's contributions have other positive externalities as well, e.g., Dr. Anji Reddy, before setting up Dr. Reddy's Laboratories, had worked for 6 years in Indian Drugs and Pharmaceuticals Limited (IDPL), a PSU (Economictimes 2013).

12.9.5 A 'Sputnik Moment' for the Biopharmaceuticals in India

Government, policy makers, biopharmaceutical industry, and all stakeholders need to realize that, as of now, it is a 'Sputnik moment' for biopharmaceuticals in India.

For the American biopharmaceutical sector, the moment was the 'Asilomar Conference' on Recombinant DNA (held in February 1975 at Asilomar State Beach, California) when discussions took place among all stakeholders on research on recombinant DNA molecules and potential concerns of such research (such as biohazards, biosafety, regulatory, ramifications, etc.) (Berg et al. 1975). This conference was held in the background that the Committee on Recombinant DNA molecules of the National Academy of Science, USA, in 1974 had concluded that an international conference was necessary to resolve the issues (of potential concerns of recombinant DNA research) and until then, scientists should halt experiments involving recombinant DNA technology. The outcome of the conference was an agreement on the guiding principles of conducting experiments using the technology safely (Berg et al. 1975). In the same manner, all stakeholders must hold such a conference/discussion to decide on a future course of action comprising the right policies, provision of a single window, process simplification, infrastructure building, etc. that can put Indian Biopharma on the global map.

12.9.6 Moving Up in the Value Chain of Biopharmaceuticals

Ascending the value chain of biopharmaceuticals is a necessity rather than a choice for the Indian Biopharma industry. As discussed above, biopharmaceuticals are going to play a dominant role in terms of global revenue for the overall industry. The next major chunk is going to come from 'biologics' and 'vaccines'. India has significant strengths in 'downstream' as evidenced by the manufacturing and supply of vaccines from India. With the launch of more than 98 biosimilar drugs in the domestic market, the Indian Biopharma industry has shown the strength to move forward in the value chain. In this way, the Indian Biopharma industry has moved away from 'reluctance' to 'acceptance'. Beyond this, the scope lies in improved biopharmaceutical products (biobetters) and novel biopharmaceutical products. Here lies the genesis for products that are not only 'Made in India' but are also 'Made for the World'.

12.9.7 Lowering of Regulatory Hurdles in Biopharmaceutical Commercialization

Regulatory hurdles in biopharmaceutical commercialization are immense. Almost 32 permissions are required for the commercialization of biopharmaceuticals in India. The number of permissions required is much more than the permissions required in other competing countries. This increases the compliance cost and time. It is like the proverbial 'holier than Pope' attitude of regulatory authorities in India. Faster approval needs to be accorded to 'breakthrough innovation'. Without making any compromise on safeguards, the regulatory pathway needs simplification, along the lines of the USA after the Asilomar Conference for rDNA technologies and public awareness of the potential of biopharmaceuticals in general. Once the cost chasm is breached (as was done in the now iconic example of hepatitis vaccines and a whole range of biogeneric products, later), a regulatory boost needs to be implemented. Modified and improved biologics with virtually the same physiological, pharmacological, and molecular mechanisms as their parent/s (read biobetters) need to be looked at more '*liberally*' instead of treating them as '*altogether new entities*', as the latter's development procedure is immensely cost and time-consuming. With the right (read competent) scrutiny—that does not neglect safety quality—this approach may turn out to be a great catalyst for launching innovative world-class products that are not entirely new in that the physiological mechanisms are largely the same as the parent prototypes, but are notably altered due to engineered advantageous additional functions/qualities.

12.9.8 Becoming a 'Bio-Science Smart Nation'

By having a maximum number of US-FDA-approved manufacturing plants outside the USA (Eximbankindia 2015), the Indian pharmaceutical industry has shown resilience through investing money in setting up facilities. By moving towards the launch of biosimilars and an edge in the manufacturing of vaccines, the Indian Biopharma industry has shown a change in the mindset from a 'Trading Nation' to a 'Trade Smart Nation'. Now is the time to move towards a 'Science Smart Nation' by becoming a pioneer through the development of biobetters and novel biopharmaceutical products. The mindset of only being a trader comes in the way when the focus does not change regarding investment in R&D and Innovation. It needs to be realized that 'we cannot become what we want to be by remaining what we are'. Results take a lot of time in the biopharmaceutical domain as evident from the journey of one molecule—Interferon, as shown in Fig. 12.3.

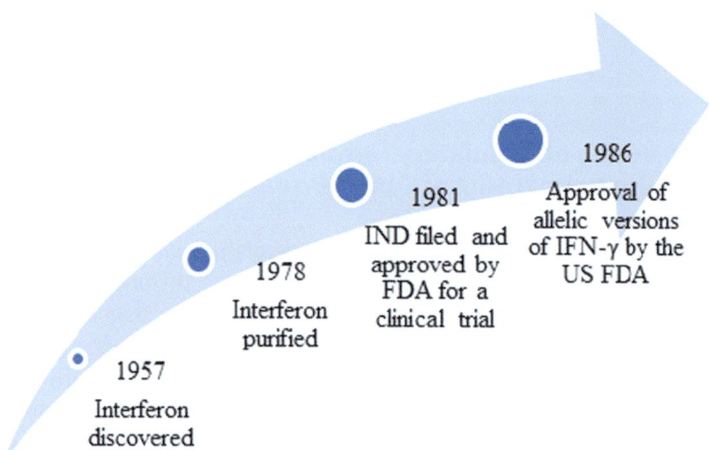

Fig. 12.3 Molecule to market: the journey of interferon

12.9.9 Need to Evolve Approach on Biopharmaceutical Drug Review for Curtailing Delays

The US experience offers lessons as the US-FDA developed four approaches, namely 'Priority Review', 'Breakthrough Therapy', 'Accelerated Approval', and 'Fast Track' (FDA 2022). The accelerated approval route, created in 1992, focusses on the 'availability of drugs', especially those drugs which are 'first available treatments', or wherein a 'drug offers advantages over existing treatments'. Fast Track route deals with a review of drugs which treat serious conditions, thereby filling an unmet medical need. Breakthrough therapy demonstrates 'substantial improvement' over available therapy. The same is relevant for 'biobetters' in India (the term 'Biobetter' was first coined by G.V. Prasad of Dr. Reddy's Labs in 2007 at a Conference on Biologics (Sharma et al. 2019)). For pricing purposes, the National Pharmaceutical Pricing Authority (NPPA) has acknowledged this through the use of 'pharmacoeconomics'; Para 5 and 15 of the Drugs (Prices Control) Order, 2013 (Nppaindia 2022) has also emphasized the application of 'principles of pharmacoeconomics' regarding new drugs, especially when the new drug is not available in the market. Hence, those biopharmaceutical products which demonstrate significant improvement in safety, and efficacy, or result in decreased mortality deserve an 'Accelerated Approval' and 'Fast Track' approval. Regulatory authorities can withdraw those drugs from the market which are found ineffective. For example, in the USA only, an Alzheimer's drug, Adulhelm of Biogen Inc., got US-FDA's accelerated approval but in May 2022, it was withdrawn due to criticism from different quarters such as doctors, researchers, and US-FDA advisers. At the same time, more than 300 drugs got approval under an accelerated approval route (Whyte 2022), thereby permitting companies to establish the benefit in confirmatory studies in a time-bound manner. Emphasis is on 'confirmatory studies in a time-bound manner' as it has been found that almost half of the drugs which got accelerated approval since 1992 are yet to show the benefit as per a study published in the *BMJ* in July 2021 (Whyte 2022). It must be noted that these were new drugs in the US market. In the Indian context, New Biological Entities (NBEs) are not common. In that sense, the above scenario does not hold true for India currently. On the other hand, in the case of distinct improvements in biobetters, whether drugs or vaccines, the same expedited approach may lead to a resurgence of R&D and invigorated outcomes instead of the current scenario where there are inordinate delays for 'biobetters', even if they have improved efficacy over their biosimilar counterparts since they are treated as New Entities.

12.10 Conclusion

The Indian pharmaceutical sector can retain its existing credentials only if it moves fast in the biopharmaceutical space. Drivers of growth in the biopharmaceutical sector should get strength from past successes in pharmaceuticals. Pharma Revolution 2.0, i.e., the leap from Pharma to Biopharma, will be possible if the Indian Pharma industry can leverage the strength of our institutional frameworks under the DBT, CSIR, and other institutions by creating mutually beneficial partnerships. At the same time, regulatory policy needs to be attuned imaginatively for taking the industry in the forward direction. Positive developments in the IPR system in India, e.g., an increase in manpower have shown an impact on the number of patents granted, and a reduction in time taken for grant of patents. The improved ranking of India in GII is another positive indication. But to nurture the biopharmaceutical sector, a lot more careful attention is required. Interventions 'upstream' of the biopharmaceutical value chain are likely to prove immensely useful, and for that purpose, the Production-linked Incentive scheme needs to be broadened to a 'Research and Production-linked Incentive scheme'. An increase in R&D expenditure for biopharmaceuticals in particular, in combination with the availability of 'risky capital', can prove to be a panacea. The biopharmaceutical industry in the private sector needs to gear up for taking up opportunities due to off-patenting and moving up in the value chain. It is a 'Sputnik moment' for biopharmaceuticals in India, and for this, regulatory authorities need to evolve for the changing times through introduction of Fast Track approaches adopted by leading nations in the area of biopharmaceuticals. By taking a similar lead in biopharmaceuticals, the Indian Biopharma industry can be catapulted to prove that India is a 'Science Smart Nation'.

References

Ahaskar A (2022) Patents granted in India tripled in last 5 years; still a fraction of China, US. Mint. https://www.techcircle.in/2022/01/31/patents-granted-in-india-tripled-in-last-5-years-still-a-fraction-of-china-us

Aninews (2021) Facility augmentation of Bharat Biotech and 3 other PSEs are being supported for, for augmented production of Covaxin. Ministry of chemicals and fertilizers. Press Information Bureau release. https://www.aninews.in/news/national/general-news/facility-augmentation-of-bharat-biotech-3-other-pses-being-supported-for-augmented-production-of-covaxin20210807065404/

Assets (2021) Indian pharmaceutical industry 2021: future is now. FICCI and Ernst & Young LLP. https://assets.ey.com/content/dam/ey-sites/ey-com/en_in/topics/health/2021/ey-ficci-indian-pharma-report-2021.pdf?download

Babu G (2021) Parliamentary committee recommends restoration of weighted tax deduction to 200% for R&D. Pharmabizcom. http://www.pharmabiz.com/NewsDetails.aspx?aid=140783&sid=1

Berg P, Baltimore D, Brenner S, Roblin RO, Singer MF (1975) Summary statement of the asilomar conference on recombinant DNA molecules. Proc Natl Acad Sci USA 72:1981–1984. https://doi.org/10.1073/pnas.72.6.1981

Bharatbiotech (n.d.) Official website of Bharat Biotech. https://www.bharatbiotech.com/covaxin.html

Biocon Annu Rep (2022). https://www.biocon.com/docs/Biocon_Annual_Report_2022.pdf

Biocon Group Fact Sheet (2024) https://www.biocon.com/docs/Fact_Sheet_Q4-FY24.pdf

Biospectrumindia, 2022. https://anyflip.com/unmb/dvta/.

BIRAC (n.d.) Official Website–BIRAC. https://www.birac.nic.in

Business (2022a) India's pharma exports touch Rs. 1.8 trillion in FY22, flat over FY21. Business Standard. https://www.business-standard.com/article/economy-policy/india-s-pharma-exports-touch-rs-1-8-trillion-in-fy22-flat-over-fy21-122050200931_1.html

Business (2022b) India supplied over 235 mn total supplies of covid vaccines to 98 countries. Business Standard. https://www.business-standard.com/article/current-affairs/india-supplied-over-235-mn-total-supplies-of-covid-vaccines-to-98-countries-122071500116_1.html

Committee (2020) Department-related Parliamentary Standing Committee on Science & Technology, Environment, Forests and Climate Change. Report no. 327. Rajya Sabha, Parliament of India. http://164.100.47.5/committee_web/ReportFile/19/126/327_2020_3_16.pdf

Conrad R, Davis K, Glos L, Liu W (2022) Estimating cost savings from new generic drug approvals in 2018, 2019, and 2020. U.S. Food and Drug Administration Report. USFDA. https://www.fda.gov/media/161540/download

CSIR (2022) CSIR labs, Ministry of Science and Technology, Govt of India. https://www.csir.res.in/csir-labs. Accessed 8 Dec 2022

Das S (2022) Expand PLI scheme for pharma to make it research-linked, says industry. Business Standard. https://www.business-standard.com/article/companies/expand-pharma-pli-scheme-to-make-it-research-linked-indian-pharma-122022500644_1.html

Differding E (2017) The drug discovery and development industry in India—two decades of proprietary small-molecule R&D. ChemMedChem 12(11):786–818. https://doi.org/10.1002/cmdc.201700043; Published online 2017 Jun 1. Weblink: https://www.ncbi.nlm.nih.gov/pmc/articles/PMC5488177/#_ffn.sectitle

DPIIT (2021) IPR reforms: evolution of the Indian IPR regime. Department for Promotion of Industry and Internal Trade (DPIIT). https://dpiit.gov.in/sites/default/files/IPR-Reforms-11June2021.pdf

Economic Times (2013) Dr. Reddy's Laboratories founder Anji Reddy passes away. Economic Times. https://economictimes.indiatimes.com/industry/healthcare/biotech/pharmaceuticals/dr-reddys-laboratories-founder-anji-reddy-passes-away/articleshow/18998140.cms

Establishment of biotechnology Park and incubation centre in Visakhapatnam (2022) Government of India, Ministry of Science & Technology, Department of Biotechnology. https://sansad.in/getFile/annex/258/AU1087.pdf?source=pqars

Eximbankindia (2015) Study on Indian pharmaceutical industry. Working Paper No. 37. Export-Import Bank of India. https://www.eximbankindia.in/Assets/Dynamic/PDF/Publication-Resources/ResearchPapers/39file.pdf

FDA (2022) Fast track, breakthrough therapy, accelerated approval, priority review, U.S. food and drug administration. https://www.fda.gov/patients/learn-about-drug-and-device-approvals/fast-track-breakthrough-therapy-accelerated-approval-priority-review. Accessed 26 Dec 2022

Ghoshal S, Piramal G, Budhiraja S (2002) World class in India. Penguin India. https://penguin.co.in/book/world-class-in-india/

Greene W (2007) The Emergence of India's Pharmaceutical Industry and Implications for the U.S. Generic Drug Market. No. 2007-05-A. U.S. International Trade Commission (USITC). https://www.usitc.gov/publications/332/EC200705A.pdf

Han K, Le Deu F, Zhang F, Zhou J (2021) The dawn of China biopharma innovation. McKinsey & Company. https://www.mckinsey.com/industries/life-sciences/our-insights/the-dawn-of-china-biopharma-innovation

The Hindu (2024) India, S. Korea, US, Japan, EU launch Biopharmaceutical Alliance. https://www.thehindu.com/business/Industry/india-s-korea-us-japan-eu-launch-biopharmaceutical-alliance/article68258492.ece

Economic Times (2024) India's Pharma Exports rise 10% to USD 27.9 bn in FY24. https://economictimes.indiatimes.com/industry/healthcare/biotech/pharmaceuticals/indias-pharmaexports-rise-10-to-usd-27-9-bn-in-fy24/articleshow/109560417.cms?from=mdr

Indiabudget (2021) Economic Survey 2021–22. Ministry of Finance, Govt. of India. https://www.indiabudget.gov.in/economicsurvey/

Indiaenvironmentportal (2021) One hundred and fifty eighth report–attracting investment in post-covid economy: challenges and opportunities for India. Rajya Sabha, Parliament of India. http://www.indiaenvironmentportal.org.in/content/469726/one-hundred-and-fifty-eighth-report-on-attracting-investment-in-post-covid-economy-challenges-and-opportunities-for-india/

Indianexpress (2022) India's biotech economy grew 8 times in last 8 years, expanding further: PM Modi. The Indian Express. https://indianexpress.com/article/india/pm-narendra-modi-biotech-startup-expo-7960647/

Indiatoday (2021) Brazil President Bolsonaro thanks PM Modi for 'sanjeevni booti' against Covid with Hanuman pic. India Today. https://www.Indiatoday.in/india/story/Brazil-president-bolsonaro-tweet-hanuman-pic-thanks-pm-modi-for-covid-vaccine-sanjeevni-booti-1761884-2021-01-22

Intellectual Property India Annual Report 2022-23. The Office of the Controller General of Patents, Designs, Trademarks and Geographical Indications. Department for Promotion of Industry & Internal Trade, Ministry of Commerce and Industry, Government of India. https://ipindia.gov.in/annual-reports-ipo.htm

Investindia (2021) Diplomacy in difficult times: India's "Vaccine Maitri" initiative, Editorial Invest India Outlook. https://www.investindia.gov.in/team-india-blogs/diplomacy-difficult-times-indias-vaccine-maitri-initiative

Investindia (2023) Pharmaceuticals: invest India, National investment promotion and facilitation agency. https://www.investindia.gov.in/sector/pharmaceuticals. Accessed 4 May 2023

Ipindia (2019) Annual report 2019–2020 of the Office of the Controller General of Patents, Designs, Trademarks, and Geographical Indications, Ministry of Commerce & Industry, Government of India. p 7. https://ipindia.gov.in/writereaddata/Portal/Images/pdf/IP_India_English_29.08.21_Final_.pdf

James TC (2020) Bulk drug industry in India: challenges and prospects. Discussion paper no. 259. Research and Information System for Developing Countries, New Delhi; https://www.ris.org.in/sites/default/files/Publication/Discussion%20Paper%20259%20T%20C%20James.pdf

Jeffrey A, Le Deu F, Li S, Zhang F, Zhou J (2022) Vision 2028: how China could impact the global biopharma industry. McKinsey & Company. https://www.mckinsey.com/industries/life-sciences/our-insights/vision-2028-how-china-could-impact-the-global-biopharma-industry

Kearney (2022) Taking India's life sciences industry to the global stage "Make In India" to fuel 4x growth in biosimilars and vaccines by 2026. KEARNEY. https://www.kearney.com/covid-19/article/%2D%2Dinsights/taking-indias-life-sciences-industry-to-the-global-stage. Accessed 24 Nov 2022

Lebedenko S (2022) The rise of Sino-Russian cooperation. Foreign Policy Research Institute. https://www.fpri.org/article/2022/05/the-rise-of-sino-russian-biotech-cooperation/

Leo L (2021) Govt aids bharat biotech with Rs. 65 crore grant, 3 PSUs' support to make Covaxin. Mint. https://www.livemint.com/companies/news/govt-aids-bharat-biotech-with-rs-65-crore-grant-3-psus-support-to-make-covaxin-11618580061242.html

Livemint (2021) Indian pharma industry to surpass $60 billion in two years. Mint. https://www.livemint.com/companies/news/indian-pharma-industry-to-surpass-60-billion-in-two-years-11629983034597.html

Mckinsey (2020) Indian pharma 2020: propelling access and acceptance, realizing true potential. McKinsey & Company. https://www.mckinsey.com/~/media/mckinsey/dotcom/client_service/Pharma%20and%20Medical%20Products/PMP%20NEW/PDFs/778886_India_Pharma_2020_Propelling_Access_and_Acceptance_Realising_True_Potential.ashx

Mondal SS, Pingali V (2017) Competition and intellectual property policies in the Indian pharmaceutical sector. Vikalpa. Indian Institute of Management, Ahmedabad. https://doi.org/10.1177/0256090917704561

Newcommittee (2013) One hundred and tenth report on FDI in pharmaceutical sector (Presented to the Rajya Sabha on 13th August, 2013; Laid on the Table of Lok Sabha on 13th August, 2013); http://164.100.47.5/newcommittee/reports/EnglishCommittees/Committee%20on%20Commerce/110.pdf

Nppaindia (2022) THE DRUGS (PRICES CONTROL) ORDER, 2013, (Notified by SO 1221 (E) dated 15.05.2013 and as amended up to vide SO 1192(E) dated 22-03-2016). https://www.nppa-india.nic.in/wp-content/uploads/2018/12/DPCO2013_03082016.pdf. Accessed 26 Dec 2022

Pharmaceuticals (2022) Notification dated July 21, 2022 on production linked incentive scheme. https://pharmaceuticals.gov.in/sites/default/files/Gazettee%20notification%20of%20bulk%20drug%20schemes_0_0.pdf

Pharmaceuticals (2023) Annual report 2021–22, Department of Pharmaceuticals, Ministry of Chemicals and Fertilizers. https://pharmaceuticals.gov.in/sites/default/files/English%20Annual%20Report%20202122%20%281%29.pdf. Accessed 4 May 2023

PIB (2022a) India's pharma exports grow by 103% since 2013–14. Ministry of Commerce & Industry. https://pib.gov.in/PressReleaseIframePage.aspx?PRID=1821747

PIB (2022b) Press Information Bureau Release. https://pib.gov.in/PressReleaseIframePage.aspx?PRID=1880666

PIB (2022c) Press Information Bureau (PIB) Release ID: 1815852 dated April 12, 2022. https://www.pib.gov.in/PressReleseDetail.aspx?PRID=1815852

Rakshit A (2016) Bengal chemicals plans to pay off debt through divestment. Business Standard. https://www.business-standard.com/article/companies/bengal-chemicals-plans-to-pay-off-debt-through-divestment-116122901146_1.html

Rasmussen ASB, Hammou A, Poulsen TF, Laursen MC, Hansen SF (2021) Definition, categorization, and environmental risk assessment of biopharmaceuticals. Sci Total Environ 789:147884. https://doi.org/10.1016/j.scitotenv.2021.147884

RBI (2021) Drivers of Indian pharmaceutical exports. RBI Bulletin. p 49–57. https://www.rbi.org.in/Scripts/BS_ViewBulletin.aspx?Id=20379

Seruminstitute (2021) Serum Institute of India (SII) re-starts COVAX exports; passes 1billion COVISHIELD dose milestone. https://www.seruminstitute.com/news/press_sii_gavi_bmgf.pdf.47

Sharma A, Kumar N, Kuppermann BD, Francesco B, Loewenstein A (2019) Biologics, biosimilars, and biobetters: different terms or different drugs? Eye 33(7):1032–1034. https://doi.org/10.1038/s41433-019-0391-5

Whyte LE (2022) FDA set to get more power to pull accelerated drugs that don't work. Wall Street J; https://www.wsj.com/articles/fda-set-to-get-more-power-to-pull-accelerated-drugs-that-dont-work-11654862400

World University Rankings 2023 by subject: life sciences. Times Higher Education (2024). https://www.timeshighereducation.com/world-university-rankings/2023/subject-ranking/life-sciences

Part V

Regulatory Framework

Genetically Modified Organisms and Products in India: History of Regulation, Promotion, and Issues

13

Prasanta Kumar Ghosh

Abstract

Following the discovery of non-sexual transfer of genetic material to an organism to add or change one or more of its genetic trait/s by Herbert Boyer and Stanley Cohen in 1973, concerns were raised by various groups about the safety from the use of such organisms, technologies, and products made therefrom. These organisms were named as recombinant DNA organisms, and products made therefrom as recombinant products. Subsequently, in an international conference by the name Asilomar Conference, held in February 1975 in California, USA, the pros and cons of the new technology were discussed globally by scientists and global opinion makers, and recommendations were made for its use on a precautionary principle basis. Every country, thereafter, came out with regulations while using the new technology. Indian rules and regulations were made, starting 1986, and over a period of 27 years up to the end of December 2023 many developments and progress have taken place. The article is a comprehensive review on how developments took place in India, and over time how the rules and regulations for the safe use of genetically modified organisms and products made therefrom, had evolved, and were implemented in the country. Several arms of the Indian government are involved in the issues. Besides the role of the Ministry of Environment and Forests & Climate Change (MOEF&CC), the role of the Department of Biotechnology (DBT) of the Ministry of Science and Technology, the Ministry of Agriculture (MOA) and its designated agencies, Ministry of Health, and Family Welfare (MOH& FW) and Food Safety and

P. K. Ghosh (✉)
Ex-Adviser, Department of Biotechnology, Ministry of Science and Technology, Government of India, New Delhi, India

Sompradip Publishers and Consultants, New Delhi, India
e-mail: pkghosh@gandipsbio.com

G. John, A. K. Panda (eds.), *Biotechnology in India - Reworking A Strategy*, https://doi.org/10.1007/978-981-97-0896-3_13

Standards Authority of India (FSSA), and Ministry of Consumer Affairs (MCA) in GMOs has also been discussed. The promotional role of RCGM of DBT has also been highlighted. Complexities about issues of labeling of GM foods in India have been underscored. Some factors requiring closer attention for speeding up new inventions in GMOs in India have been identified.

13.1 Introduction

India has made phenomenal progress in biotechnology in its broad spectra of applications. The Department of Biotechnology (DBT) of the Government of India had played a key role in the endeavor by putting in place appropriate rules and procedures, and by disseminating the areas of research and applications during the earlier years, starting Feb 1986 under the leadership of Dr. S. Ramachandran, the founder Secretary of DBT. The initial hard labor fructified into the present modern Indian biotech industry. Indian biotech sector comprising five main sub-sectors, namely biotech pharmaceuticals; agriculture biotechnology; industrial biotechnology including effluent management projects; bioinformatics; and bio-services consisting of contract and clinical research organizations, as well as contract manufacture has grown tremendously over the years. It has been estimated that the Indian biotech sector in 2023 was valued at US $ 137 billion which was 3.65% of the Indian GDP during this year (INVESTINDIA 2024).

When the direct, non-sexual transfer of genetic material to an organism in order to add or change one or more of its genetic trait/s was discovered by Herbert Boyer and Stanley Cohen in 1973, concerns were raised by various groups including the scientists, the lawyers, the biologists, and the powerful opinion makers about the safety from the use of such products. Through one international conference by the name, "The Asilomar Conference on Recombinant DNA" the issues of potential biohazards were discussed where the Nobel Laureates like Paul Berg and Maxine Singer were present among others of about 140 professionals from various disciplines. The conference (NCBI 2022) was held in February 1975 at Asilomar State Beach, California. It was concluded at that time that there was a need for the regulation of research and experimental work on the new technology. The participants felt that scientific research on the new technology, designated as recombinant DNA technology (rDNA technology), generating genetically modified organisms(GMOs) should be brought into the public domain, so that the public are aware about the safety of new recombinant DNA products (rDNA products) emanating from such research. Research should be carried out in a transparent manner on a precautionary principle. It was therefore necessary to draw up guidelines to ensure the safe use of this technology.

Following the recommendations of the Asilomar Conference, different countries came up with their rules and regulations to deal with this new technology. The Indian Government propounded the Environment Protection Laws in 1986 and Rules in 1989 from its Ministry of Environment and Forests & Climate Change (MOEF&CC) [then known as the Ministry of Environment and Forests (MOE&F)] to deal with environmental safety issues as well as safety to human and animal health from the use of rDNA based technologies (THE ENVIRONMENT (PROTECTION) ACT 1986; Biosafety 2022).

Even before the Asilomar Conference was held to discuss on safety related issues emanating from the newly invented rDNA technology in 1975, there was the Stockholm Conference of the United Nations Conference on the Human Environment(UNCHE), held at Stockholm in June, 1972 to discuss actions to be taken on ecological damages created by human activities. India participated in the conference. The then Prime Minister of India Mrs. Indira Gandhi in her speech put forward the connection between ecological management and poverty alleviation (Wikipedia Contributors 2022). The roots of the future enactment of the Environment Protection Act (EPA) were indeed seeded in that conference, even though the recombinant DNA technology of Herbert Boyer and Stanley Cohen was not yet discovered; the new technology was discovered in 1973 and had potential of damaging the environment as well as human and animal health.

Coming back to Indian developments on the creation of Environment Protection Laws, since industrial activities of handling, manufacture, packaging, sale, and distributions take place from industries, there was a need to keep a watch on the industries, which were involved in biotechnologies including the rDNA based products and services. Consequently, there was a need to identify a government department/ministry to provide oversight to such industries. Furthermore, since rDNA products have wide use in multiple sectors of human activities such as in agriculture including horticulture, animal husbandry, and fisheries; food, feed, and process activities; medicines; as well as industrial activities utilizing recombinant products and organisms, all these sectors regulated by the government also were required to be vigilant to deal with such products and services. Accordingly, the safety laws dealing with rDNA products and services had to have linkages with the main law of safety, conceptualized and propagated by the government from the MOEF& CC, while the scientific knowledge in overseeing, implementing, and promoting the rDNA technologies was entrusted to the DBT of the Ministry of Science & Technology (MOS&T).

13.2 Creation of the Department of Biotechnology (DBT)

The Department of Biotechnology (DBT), one of the Departments in the Ministry of Science and Technology, was created in February 1986 with the approval of the then Prime Minister, late Rajiv Gandhi. The DBT on its website at its description of "Mission" has stated that it shall devote wholly to achieve excellence in the promotion of biotechnology in the country within the ambit of the Allocation of Business Rules, 1961 as enumerated through the Government of India notifications. The first two notifications were no. CD-172/86 dated 27.2.86 and No. CD-87/87 dated 31.1.87. All the mandates of DBT are expressed and published in the Government of India (Allocation of Business) Rules Book, which are amended and published by the Government from time to time.

In India, all Central Government activities are carried out through Ministries and Departments, created through, and notified in the Allocation of Business Rules, 1961. The Rules emanate, in exercise of the powers conferred by Clause (3) of Article 77 of the Indian Constitution (Legislative 2022a). The Government of India (Allocation of Business) Rules, 1961 was framed and notified on January 14, 1961. The Rules have been modified from time to time, and according to the latest printed version, as amended up to 31st January 2017, the Department of Biotechnology (DBT) was allocated its mandate of activities (Government of India (Allocation of Business) Rules 1961). This mandate was later revised but was not notified; the revised updated mandate was quoted in the Parliamentary Standing Committee Report of February 2018. The Department-related Parliamentary Standing Committee on Science & Technology, Environment & Forests considered the "Demands for Grants 2018–2019" of the DBT in its meeting held on 13th February 2018, and the ensuing Report contained the latest mandate of DBT (Committee 2022), which is at Annexure-I in Table 13.1.

One of the important tasks of promoting the inclusion of, "Establishment of National Biotechnology Regulatory Authority" in the DBT mandate (Table 13.1) emanates from safety issues originating from the signing and adoption of the International Treaty entitled, "Cartagena Protocol on Biosafety" by India for transboundary movement of Living Modified Organisms (LMOs), discussed later. LMOs are living modified modern biotech organisms, which are genetically modified organisms (GMOs) that are multipliable, and can be used under *contained conditions* (such as genetically modified microorganisms or cells and tissues, deployed for the manufacture of rDNA based medicines and other products such as rDNA based enzymes, and other substances). LMOs can also be used and multiplied in the *open environment* such as in agriculture as seeds and planting materials, or as genetically modified animals, deployed in contained or open environment.

13.3 Licensing Requirements for Industrial Activities in GMOs and Products Therefrom in India

The mandate of the DBT describes multiple functions which are leaned toward carrying out research and development activities in biotechnology through programs, activities, human resource development, institution building, and other administrative activities linked with genetically manipulated products. In particular, the mandate includes besides other work-functions, the following:

(a) To serve as the administrative and implementing department for fulfilling the national objectives in biotechnology.
(b) To manufacture and ensure application of recombinant, cell-based and DNA vaccines, diagnostics, and other biotechnological products (which include all rDNA products).

Table 13.1 Mandate of DBT

1. To evolve policies and integrated programs in biotechnology and ensure their implementation and monitoring
2. To identify specific programs of Research and Development and manufacturing in biologicals and biotechnology and oversee the initiation and pursuit of related research and manufacturing activities
3. To identify, set up and support centers of excellence for Research and Development in biotechnology and ensure proper dovetailing of their activities as per national priorities and objectives
4. To act as a screening, advising, and approving agent of the government with regard to import and transfer of new technologies for the manufacture of biologicals, biotechnological products and their intermediates
5. Evolve safety guidelines for biotechnology Research and Development and manufacturing in India
6. To act as the central agency for the import of genetically manipulated materials, culture, cells, specimens, tissues and biotech products including DNA and RNA of any type or size and for promoting their production in the country
7. Serve as the inter-ministerial and interagency nodal point for all specific international bilateral and multilateral Research and Development collaborations and agreements in the area of biotechnology, act as the nodal point for all technology transfers in the area of biotechnology
8. Manufacture and ensure application of recombinant, cell-based and DNA vaccines, diagnostics, and other biotechnological products
9. To evolve programs for human resource development in the area of biotechnology
10. Serve as an administrative and implementing department of agencies, commissions, boards, etc. specifically formed by the government for fulfilling the national objectives in biotechnology and also to serve as the nodal point for bioinformatics including training and creation of infrastructure, collection, dissemination and exchange of information relating to biotechnology
11. Establishment of National Biotechnology Regulatory Authority
12. Matters relating to:
(a) International Centre for Genetic Engineering and Biotechnology (ICGEB), New Delhi
(b) National Institute of Immunology (NII), New Delhi
(c) National Centre for Cell Sciences (NCCS), Pune
(d) Centre for DNA Fingerprinting & Diagnostics (CDFD), Hyderabad
(e) National Institute for Plant Genome Research (NIPGR), New Delhi
(f) National Brain Research Centre (NBRC), Manesar, Gurgaon
(g) Institute for Bioresource and Sustainable Development (IBSD), Imphal
(h) Institute of Life Sciences (ILS), Bhubaneshwar
(i) Translational Health Science and Technology Institute, (THSTI), Faridabad
(j) Rajiv Gandhi Centre for Biotechnology, (RGCB), Thiruvananthapuram
(k) UNESCO Regional Centre for Education, Training & Innovation in Biotechnology, Faridabad
(l) National Institute of Biomedical Genomics, (NIBMG) Kolkata
(m) Institute for Stem Cell Science and Regenerative Medicine, Bengaluru
(n) National Agri-Food Biotechnology Institute Mohali, (NABI), Punjab
(o) National Institute of Animal Biotechnology, (NIAB), Hyderabad
(p) Center of Innovative and Applied Bioprocessing (CIAB), Mohali

(continued)

Table 13.1 (continued)

13. Matters relating to:
(a) Bharat Immunologicals and Biologicals Corporation Limited (BIBCOL), Bulandshahr, UP
(b) Indian Vaccine Corporation Limited, (IVCOL)
(c) Biotechnology Industry Research Assistance Council (BIRAC), New Delhi

(c) To act as the central agency for the import of genetically manipulated materials, culture, cells, specimens, tissues, and biotech products including DNA and RNA of any type or size and for promoting their production in the country.

It was felt that there was a need for assuming the administrative responsibility by the department for safe use of technologies requiring to deploy GMOs and products derived therefrom. Industrial activities that required using GMOs should be pursued on a precautionary principle. Discharged effluents from such manufacturing plants are anticipated to contain GMOs in the wastewater if not properly treated and if GMOs are not fully deactivated; the risks, if any need to be ascertained in advance to prevent any untoward effects to the environment and to human and animal health from the living GMOs discharged. A policy paper was therefore prepared by the author under the guidance of the Secretary, and was submitted to the Ministry of Industry (MOI), seeking approval /consent of the latter for the DBT to work as the Administrative Ministry (AM) for industries and establishments in the country, which would be dealing with GMOs.

The Ministry of Industry is the administrative head for all matters relating to Industries (Development and Regulation) Act in the country, and it allocates different businesses of industrial activities to other Ministries and Departments, while coordinating all activities relating to the Act. The Ministry of Industry remains the chief one for industrial activities in the country, working and authorizing activities through the administrative departments and ministries, the companies and Industries in the country to carry out different business activities in India and coordinates the activities through the assistance of other Ministries [through their expertise], making them the Administrative Ministry, and empowering them to develop and regulate the growth of the business in a healthy and rational manner within the provisions of the Indian industries under Industries (Development & Regulations) Act [I (D&R) Act].There was no notified Department of the Government in 1987 for carrying out the administrative work, relating to biotechnology-based industries. Indeed, there was no clear-cut definition on which activities constitute biotech industries.

During those periods in 1987 and earlier, India excelled in the production of fermentation-based antibiotics, certain vitamins, microbial transformation of a few drug substances, alcohol production from molasses, microbial enzymes, extraction of active substances from plant parts, extraction of glandular products like insulin from bovine and porcine pancreas, liver extracts from slaughtered domestic animals, and the like. There was no modern biotechnology industry yet in the country. One of the earliest papers depicting the status of Indian biotechnology products was published in 1993, where the then range of biotech products used in India were described (Ghosh 1993).

In the proposal to the Ministry of Industry, the DBT used the term "biotechnology," based broadly on the definition of Károly Ereky (1919), the Hungarian agricultural engineer, who coined the term "biotechnology" in 1919 for any industrial activity utilizing living organisms. Based on the submissions made by the DBT, the department was accepted as the administrative ministry for biotech products and processes in the country. This event ushered in a new era, starting 1987, and DBT got involved in the approval, promotion and furtherance of Indian Biotech industries, which included the conventional as well as the modern biotechnology industries (based on the use of GMOs and products thereof). In DBT, a small cell was created. With this set up, technical assessment was made on the licensing proposals submitted to the government. Subsequently, in 1991 when the process of liberalization set forth by the government, required reduction in the scope of enforcing non-tariff barriers to imports, the Industrial Licensing Policy was amended and modified by the government, and multiple steps of deregulation and relaxation were announced. The new I (D&R) Act reduced the scope of technical audit. However, because of environmental safety as well as, human and animal health safety implications from rDNA products, certain industrial activities such as (a) bulk drugs produced by the use of recombinant DNA technology, (b) bulk drugs requiring in-vivo use of nucleic acids as the active principles, and (c) specific cell/tissue targeted formulations were restricted, and such activities required an Industrial License (Pharmaceutical Policy 2002) under the Indian I (D&R) Act before they could be pursued in the country. This system provided government vigilance on the activities carried out in India by the modern biotech industries.

13.4 International Pressure on Equitable Use of Global Biological Diversity and Safe Use of Modern Biotech Products

13.4.1 Convention on Biological Diversity and the Genesis of GMOs and LMOs

The world community realized that natural resources are not infinite and therefore it is necessary to take measures on a philosophy that sets out its sustainable use. In the year 1988, at a United Nations Environment Programme (UNEP), an Ad Hoc Working Group of Experts on Biological Diversity was set up. Subsequently, through an Ad Hoc Working Group of Technical and Legal Experts constituted by the Ad Hoc Working Group, a legal text was drafted out, which addressed the conservation and sustainable use of biological diversity, and the sharing of benefits arising from their utilization with countries (sovereign states that have a defined territory and that administers its own government) and local communities. Later in 1991, an intergovernmental negotiating committee was established for finalizing and adopting the text (History of the Convention 2022). Thereafter, through several international meetings and discussions, procedures for safe handling of GMOs and

LMOs were framed and a global Treaty was signed by the Member Countries. A note depicting the chronological events is at Annexure-II.

13.5 Indian Acts, Rules, and Regulations on the Use of GMOs and LMOs

13.5.1 Ministry of Environment and Forests (MOE & F) [Presently Renamed as Ministry of Environment, Forests, and Climate Change (MOEF & CC)] and the Department of Biotechnology (DBT)

The Indian Environment (Protection) Act (Act) was enacted in 1986. The act was enacted under Article 253 of the Indian Constitution (Legislative 2022b); in other two Articles, namely Article 48A and Article 51A, the responsibilities of the States in India to protect the environment including safeguarding the forests and wild lives of the country, and the duties of every citizen to protect the environment respectively are elaborated.

The objectives of the Act were to establish and provide protection of the environment. The Act empowers the Central government to establish authorities charged with the mandate of preventing environmental pollution in all its forms and to tackle specific environmental problems that are peculiar to different parts of the country. The Act is a comprehensive legislation for preventing environmental pollution emanating from all kinds of substances used and handled in the country. The Act is described in four chapters: (a) Chapter I Preliminary; (b) Chapter II General Powers of the Central Government; (c) Chapter III Prevention, Control and Abatement of Environmental Pollution; and (d) Chapter IV Miscellaneous. In these four Chapters, there are in total 26 Sections dealing with different matters and aspects of the Act. The Act is implemented through a set of rules, which are popularly known as Rules 1989. Before further elaborations are made on the Rules 1989, some more amplification, explication, and interpretations are necessary on the Act, as this is an extremely important Act for environmental protection. The Act provides powers to the Central Government through its various paragraphs. These are briefly discussed and placed at Annexure-III. In no part of the Act the words such as genetically engineered organisms or cells are mentioned.

Using the powers conferred upon the Central Government in the three Sections mentioned in the Act, [Sections 6, 8, and 25] the Central Government at a later date came out with Rules for the manufacture, use, import, export, and storage of hazardous microorganisms and genetically engineered organisms or cells (Rules 1989); the Rules were notified on December 05, 1989. The Rules 1989 became effective (Geacindia 1993) from 01.10.1993. Highlights of the Rules-1989 can be seen at Annexure-IV.

The rules for the manufacture, use, import, export, and storage of hazardous microorganisms, genetically engineered organisms or cells, 1989 (Rules–1989) are implemented by the Ministry of Environment, Forest and Climate Change (MOEF&CC), Department of Biotechnology (DBT) and State Governments (SGs).

Following the Rules—1989, a series of guidelines on contained research, biologics, confined field trials, food safety assessment, and environmental risk assessment of genetically modified organisms and products thereof have been formulated and notified, based on which the use of such substances are regulated in the country. The new genome engineering technologies including gene editing technologies like CRISPR/Cas9 are also covered under the modified rules (Chimata and Bharti 2019).

The DBT came out with new guidelines (Dbtindia 2022) in May 2022, wherein three kinds of transgenic plants had been defined, which would be created by employing gene editing techniques such as Zinc Finger Nucleases (ZFNs), Transcription Activator-Like Effector Nucleases (TALENs), and Clustered Regulatory Interspaced Short Palindromic Repeats (CRISPR). Three kinds of transgenic plants may be developed by using these techniques, which have been defined in the guidelines as those produced by (a) site-directed mutagenesis without using a DNA sequence template, which plants shall be known as Site-Directed Nuclease(SDN)-1 products, (b) site-directed mutagenesis using a DNA sequence template, which plants shall be known as Site-Directed Nuclease (SDN)-2 products, and (c) site-directed insertion of gene/ large DNA sequence using a DNA sequence template, which plants shall be known as Site-Directed Nuclease (SDN)-3 products. The SDN-1, SDN-2 plants shall be exempted from the application of Clauses 7–11 of Rule-1989. The SDN-3 plants would be regulated as per the Rule-1989.

MOEF&CC along with the Department of Biotechnology (DBT), Ministry of Science & Technology and the State Governments carry out various functions of implementation of the Act and the Rules-1989. There are essentially six committees, the functions of which are elaborated in the Rules-1989, and through these six committees, the Act and the Rules-1989 are implemented. The details of the six committees and the manner of functioning are elaborated briefly at Annexure-V.

13.6 Ministry of Agriculture

The Indian Seeds Act, 1966 and the Seeds (Control) Order 1983, all of which are implemented by the Ministry of Agriculture (MOA) and its designated agencies, did not have any entry of genetically modified seeds, as such seeds were not known in India at that time. Therefore, the MOE&F in its multiple notifications (Geacindia 2022) bearing numbers GSR 584 (E), 585 (E), GSR 586 (E), 587 (E), GSR 588 (E), and 589(E), all dated 01.09.2006, provided authorization to the implementers of Seed Act and Order for their enabling them [MOA and its designated agencies] to appoint Seeds Inspectors under Section 13 of the Seeds (THE SEEDS ACT 1966) Act 1966, and under Section 12 of the Seeds (Control) (Seed Control Order (Amendment) 2006) Order 1983. A notification of MOA, covering the rules and regulation for the import of germplasm /GMOs/transgenic plant material for research purpose was issued. The notification also covered procedures for regulating the import into India of agricultural articles such as live insects and microbial cultures, and plants and plant products for use for other purposes also as per Plant Quarantine Order 2003 (Plant Quarantine Order 2003).

Also, liberty was extended for the setting up of testing laboratories for genetically modified seeds (GM seeds), persons to be appointed for testing, etc., and even for revoking orders already granted for using GM seeds, if necessary.

Following the MOE&F notifications on GM seeds, the Ministry of Agriculture had amended (Seeds (Control) Amendment Order 2019) their Seeds Control Order 2006 by two notifications incorporating the above amendments, for dealing with GM seeds.

13.7 Ministry of Health and Family Welfare and Food Safety and Standards Authority of India

With the invention and introduction of genetically engineered plants including crop plants, concerns were raised globally to determine the safety aspects from the consumption of such substances as food, feed, and fodder. The Cartagena Protocol on Biosafety (Protocol) is the first International Treaty, which was adopted in 2000, to deal with the safety of such foods, and regulations were advocated for the transboundary movement of LMOs besides the setting up of statutory regulatory infrastructure in member countries for preparing documents based on the laws of each country for safe use of LMOs. Following the adoption and being a signatory to the Protocol, India took steps for the use of such substances. In the meantime, International Regulatory Bodies like Codex Alimentarius Commission prepared documents for determining standards of foods derived from recombinant DNA plants for consumption and finalized its first Report (FAO n.d.) in 2003. The Indian Council of Medical Research (ICMR) came out (Dbtindia n.d.) with "Guidelines for the Safety Assessment of Foods Derived from Genetically Engineered Plants" in 2008, which was updated in 2012 (ICMR-Guidelines). The Indian guidelines take into consideration the contents in the Protocol and the recommendations of the Codex Alimentarius. The Food Safety and Standards Authority of India (FSSAI) under the Ministry of Health & Family Welfare, Government of India (MOH&FP) is the food regulatory authority of India that sets standards for foods. FSSAI was set up (FSSAI n.d.-a; FOOD SAFETY AND STANDARDS ACT 2006) in 2006; it initially operated through Food Safety and Standards Act, 2006 (FSAS-2006), which Act was later modified as Food Safety and Standards (Amendment) Act, 2008 (FSAS-Amendment-2008) with certain modifications, and concurrently from time to time multiple "Enforcement Notifications" were issued. The FSAS-2006, with its modifications and enforcement notifications, regulates and checks for safe food manufacture, storage and handling practices of Food Business Operators in India. The FSAS-2006 (Act No. 34 of 2006) received the assent of the President of India on 23rd August 2006 and was published in the Gazette of India (Extraordinary) Part I, Section 1 dated 24th August 2006.

Foods, according to FSAS-2006 include those derived and manufactured from genetically modified substances (Indiaenvironmentportal n.d.). A license is issued to the Food Business Operators by the FSSAI and a compliance badge, if the foods are manufactured by the operators in accordance with the standards. FSSAI had set up standards for foods manufactured from genetically engineered plants, utilizing the ICMR-Guidelines. FSSAI notified (FSSAI 2020) on August 21, 2020, under Section 22 of the FSAS-2006 that effective January 1, 2021, specific food product imports into India identified in Annex-I of the notification must be accompanied by a certificate from the exporting country, declaring that the product is of non-GM origin and does not contain genetically modified organisms and are also not genetically modified.

The issue of labeling or not labeling food, produced from genetically modified organism not containing any living genetically modified organism therein, is a controversial matter the world over. In India, the issues of identification and labeling are supposed to be legally dealt with under the Prevention of Food Adulteration Act & Rules (PFAA& R) (FSSAI n.d.-b) administered by the Food Safety and Standards Authority of India (FSSAI) under the Ministry of Health & Family Welfare, Government of India (MOH&FW). MOH&FW had proposed a mandatory labeling policy for all GM foods in 2006, and had drafted a rule, which was to be included as a new para in the PFAA&R as, "37-E Labeling of Genetically Modified Food," and was to be inserted after 37-D in the PFAA&R with details to conform to labeling requirements as follows:

(a) GM food, derived therefrom, whether it is primary or processed or any ingredient of food, food additives or any food product that may contain GM material shall be compulsorily labeled, without any exception.
(b) The label of all packages(s) of GM food(s) or foods containing ingredients, derived from biotechnology, or bioengineering or food additives or any food product that may contain GM materials shall indicate that they have been subject to genetic modification. These provisions will be applicable to all such products both imported or domestically produced; and.
(c) The label of imported GM food or derived therefrom, whether it is primary or processed or any ingredient of food, food additive or any food product that may contain GM material shall also indicate that the product has been cleared for marketing and use in the country of origin so that the verification, if needed can be taken up with that country without having to resort to testing.

Imports of GM foods of various kinds are taking place in India, and such foods are most often non-labeled. In certain cases, these foods are labeled as "may contain GM ingredients." However, up to August 2018, FSSAI had neither put the regulations for approvals for GM foods in place nor taken any measures to stop the imports of such products into the country. FSSAI has, however, stated that they have not approved the import or marketing of GM foods in India. An Indian Parliament Report of August 2018 has the details (PARLIAMENT OF INDIA RAJYA SABHA n.d.). On 18th November 2022, FSSAI published a draft regulation (FSSAI 2022), through a notification on Food Safety and Standards (Genetically Modifies Foods) Regulations, 2022. These regulations were to apply to GMOs intended for food use, food ingredients produced from GMOs that contain modified DNA, and also those that do not contain modified DNA. There is no further notification on whether the contents of the draft notification are presently an enforceable law of the country on GM foods.

13.8 Ministry of Consumer Affairs

The Ministry of Consumer Affairs, which is responsible for empowering the consumers through awareness and education and for enhancing consumer protection and safety through progressive legislations and for effective grievance redressal besides others, brought out a notification (Indiagminfo 2012) for bringing in a GM food labeling policy in the country in the year 2012. According to this notification it was mandatory to label every food package containing GM food from January 1, 2013. This policy is yet to be implemented in the country.

13.9 Emerging Issues on Biosafety Regulations in India and Way Forward

Stem cell research in germ lines, in vivo and in vitro genome editing technologies, xenotransplantation procedures, in-vitro 3D bioprinting technologies for producing human organs and use of artificial intelligence (AI) technology in multiple facets of advanced biotechnology in the whole spectra of usage including biopharmaceuticals, agriculture, industrial products, and environment management issues are making rapid advances in some parts of the world. Such advancements are also going to be percolating in Indian bio economies. Novel biosafety rules and procedures would have to be evolved for dealing with these developments.

13.10 Discussion

13.10.1 The Need for Regulation of GMOs and Products Therefrom Globally

After the invention of new biology products, where methods were invented for the development and use of organisms manipulated by genetic modification or genetic engineering methods of direct, non-sexual transfer of genetic material to an organism in order to add or change one or more of its genetic trait/s, in early 1970s people all over the world became suspicious about the safety from such products. The invention of such a technology changed the course of development of biotechnology industry in the world. Since the technology was a revolutionary one in industrial development, a note is inserted to provide a historical flavor of the invention of this technology at Annexure-VI.

13.10.2 Regulations of GMOs in India and Certain Innovative Review Committee on Genetic Manipulation (RCGM) Steps

The Indian Environment (Protection) Act, enacted in 1986 and the Rules for the manufacture, use, import, export, and storage of hazardous microorganisms, genetically engineered organisms or cells, published in 1989, followed by a series of guidelines on contained research on biological material, confined field trials, food safety assessment and environmental risk assessment of genetically modified organisms and products thereof have been implemented by the MOEF&CC; the DBT; and, where required, by involving the State Governments.

For the drugs and pharmaceutical substances requiring the use of GMOs, the environmental impact from the use of such GMOs is first assessed by the RCGM of DBT. This is followed by authorization for the use of such substances by the GEAC, which is based on the recommendations and clearance by the RCGM. The RCGM also authorized the conduct of experiments on rDNA drugs in India, which were not yet approved anywhere. Only after the authorization for use under Indian EPA, the drugs and pharmaceutical substances requiring the use of GMOs can move to the next stages of statutory approvals. These measures set into motion the actions for the authorization of such drugs and pharmaceuticals, under the Indian Drugs Act, where safety and efficacy data on rDNA drugs are generated on human subjects. The GEAC clearance requires generation of safety data on relevant two animal models at the minimum.

One major decision taken in early 1990s by the RCGM was that it would authorize under the Indian EPA, the conduct of R&D in GMOs used for producing active drug substances, where the GMOs have not yet been approved anywhere in the world for use for producing recombinant drug substances, which are to be deployed on human subjects. RCGM would authorize the generation of safety data on relevant lab animals. Deploying such drugs on human subjects is based on satisfactory compliance of the provisions under the Indian Drugs Act.

RCGM authorization under the EPA was to generate animal toxicity and allergenicity data on *relevant animal species*. If the experimentally determined animal toxicity and allergenicity data were satisfactory, then based on RCGM recommendations, the GEAC would clear the GMO for further evaluation of the rDNA drugs for use in generating safety and efficacy data on human subjects, under the Indian Drugs Act. The GMOs cannot be used for producing drug substances under the Drugs Act, until and unless these are cleared under the Indian EPA. The contributions of RCGM toward the advent of modern biotech products in India have been commendable from 1992 onwards and is summarized in Annexure-VII.

13.10.3 Labeling Issues

Within the Clause 2 of Rules –1989 are embedded the intentions of the government that the GMOs and the products thereof are to be identified, whenever these are offered for sale. Para 4 of Clause 2 of the Rules –1989 has the details. A note on Para 4 of Clause 2 of the Rules and the issue of labeling of food products in India and the consequential issues to be tackled by the government is at Annexure-VIII.

All GMOs and products thereof for sale in the areas of drugs and pharmaceuticals, Industrial enzymes and Reagents and diagnostics are sold with the declaration in the accompanying literature or in labels that these products either GMOs [when sold as living entities] or bulk substances or formulations, the active substances of which are manufactured by using recombinant DNA technology, or genetic engineering technology or GM technology. However, labeling of agricultural products of GM origin have not yet been done in India. Imported packaged foods sold in the country have sometimes been labeled by using such words as "may contain ingredients produced by GMOs," but often these have not been labeled. Bt-cotton seeds produced in India are used for extracting oil there from, but the oil is not labeled as derived from GMO-based cotton seeds. The cotton seed cake after oil extraction is used for making feed formulations for animals. But the feed formulations are not labeled. India is also importing sizable quantities of edible oil produced from GM rape seeds and GM soya beans; however, such oils are not labeled as of GM origin. As has been mentioned in the text, the Ministry of Health and Family Welfare and Food Safety and Standards Authority of India as well as the Ministry of Consumer Affairs are in the process of implementing labeling statutes of GM products.

13.11 Schematic Flow of Events for Dealing with rDNA Products: Present and Future

A schematic flow of events for dealing with rDNA products in India involving the history of development of Indian EPA, and the Ministries, Departments, and Regulations involved in the process as also the issues handled legally is depicted in Fig. 13.1. The figure includes anticipated needs for the creation of newer ministries for handling newer biological products and processes.

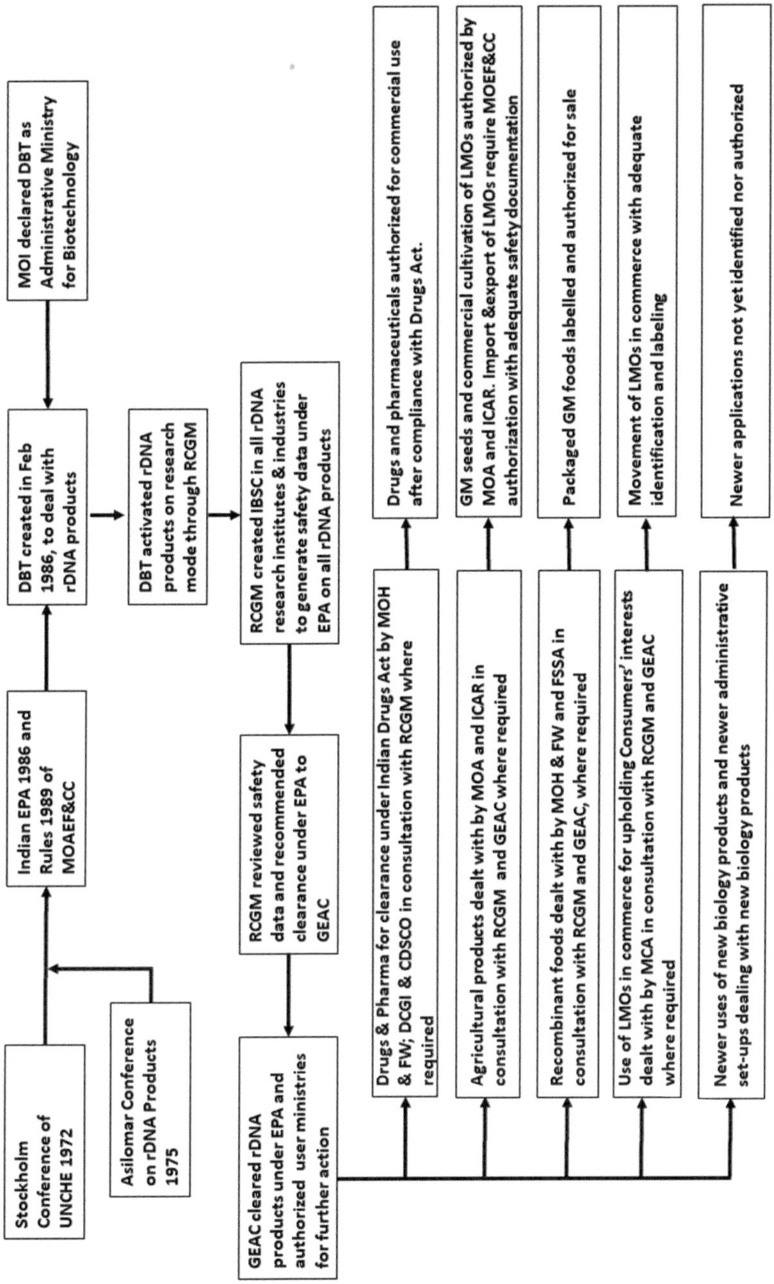

Fig. 13.1 Schematic Flow of events for dealing with rDNA products in India: Ministries, Departments and Regulations

13.12 Recombinant DNA Based New Products

It is anticipated that human abilities to generate and analyze biological data by using new algorithms for computational analysis, using genomic information and imaging techniques of biological processes will expand phenomenally. Use of patient's genes and other biological materials including cells and tissues for diagnosis, treatment, and prevention (from future disease) is another new area which is expanding rapidly. The 3D bioprinting technology on the horizon is anticipated to generate natural human tissues and organs. Stem cell research in germ lines, in vivo and in vitro genome editing technologies, xenotransplantation procedures, and use of artificial intelligence (AI) technology in multiple facets of advanced biotechnology are evolving fast. For all such evolving technologies, it is anticipated that the new safety and efficacy issues shall come up, and for effectively regulating such technologies, products, processes, and materials newer regulatory bodies and ministries may have to be created during the future years.

13.13 Concluding Remarks and Way Forward

India has instituted sound and powerful environment protection acts and rules to deal with genetically modified organisms and products manufactured and used there from. The rules have been updated from time to time, taking into consideration the scientific developments. By implementing the acts and rules, India has made safe use of GMOs and products made there from. This has resulted in the development of a sound and competitive modern biotech industry in the country. Policies on industrial licensing procedures were also established to keep a watch on industries involved in the manufacture and use of GMOs. Several arms, departments, and committees constituted by the government had done commendable work to promote the applications of modern biotechnology, on a precautionary principle. The progress made by India is also remarkable in multiple aspects such as in the establishment of a sound knowledge based biotech industry, conduct of research and development, infrastructure establishment, skilled man-power development, and instituting national and international collaborations. There are, however, issues, especially in the introduction of GM agricultural food products including large-scale cultivation of GM food crops, where identification, segregation, harvesting, and sale, followed by use for manufacturing processed and packaged foods may need identification and labeling of GM foods. There is a need to come out with more policies acceptable to most sections of Indian population to attract the implementation of GM technologies entering human food chain directly and indirectly.

Soon after Stanley N. Cohen and Herbert W. Boyer invented the rDNA technology in 1973, the speed of progress in its applications during the last 50 years had been phenomenal in multiple areas, but especially in human medicines, and in certain countries which have benefitted more than many others.

Access to appropriate scientific material including chemicals, biochemicals, experimental animals, high-tech separation and purification equipment, sophisticated instruments for precise and delicate measurements is the bedrock of research in biotechnology. The deployment of skilled young people from developing countries, conducting laborious high class and risky R&D; protection of new knowledge based on stringent IPR laws add to the core effort. Conduct of clinical research in lab animals, followed by human clinical trials; with fast tracking of regulatory approvals; investment both by the government as well as the private sector in research and development; promotion of innovative research and creating start up industries, etc., are the main factors for breakthrough in scientific research in modern biology.

While deploying the new technologies on a precautionary principle, the developed countries have made faster progress where the regulators have continuously increased their speed of understanding of human biology. The speed of understanding and relaxing rules in newer experiments resulted in faster approval, and application of new products for human benefits. A deeper understanding of how the human immune system responds to disease causing microbes and approving the intervening agents fast for human use, on a precautionary principle, enabled the development of life saving vaccines. Knowledge of genomic sequencing and identification of aberrations in nucleotide sequences and its translated proteins have led to the development of new and safer small molecules, where the regulators have played a significant role in speeding up the application process.

Most developing nations face challenges in competing with developed countries unless equal level playing ground is in place. As breakthrough inventions are becoming difficult at the global level, developing countries like India can take advantage from the observed natural phenomenon of slow speed of invention of new substances in modern biotechnology, especially in the new drug development arena by picking up and shortlisting products/processes and planning out to conduct in-house/in-country research on the chosen hot areas/products/processes to understand the mechanism of action of the new substances (where relevant) in animal model or in cell lines or on in vitro 3D-bioprinted devices, if such new substances are not yet approved but are available from non-IPR protected sources. Sound knowledge in target validation, identifying the druggable biological targets and understanding the biological mechanism of action between the ligands and the receptors, toxicity data generation by using the ligands, proficiency in synthetic biology, knowledge in a wide spectra of chemical synthesis, and identification of the highly active small molecules from the multiple ones represented by a general chemical structure with multiple functional radicals, followed by working on various synthetic methods not described in the IPR are some essential requirements of expertise. On large biological molecules, biosynthetic pathways along with the choice of suitable expression host need to be worked upon and expertise needs to be developed. In complementation, updating and applying the tools of AI also need to be mastered. Conducting research and development work by using any substance, protected or unprotected, is legal. However, use of such substances in trade is guided by the provisions of IPR under the provisions of the trade-related aspects of intellectual property rights (TRIPS) of the World Trade Organization (WTO). Advancing research and developmental work on new substances/processes can speed up the market introduction faster after the expiry of the IPR on the protected substances/process. Taking proactive steps through intensive research is thought to be beneficial. This can be achieved by speeding up the regulatory oversight and also by assisting the front-line drug development researchers to reach the conclusive steps much faster. At the present time, speed at all levels would ease survival in the global context and is the essence of success.

In the context of India, a lot of factors contribute to the slowness in the implementation of regulatory dictum for new biology products and processes and this requires significant updating and catching up. Some of these major contributing factors are: inadequate funds for carrying out R&D in front-line research, especially from the private sector entrepreneurs; slowness in the conception of road map in S&T development in biological research; unavailability of multiple chemical, biochemicals and biological substances required in research and applications; considerable restrictions in import/export/exchange/transfer of living genetic materials; too much weightage to the opinion makers against front-line scientific experimentation; fear of losing the voting support of less educated and biased public, and some more.

Slowness in the implementation of regulatory dictum affects the enthusiasm of the entrepreneurs who have invested in generating new knowledge to derive benefits there from. Delays are detrimental to catching up with progress globally. Therefore, for better progress, regulators need to adapt to fast decisions, lessen their knowledge gap and increase skill capabilities to match their scientific peers from developed countries. Scientific teams consisting of highly talented scientists, technologists, medical professionals, and administrators can guide the regulators continuously. The regulators need to be intermittently trained to remain up to date in latest scientific developments. In line with the above factors, there is a need to periodically review and strengthen the existing regulatory set up in India.

The "Grassroots Innovators of India" are being promoted to a considerable extent presently by creating BIRAC, a feat that DBT accomplished in 2012. Earlier DBT had promoted the transfer of institutionally developed technologies, starting in 1988, which were developed in Indian public funded institutions. Most of these technologies could not sustain the fierce market competition and market needs. A success story was the Mw vaccine, an immunomodulator used in the treatment of leprosy. It was not used extensively for leprosy treatment because better therapeutic substances were available to contain and treat leprosy. The Mw based vaccine was further improved and developed by the industry, where it was found to impart innate immunity to some extent in multiple situations of immune losses, such as in certain types of cancer, tuberculosis, COVID-19, flu, etc. Besides Mw vaccine, there are a few other institutionally developed innovative technologies, especially in the disease diagnosis area, that have been industrially successful. Promotion of BIRAC by DBT, has led to benefit the small Indian incubators and start-ups, and the efforts may promote in future, the availability of cheaper, already-know modern biotech products in the marketplace. For coming out with globally competitive pioneering biotech products, more innovative approaches are required to identify better funding opportunities, improved infrastructure, more international

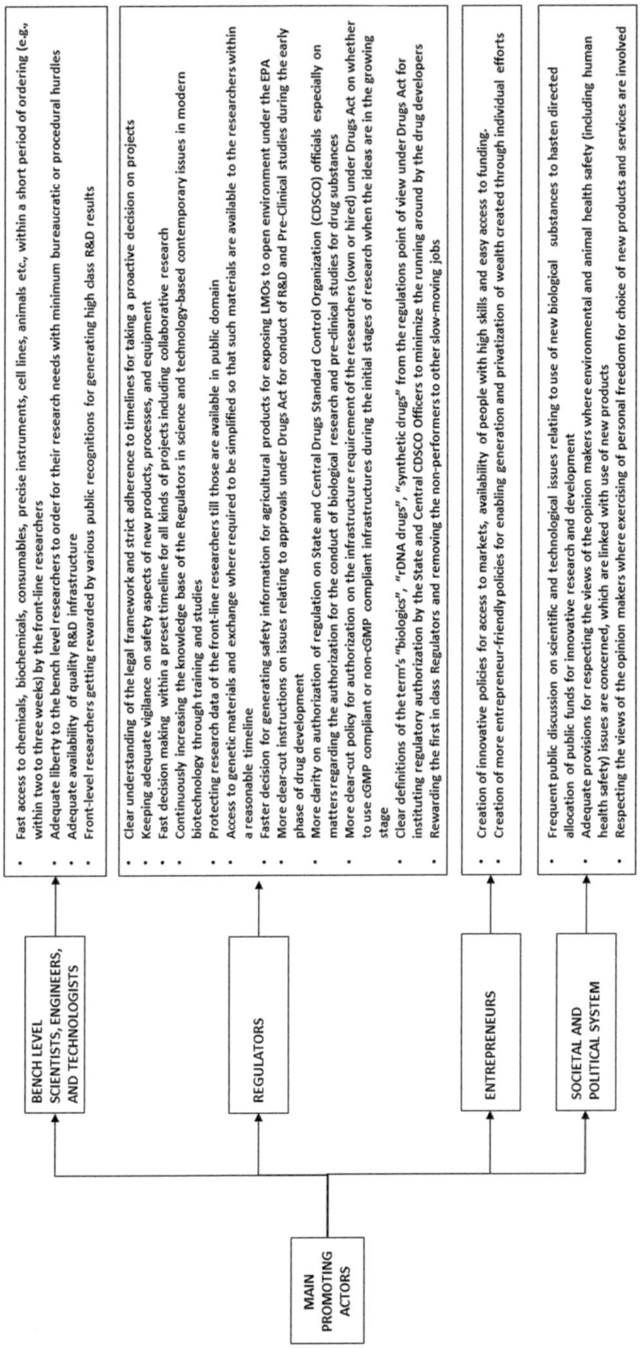

Fig. 13.2 Main factors & linked actors for fast promotion of modern biotech products in the market in India

collaborations, and more flexible public and private participation, with the support of the government.

The main actors and contributors who promote the fast applications of modern biotech-based products and services include the several front-level forerunners such as the bench level scientists, engineers, and technologists involved in R&D work; the regulators monitoring the conduct of the experiments and developmental work requiring vigilance because of linked environmental and human health safety issues; the entrepreneurs involved in generating own personal wealth; and the societal and the political system governing the broad directions of progress in the society through the use of innovations and new discoveries in the multiple areas of applications. For each of these actors, there are presently constraints in making fast progress. In the Indian context, main factors requiring closer attention and intervention for speeding up the invention and use of application-oriented modern biotechnologies have been broadly identified and depicted in Fig. 13.2.

Acknowledgement Dr. Satish Kumar Gupta, PhD former Senior Scientist, NII, New Delhi and presently an ICMR Emeritus Scientist, and Dr. Renu Swarup, PhD, Former Secretary, DBT, Government of India had read the manuscript and had made useful suggestions. Their contributions are gratefully acknowledged.

Authors' Contributions Prasanta Kumar Ghosh is responsible for conceptualization, literature search, methodology, concept visualization and integration, writing, reviewing, and editing of the article.

Annexure-I

(Mandate of the Department of Biotechnology, Government of India in Table 13.1)

Annexure-II

(Chronological events following international pressure on equitable use of global biological diversity and safe use of modern biotech products)

1. The Nairobi conference of 1992.

   A global conference was held in Nairobi, Kenya, in 1992 for the adoption of the "Agreed Text of the Convention on Biological Diversity" and its conclusions were finalized in the text of Nairobi Final Act (CBD 2022). The text was opened for signature on 5 June 1992 at the United Nations Conference on Environment and Development (the Rio "Earth Summit"). India signed (Treaties 2022) the Text of the Convention on Biological Diversity Text on 5 Jun 1992, and this was ratified on 18 Feb 1994. The Convention aims at (a) the conservation of biological diversity (or biodiversity); (b) the sustainable use of its components; (c) and a fair and equitable sharing of benefits arising from genetic resources. Since at that time the rDNA technology was already invented and was being practiced in certain parts of the world, many countries were apprehensive, were afraid and suspicious about issues of safety from the use of such products and the technology in their territories. Their fear was that the products and the technology may be detrimental to the environment and human health. A clause was therefore incorporated (Convention on Biological Diversity 1992) at Article 19(3) of the Convention, which reads as under:

"Article 19(3): The Parties shall consider the need for and modalities of a protocol setting out appropriate procedures, including, in particular, advance informed agreement, in the field of the safe transfer, handling and use of any living modified organism resulting from biotechnology that may have adverse effect on the conservation and sustainable use of biological diversity."

2. The Genesis of GMOs and LMOs.

   In pursuance of Article 19(3) of CBD for safe handling of modern biotech products, the Parties to the CBD recommended the setting up of an Open-ended Ad Hoc Working Group on Biosafety in order to prepare its recommendations on the use and movement of LMOs across countries. The Open-ended Ad Hoc Working Group on Biosafety of the CBD had met six times between July 1996 and February 1999 and came out with a text which was adopted by the Parties (Countries) to the document, on Jan 29, 2000. The adopted document was the Cartagena Protocol on Biosafety (Protocol), which was the document on the genesis for the terms GMOs and LMOs.

3. Cartagena Protocol on Biosafety (Protocol) and Biosafety Clearing House.

   Modern biotechnology, according to the Cartagena Protocol, refers to development and use of organisms modified by genetic modification or genetic engineering methods of direct, non-sexual transfer of genetic material to an organism in order to add or change one or more of its genetic trait/s; the resulting organisms are known in short as GMOs. When such modified organisms are living, these are designated as Living Modified Organisms or LMOs in short. The GMOs as such or the products manufactured by using them refer to as modern biotech products. This definition is also being used in India now when referring to modern biotech products.

   The Protocol prepared by the Open-ended Ad Hoc Working Group on Biosafety was adopted on 29 January 2000 at Cartagena, Columbia, and it entered into force on 11 September 2003. The author was a Member of the Indian Delegation, and he was present at Cartagena on 29 January 2000. He had been nominated to participate in the Open-ended Ad Hoc Working Group on Biosafety as one of the Indian representatives. He attended the first meeting held at Aarhus, Denmark, during 22–26 July 1996. He was nominated as the Indian Representative to participate in all the meetings later, as the subject matter was highly technical, and the author had a background to understand the issues. The Ministry of Environment & Forests nominated the author, and DBT consented to the nomination.

   In framing the Protocol, an Intergovernmental Committee for the Cartagena Protocol on Biosafety (ICCP) was established later, following the initial meetings of the Ad Hoc Working Group, with ten Members chosen from different countries, and the author was selected at a later stage as one of the ten Members of ICCP (CBD 2001, 2002). The author had contributed to the technical part of the Protocol, which was immensely appreciated. India is a signatory to the Protocol (BCH n.d.) and India ratified it on January 23, 2003.

According to Article 11 of the Protocol, Member countries are required to set up a Biotechnology Regulatory Authority. The Indian National Biotechnology Regulatory Authority has not yet been established. India continues to use its Environment (Protection) Act 1986 and Rules 1989 for handling issues relating to the use of rDNA products in the country.

As regards modern biotechnology regulatory authority for the introduction of genetically modified substances, meant for use in research or applications in Indian establishments including in institutions, universities, all legal entities in India including industries, the DBT was using the Environment (Protection) Act 1986 (Act) and Rules 1989 for handling rDNA products in the country (Biosafety 2022; Wikipedia Contributors 2022). These handlings were connected to I(D&R) Act only for the industries, as a prerequisite for establishing an industrial outfit/ infrastructure, and the need to inspect, where necessary, whether the industrial activities were carried out, complying the provisions of safety as enumerated in the Indian Environment (Protection) Act 1986 (Act) and Rules 1989. The Cartagena Protocol on Biosafety, [in short, the Protocol], clearly spent out the need for the setting up of a Biosafety Clearing House [the term used in India for such an establishment was termed as the National Biotechnology Regulatory Authority].

The National Biotechnology Regulatory Authority was to be established by the Indian Government to deal with such substances, following certain clauses of the Protocol. National Biotechnology Regulatory Authority was to work among others, as the "Biosafety Clearing House" of the country for the LMOs. The words "Biosafety Clearing House" is referred to in multiple Articles (clauses) of the Protocol such as in 6, 10, 11, 12, 13, 14, 18, 20, 23, and 24. According to Article 11 of the Protocol, each Party [Country] is to create its Biosafety Clearing House.

Annexure-III

(Some details of the relevant sections of the Environment (Protection) Act 1986 empowering the government to implement it through promulgated Rules-1989)

Relevant Sections of the Act-1986.

Section 6 of the Act is on rules to regulate environmental pollution, and reads as under:

"Rules to regulate environmental pollution. (1) The Central Government may, by notification in the Official Gazette, make rules in respect of all or any of the matters referred to in section 3. (2) In particular, and without prejudice to the generality of the foregoing power, such rules may provide for all or any of the following matters, namely: (a) the standards of quality of air, water or soil for various areas and purposes; (b) the maximum allowable limits of concentration of various environmental pollutants (including noise) for different areas; (c) the procedures and safeguards for the handling of hazardous substances; (d) the prohibition and restrictions on the handling of hazardous substances in different areas; (e) the prohibition and restrictions on the location of industries and the carrying on of processes and operations in different areas; (f) the procedures and safeguards for the prevention of accidents which may cause environmental pollution and for providing for remedial measures for such accidents".

Section 8 of the Act is about persons handling hazardous substances, who are to comply with procedural safeguards.

This section authorizes and dictates that, "No person shall handle or cause to be handled any hazardous substance except in accordance with such procedures and after complying with such safeguards as may be prescribed."

Section 25 of the Act is about the power of the Central Government under the Act to make Rule.

Section 25 of the Act reads as under: "(1) The Central Government may, by notification in the Official Gazette, make rules for carrying out the purposes of this Act. (2) In particular, and without prejudice to the generality of the foregoing power, such rules may provide for all or any of the following matters, namely: (a) the standards in excess of which environmental pollutants shall not be discharged or emitted under section 7; (b) the procedure in accordance with and the safeguards in compliance with which hazardous substances shall be handled or cause to be handled under section 8; (c) the authorities or agencies to which intimation of the fact of occurrence or apprehension of occurrence of the discharge of any environmental pollutant in excess of the prescribed standards shall be given and to whom all assistance shall be bound to be rendered under sub-section (1) of section 9; (d) the manner in which samples of air, water, soil or other substance for the purpose of analysis shall be taken under sub-section (1) of section 11; (e) the form in which notice of intention to have a sample analyzed shall be served under clause (a) of sub-section (3) of section 11; 9 (f) the functions of the environmental laboratories, the procedure for the submission to such laboratories of samples of air, water, soil and other substances for analysis or test; the form of laboratory report; the fees payable for such report and other matters to enable such laboratories to carry out their functions under sub-section (2) of section 12; (g) the qualifications of Government Analyst appointed or recognized for the purpose of analysis of samples of air, water, soil or other substances under section 13; (h) the manner in which notice of the offence and of the intention to make a complaint to the Central Government shall be given under clause (b) of section 19; (i) the authority or officer to whom any reports, returns, statistics accounts and other information shall be furnished under section 20; (j) any other matter which is required to be, or may be, prescribed".

The Act provides powers to the Central Government to take all such measures as it deems necessary or expedient for the purpose of protecting the environment in coordination with the State Governments. It empowers the Central Government to (a) plan and execute a nation-wide program for the prevention, control, and abatement of environmental pollution; (b) lay down standards for the quality of environment in its various aspects; (c) lay down standards for emission or discharge of environmental pollutants from various sources; (d) restrict areas in which any industries, operations or processes or class of industries, operations or processes shall or shall not be carried out subject to certain and specific safeguards; (d) can appoint officers under this Act for various purposes and entrust them with the corresponding powers and functions; (e) has the powers to direct the closure, prohibition or regulation of any industry, operation or process, and even to stop or regulate the supply of electricity or water or any other service to such establishments; (f) restrict individuals or organizations on the discharge or emission or permission to discharge/emit any environmental pollutant in excess of the prescribed standards;(g) ordering for compliance with procedural safeguards, where individuals can be ordered not to handle or causing not to handle any hazardous substance except in accordance with the procedure and safeguards, as prescribed.

In addition, the Act provides authorization to the Central Government powers of entry and inspection, and for this purpose, any person can be empowered by the Central Government to have a right to enter (with the assistance deemed necessary) at any place for the inspection of compliance of any orders, notifications and directions given under the Act; and for the purpose of examining (and where required seizing) any equipment, industrial plant, record, register, document or any other material object may furnish evidence of the commission of an offence punishable under this Act.

As per the Act, the Central Government is entitled to (a) establish environmental laboratories; (b) recognize any laboratory or institute as environmental laboratories to carry out the functions entrusted to such a laboratory; (c) is also entitled to make rules specifying the functions of environmental laboratories; and (d) a Government Analyst can be appointed by the Central Government for the analyzing the samples of air, water, soil or other substance, which may have to be sent to a recognized environmental laboratory for analysis.

Certain kinds of penalties have been included in the Act and non-compliance or contravention to any of the provisions of the Act is considered as an offence. If an offence under this Act is committed by a company, every person directly in charge of the company, at the time of the commitment of offence, is deemed to be guilty unless proven otherwise. If an offence under this Act has been committed by any Department of Government, the Head of the Department (HOD) shall be deemed to be guilty of the offence unless proven otherwise. Any officer, other than HOD, if proven guilty, shall also be liable to be proceeded against and punished accordingly.

As regards cognizance of offences, no court shall take cognizance of any offence under this Act except on a complaint made by (a) the Central Government or any authority on behalf of the former, or (b) a person who has approached the Courts after a 60-day notice has been furnished to the Central Government or the authority on its behalf.

Annexure-IV

(Highlights of the Rules-1989)

The Rules-1989 are described in **20 clauses,** namely (1) Short Title, Extent and Commencement; (2) Application; (3) Definition; (4) Competent Authorities; (5) Classification of Microorganisms or Genetically Engineered Product; (6) Animal Pathogens and Plant Pests; (7) Approval and Provisions, etc.; (8) Production; (9) Deliberate or Unintentional Release; (10) Permission and Approval of Certain Substances; (11) Permission and Approval of Food Stuffs; (12) Guidelines; (13) Grant of Approval; (14) Supervision; (15) Penalties; (16) Responsibility to Notify Interpretations or Accidents; (17) Preparation off-site Emergency Plan by the DLC; (18) Inspection and Information Regarding Finance; (19) Appeal; and (20) Exemption. Each clause depicts specific aspects of the Rules-1989. Essentially, the keepers and the implementers of the Rules-1989, especially the MOEF & CC and the DBT have, the following activities to be watched and regulated continuously:

1. Synopsis of Clause 7: Approval and Prohibitions, etc. on GE organisms/ substances.
   (a) No person shall import, export, transport, manufacture, process, use or sell any genetically engineered organisms/substances or cells except with the approval of the GEAC.
   (b) Use of any genetically engineered organisms or cells for the purpose of research shall only be allowed in laboratories or inside laboratory area notified by MOEF & CC.
   (c) The GEAC shall give directions to the occupier to determine or take measures concerning the discharge of genetically engineered organisms or cells mentioned in the Schedule of the Rules-1989 from the laboratories, hospitals and other areas including prohibition of such discharges and laying down measures to be taken to prevent such discharges.
   (d) Any person operating or using genetically engineered organisms mentioned in the schedule for scale up or pilot operations shall have to obtain license issued by the GEAC for any such activity. The possessor shall have to apply for license in prescribed proforma.
   (e) Certain experiments for the purpose of education within the field of gene technology or microorganism may be carried out outside the laboratories and laboratory areas mentioned in sub-rule (2) and will be looked after by the Institutional Biosafety Committee.
2. Synopsis of Clause 8: Production.

   Production or manufacturing operations in which genetically engineered organisms or cells or microorganisms are generated or used shall not be commenced except with the consent of the GEAC with respect of discharge of genetically engineered organisms or cells into the environment. This shall also apply to production taking place in connection with development, testing and experiments where such production, etc., is not subject to rule 7.
3. Synopsis of Clause 9: Deliberate or unintentional release.

   (a) Deliberate or unintentional release of genetically engineered organisms/ hazardous microorganisms or cells, including deliberate release for the purpose of experiment shall not be allowed.
   (b) The GEAC may in special cases give approval of deliberate release.
4. Synopsis of Clause 10: Permission and approval for certain substances.

   Substances and products, containing GMOs shall not be produced, sold, imported or used except with the approval of GEAC
5. Synopsis of Clause 11: GM Food Stuff.

   Food stuffs, ingredients in food stuffs and additives including processing and containing or consisting of genetically engineered organisms or cells, shall not be produced, sold, imported or used except with the approval of the GEAC.
6. Synopsis of Clause 12: Guidelines.
   (a) Any person who applies for approval under rules 8–11 shall, as determined by GEAC submit information and make examinations or cause examinations to be made to eradicate the case, including examinations according to specific directions and at specific laboratories. He/ she shall also make available an on-site emergency plan to GEAC before obtaining the approval. If the authority makes examination itself, it may order the applicant to delay the expenses incurred by it in so doing.
   (b) Any person to whom an approval has been granted under rules 8–11 shall notify the GEAC of any change in or addition to the information already submitted.
7. Synopsis of Clause 13. Grant of approval.
   (a) All approvals of the GEAC are conditional, with terms and conditions shall be stipulated, and information on the implementation of stipulated conditions are to be furnished by the applicants to the State Biotechnology Coordination Committee or to the District Level Committee.
   (b) All approvals of the GEAC shall be for a specific period not exceeding 4 years at the first instance renewable for 2 years at a time. The GEAC shall have powers to revoke such approval in situations such as new information about the harmful effects on human and animal health, damage to the environment, or non-compliance of any condition stipulated by GEAC.

The Rules-1989 have subsequently been revised in several of its elements a few times, and a few notifications were therefore issued by the government.

First, the effective date for the implementation (Geacindia 1993) of the Rules-1989 was notified on September 1993 as 01.10.1993. Thereafter, the Clause 4 of Rules-1989 concerning the officials of the Genetic Engineering Approval Committee were amended (Geacindia 2005) on July 14, 2005.

In the original Rules-1989, genetically modified substances including the genetically modified seeds and products thereof were regulated through the provisions of the Rules-1989. Later, it was decided by the government in 2007 that only the living genetically modified substances need be controlled under the Act and the Rules-1989. Accordingly, processed food stuffs and additives including processing aids derived from genetically modified substances where the end products were not living any more (because of processing), were exempted (IBKP 2007) from the applicability of Clause 11 of Rules-1989.

Annexure-V

(The names of the six committees and their operational procedures in brief)

The six committees through which the Rules-1989 are implemented are as under:

1. Genetic Engineering Appraisal Committee (GEAC) [GEAC was named as Genetic Engineering Approval Committee].
2. Review Committee on Genetic Manipulation (RCGM).
3. Institutional Biosafety Committee (IBSC).
4. State Biotechnology Coordination committee (SBCC).
5. District Level Committee (DLC).
6. Recombinant DNA Advisory Committee (RDAC).

The GEAC and the RCGM are the key drivers, responsible for regulating the functions of the Act and the Rules-1989. The RCGM through the constitution of the IBSC at every establishment that conducts research utilizing genetically modified organisms, keeps vigilance of the safety aspects emanating from the use of GMO to the environment and to human and animal health. Through the SBCC and DLC the progress of monitoring of experiments utilizing GMOs are obtained; these committees also keep a watch on hazards, if any from the use of GMOs in territory, and ensures safe use of GMOs.

The RCGM functioning at DBT is mandated to monitor the safety related aspect in respect of on-going research projects and activities involving genetically engineered organisms/hazardous microorganisms. RCGM shall bring out Manuals of guidelines specifying procedure for regulatory process with respect to activities involving genetically engineered organisms in research use and applications including industry with a view to ensure environmental safety. All ongoing projects involving high risk category and controlled field experiments shall be reviewed to ensure that adequate precautions and containment conditions are followed as per the guidelines. The RCGM shall lay down procedures restricting or prohibiting production, sale, importation and use of such genetically engineered organisms or cells as mentioned in the Schedule of the Act and Rules-1989. The RCGM operates with the scientific advice from experts, who are selected from various organizations besides experts from the DBT. RCGM is the main driver in the development and application of GMOs and products thereof in the country, as most such products are research- based, and invented elsewhere where the IPR on such products has expired or does not exist or are developed in India based on newer innovative processes, and where the questions of environmental safety as well as safety to human and animal health under the Act and Rules-1989 have not yet been resolved. The RCGM is required to develop scientific data for presentation to the GEAC, seeking clearance of such products and processes under the Act and Rules-1989 from the GEAC. RCGM does it through the IBSC, the SBCC and the DLC, besides appointing subgroups and seeking their advice. For recombinant DNA based drugs and formulations thereof, the products developed in India cannot be tested under the Indian Drugs Act until and unless such products are cleared by the GEAC, based on the Reports and evaluations, generated and submitted by the RCGM to the GEAC.

Annexure-VI

(A note on the historical background of the invention of rDNA technology and earlier period of industrial pollutions)

Stanley N. Cohen and Herbert W. Boyer are regarded as the first group who invented in 1973, methods to insert recombined genes into bacterial cells, and produce recombinant life forms, which were non-existent in nature. However, this invention was possible because Werner Arber discovered restriction enzymes (type-I) in 1968, followed by the discovery and purification of type-II restriction enzymes by Hamilton O. Smith in 1970. Type II restriction enzymes were found to be essential to genetic engineering, because of their ability to cleave specific sites within the DNA. Daniel Nathans used type II restriction enzymes, and his work during 1970–1971 helped advancement of the technique of DNA recombination. Almost during the same period, Paul Berg developed methods for splitting DNA molecules at selected sites. He further developed methods for attaching such segments of the molecule to the DNA of a virus or plasmids, which could then enter bacterial or animal cells. All these scientific works created the foundations, based on which the recombinant DNA technology was invented (Britannica 2018; Pingoud et al. 2014).

Concerns developed in the minds of sections of the global society on issues relating to environmental safety from the release of GM organisms in the environment as also safety to human and animal health resulting from use of such substances.

In the meantime, during the 1960s, with human activities and industrialization the world over, it was becoming evident that these had caused ecological disturbance which brought about changes in the spatial patterning in the ecosystems of the habitats of living organisms including animals, plants, and smaller life forms including different kinds of microorganism, and environmental pollutions were rising, causing depletion of natural resources. Consequently, public the world over started raising concerns, insisting the governments to bring in laws and regulations to prevent people from destroying/degrading the environment. In the USA, in particular, the enactment of the Clean Water Act of 1960 and the Water Quality Act of 1965; the Clean Air Acts of 1963 and 1967, etc. are the initial reactions from humankind and governments, showing concerns on polluting natural resources from human actions. These measures followed the creation of federal legislation in 1970, when USA enacted the first National Environmental Policy Act, which required Environmental Impact Statement of all major federal actions that affected the quality of human environment. The USA set up its Environmental Protection Agency in

1970 to enforce the federal regulations (Environmental Movement 2022). The actions of this agency followed the enactment of Clean Air Act of 1970; the Pesticides Control Act of 1972; the Ocean Dumping Act of 1972; and the Water Pollution Control Act Amendment of 1972, etc. As the people in USA started getting concerned from human activities that created or had potential for creating substantial environmental disturbances, such concerns quickly spread to other countries.

Therefore, even before the Asilomar Conference was held to discuss on the safety issues emanating from the newly invented related recombinant DNA technology, concerns were raised and issues on ecological damage created by human activities were discussed globally in the Stockholm Conference of the United Nations Conference on the Human Environment, held in 1972, as already mentioned.

Annexure-VII

(A brief on the induction and progress in the use of rDNA products in India)

The RCGM authorized a recombinant species of methylotrophic yeast *Pichia pastoris* for producing recombinant DNA products such as hepatitis B vaccine, insulin, etc. as early as in 1993. RCGM was in fact authorizing research on rDNA products, using multiple kinds of yeast strain such as *Saccharomyces cerevisiae*, *Hansenula polymorpha, Pichia pastoris,* and *Schizosaccharomyces pombe*.

The recombinant yeast species *Pichia pastoris,* using methanol as the carbon source for maximizing the expression of recombinant DNA products was yet not considered a safe host for producing human medicines elsewhere in the world in 1992–1993. The US FDA granted (Reyes et al. 2021) as "a no questions" status to *P. pastoris*-derived enzyme preparation only in 2006, and an approval to a recombinant pharmaceutical product Kalbitor® produced in recombinant *P. pastoris* thereafter in 2010.

In 1992, Dr. K. I. Varaprasad Reddy, an Electrical Engineer along with Dr. Gita Sharma, a Professor of Microbiology, Osmania University, Hyderabad visited the author in DBT, when Dr. Reddy was exploring new ideas in biotechnology for making investments. They were advised on multiple biotech products including setting up of a hepatitis B vaccine manufacturing unit (Ghosh 1992). Dr. Reddy decided to invest on developing a technology and a manufacturing unit to produce recombinant DNA based hepatitis B vaccine, initially in collaboration with Dr. Gita Sharma, and later involving many other agencies and people, and promoted Shantha Biotechniques, Hyderabad in 1993. The expression host, a recombinant *P. pastoris* yeast was obtained by Shantha from Phillips Petroleum Company in Bartlesville and Research Corporation Technologies (RCT), USA in 1993. Pichia system produces recombinant proteins and grows efficiently in media containing methanol. Phillips and RCT made available the *P. Pastoris* expression system to Shantha for research use initially and subsequently for commercial use. RCGM authorized Shantha to generate safety and efficacy data on recombinant *P. pastoris*-derived recombinant proteins under the Indian EPA in 1993.

The decision of the RCGM to allow generating safety and efficacy data under the Indian EPA, on recombinant *P. pastoris*, and for the rDNA drugs produced in recombinant *P. pastoris* as early as in 1993, enabled India to use such substances for evaluation in India on human subjects under the Indian Drugs Act. This enabled gaining substantial time advantage in the early development and marketing of rDNA based drugs in India.

There are several such kinds of other decisions on the requirement of safety evaluation of rDNA drugs on primates, where RCGM advised avoiding generation of safety data as RCGM felt that near-similar knowledge could be gathered from information generated in the right kind of small animals. The HBV virus challenge test for testing the efficacy of rDNA vaccine was carried out in mouse model with certain modifications. These advises, based on sound scientific logic ushered the Indian modern biotech industry to obtain time advantage in introducing rDNA drugs in Indian market.

Furthermore, earlier several irregularities were found in the generation of human safety and efficacy data, which were developed by companies under the Indian Drugs Act (Naik n.d.), which could be minimized while in the process of introducing rDNA products in Indian market, because of high levels of vigilance on safety, instituted by the RCGM on such products.

In 1991, Indian government got initial acquaintance with recombinant DNA based Bt-cotton, which is a genetically modified pest resistant plant cotton variety. Bt-cotton plant produces an insecticide protein (a toxin to boll worms) in its tissues while growing. The genes of the protein are derived from bacterium *Bacillus thuringiensis* and are inserted into the genome of the Bt-cotton plant variety. Monsanto Inc., USA who developed the technology, was in negotiations with India for its sale, in 1991. Bt-cotton was not then approved anywhere in the world, not even in USA; field trials were conducted in USA for the first time in1993 and was thereafter approved for commercial use in that country in 1995.

Indian Government from the DBT could proficiently conduct the negotiations for the procurement of the Bt-cotton technology in 1991, because the Indian Act and Rules-1989 were in place. Though the negotiations did not fructify into an agreement, Indian officials learnt a lot, and used such knowledge through RCGM to frame guidelines for confined field trials, food safety assessment and environmental risk assessment of genetically modified agricultural substance. Later, when Monsanto came up in collaboration with Maharashtra Hybrid Seeds Company Ltd., (MAHYCO), Mumbai in 1994 seeking permission of the Indian Government to use the Bt-cotton varieties in Indian agriculture, it was easier to deal with the project by the Indian Government. DBT authorized the initial imports of Bt-cotton seeds to MAHYCO in September 1994, and the RGGM along with Indian Council of Agricultural Research (ICAR) through the authorization of GEAC, effected the generation of safety and efficacy data on Bt-cotton in India.

For assessing the of GM agricultural products like Bt-cotton technology, the RCGM had designed a comparatively safe Green House with necessary barriers and a suitable environment for growth of transgenic plants. This enabled generating large initial scientific data. Based on such data, decisions for limited open-field trials could be taken. No other country had used such low-cost Green House for generating initial safety data on transgenic plants. Use of such facilities enabled considerable reduction in costs, and time saving.

After being satisfied with the data generated in the RCGM-designed Green House, limited open-field experimentation were being authorized for the conduct of experiments using transgenic plants to generate information on the extent of expression of the transgenes (e.g., Bt-cotton plants for Bt-proteins) in the plant parts, GM-pollen flow and trans-pollination with near-relative plants, creation of weediness or invasive properties in other plants resulting from pollination with GM-pollens, development of soil toxicity, if any, by the cultivation of the GM plants, yield difference/change, intensity of use of water, fertilizers, pesticides, etc., and to compare the behavior and properties with the non-transgenic counterpart plants. Well-designed, scientifically valid experiments were conducted in the field and the trials were continuously monitored by a team of Experts authorized by the RCGM. The representatives from the ICAR were always inducted in such Expert Teams.

In the original Rules-1989, genetically modified substances including the genetically modified seeds and products there of were regulated through the provisions of the Rules-1989 under the Act. The Rules-1989 require that all GM organisms and products thereof are introduced only after these are found to be environmentally safe as well as safe for use as food for human and animals. GM food can be non-propagating substances of GM origin which are not "living" or GM organisms themselves such as GM edible yeast or lactobacillus or a host of agricultural produces such GM corn, potato, tomato, soybean, wheat, rice, mustard/rape seed, etc. Bt-cotton was the first GM crop that was approved for release into the open environment in India in 2002 after exhaustive safety assessment, including food safety assessment of the Bt-cotton seeds on animal feeding trials (Ghosh and Ghosh 2007). Before approval of Bt-cotton technology by the GEAC, extensive environmental safety data as well as food safety data were generated. Multi-centric field trials were conducted under the authorization of RCGM. The soil of the experimental lands was evaluated post harvesting of the Bt-crop to assess if any toxicity had developed in the soil. For food safety data, feeding experiments were carried out, starting from small lab animals to goats, cows, buffaloes, fish, and birds. The viscera of sacrificed goats including the lungs, the heart, and the organs of the digestive, excretory, and reproductive systems were examined after feeding 20 animals with meal rich in Bt-cotton seeds and compared with controls. Only, and after being satisfied about the safety of the GM product, commercial release was authorized in March 2002. This approval enabled India to be the world's No. 1 producer of cotton and second largest exporter, as cotton production jumped several folds.

Indian authorization for the commercial cultivation of Bt-cotton was accorded, despite a Writ Petition filed by activists in 1999 [Writ Petition No 71 of 1999 in the Supreme Court], which in substance was a prayer by the Petitioner to disallow the introduction of Bt-cotton technology in Indian agriculture and to stop the implementation of the provisions of the biosafety laws (Lifegate n.d.). The case was not concluded yet in 2002, when the approval was accorded. Thereafter, from March 2002 to October 2022, use of GM technology in Indian agriculture had gone through multiple stages of assessment, protest from non-government organization (NGO) activists, and as a consequence the technology took a back seat in the economic development in the country.

In the meantime, several Bt- genes have been utilized in India for producing transgenic Bt-cotton hybrid seeds such as Mahyco-Monsanto (cry1Ac and cry1Ac plus cry1Ac + 2Ab, Monsanto, USA technology); Nath seeds(cry1Ab plus Ac fusion, technology from China); JK seeds(cry1Ac modified, technology developed at IIT Kharagpur); Syngenta (vip3A + cry1 Ab, technology from Syngenta Switzerland); Dow Agri. Science(cry1 Ac + cry 1F,Dow Agri., USA); Metahelix (cry1 C, technology developed at Bangalore); ICAR(cry1 Aa3,cry1F,cry1 Ia5,cry1 Ab); and NBRI(cry1 Ec) (Agritech n.d.). Among the marketed transgenic cotton seeds, the sale of maximum quantities was from Mahyco-Monsanto supplies. Introduction of Bt-cotton seeds caused a rapid increase in the production of cotton in India.

Following the use of the first Bt-cotton technology, and over the years realizing the potential of GMOs in agriculture, the RCGM of DBT had allowed experimentation for the generation of environmental safety data in labs or in limited field trials, on nine GM crops, namely different kinds of other Bt-cotton as mentioned above, and varieties of transgenic potato, tomato, rice and mustard (IATP 2022a). There was no commercial release of any transgenic plants other than the Bt-cotton till October 2022.

In October 2022, 20 years after Bt-cotton approval, Government of India approved the commercial cultivation of genetically modified (GM) mustard seeds (Reuters n.d.). The transgenic mustard (Dhara Mustard Hybrid-11 or DMH-11) is a genetically modified hybrid variety of the mustard species *Brassica juncea*. The team under the leadership of Professor Deepak Pental from the Centre for Genetic Manipulation of Crop Plants, University of Delhi developed it. DHM-11 involves the use of the *bar, barnase,* and *barstar* chimeric gene system. DMH-11 is a hybrid mustard seed which is produced by crossing two parental lines Varuna bn3.6 and EH-2 modbs 2.99, which contain the *bar, barnase,* and *barstar* genes. DMH-11 seeds are fully fertile.

In *bar barnase barstar* system, rapeseed plants are generated in two distinct types. In one type, rapeseed plants containing a chimeric dominant gene *bar barnase* driven by a tapetum-specific promoter is created. The promoter causes the expression of the *barnase* gene specifically to anther tapetal cells, causing selective destruction of the tapetal cell layer which surrounds the pollen sac. The *bar* gene is driven by the CaMV 35S promoter and is expressed from the seedling stage itself of the plant. *Bar* genes express phosphinothricin acetyltransferase, an enzyme which confers resistance to glufosinate by transforming the herbicide into N-acetyl-l-glufosinate (NAC). NAC is non-toxic to plants. Glufosinate is available as its ammonium salt and is sold as Basta (trade name) herbicide the world over. Basta is spread at seedling stage of the crop to weed out susceptible plants from the field for ensuring the production of male sterile plants only. In the other type, a chimeric gene encoding the *barnase*-specific RNase inhibitor called *barstar* gene under the control of a strong promoter is generated. When the two types are crossed, the F1 progeny expresses both the genes in the anther, and therefore the fertility is restored, which is because of the formation of tapetal cell specific barnase/barstar protein complexes; these protein complexes inactivate the *barnase* expressed protein. Initially the male fertile lines are generated from the above two lines, when a homozygous *barstar-bar/barstar-bar* restorer line is formed. This line is crossed with the male sterile plants containing the chimeric gene *barnase-bar* genes. The F1 lines here are created from the pollens from male fertile plants to male sterile plants, pollinating the male sterile plants. Care must be taken not to allow the seeds from the restorer line to get mixed with the F-1 hybrids, and therefore the seeds from the restorer line are to be harvested first and used separately. In this high yielding F1 hybrids are generated (Chand et al. 2018). In India now, with the approval of DMH-11 commercial cultivation, the production of mustard oil in India is anticipated to increase sizably over a period from the cultivation and sale of hybrid F1 brassica seeds.

Coming back to the Rules-1989, the operating contents were later amended, and in the amendment, it was decided by the government in 2007 that only the *living* genetically modified substances need be controlled under the Act and the Rules-1989. Consequently, processed food stuffs and additives including processing aids derived from genetically modified substances where the end products were not *"living"* any more (especially because of life-inactivating processing operations), were exempted (IBKP 2007) from the applicability of Clause 11 of Rules-1989.

Even though India has been enormously benefitted (Jonas and Matin 2012) and have presently become the largest producer of cotton in the world (PIB 2022), there have been many critics (IATP 2022b; Downtoearth 2022) who expressed adverse views on the technology. Opinion makers exist in every society, and their views become stronger in democratic settings, where everyone has the right to speak. It is such a group that influenced political opinion to move in the negative direction, trying to prevent the use of other newer GM technologies in Indian agriculture, after the Bt-cotton technology was already inducted in the country.

The anti-GM crops activists filed a writ petition in the Supreme Court of India in 2005, seeking a moratorium on the release of all kinds of GMOs in Indian agricultural environment, pending evolution and application of comprehensive, transparent and rigorous biosafety test protocols. The Supreme Court consequently set up a Technical Expert Committee (TEC) of five scientists, namely (1) Dr. Imran Siddiqui, plant development biology scientist & group leader at the Hyderabad-based Center for Cellular & Molecular Biology; (2) Dr. P S Ramakrishnan, emeritus professor of environmental sciences and biodiversity from Delhi's Jawaharlal Nehru University; (3) Dr. P S Chauhan, genetics toxicology and food safety expert; (4) Dr. P C Kesavan, former BARC scientist noted for his work on genetic toxicology and radiation biology who is currently distinguished fellow; M S Swaminathan Research Foundation, Chennai; and (5) Dr. B Sivakumar, former director of National Institute of Nutrition, Hyderabad. Later, Dr. R. S. Paroda, former Director General of Indian Council of Agricultural Research (ICAR) was also inducted in the TEC. The TEC was to review GMO related concerns and issues. The first five Members of TEC submitted their Report in July 2013, which was not signed by Dr. R. S. Paroda, who submitted a separate Report. Later, on direction of the Supreme Court, the Report prepared by Dr. R. S. Paroda, was made public. The TEC had in its recommendations (Business 2022) suggested a blanket ban of 10 years on field trials of transgenic food crops as also on crops where India is considered as the a center of origin or diversity of a species. Dr. Paroda did not agree to a blanket ban on field trials. Interestingly, neither the Supreme Court nor the Supreme Court Panel released issued any verdict on the TEC Reports; the Supreme Court Panel however suggested a moratorium on GM trials. In the meantime, in 2012, the Parliamentary Standing Committee on Agriculture in its 37th Report suggested and asked for an end to all GM field trials in India (PARLIAMENT OF INDIA RAJYA SABHA 2022). Following the Supreme Court Panel suggestion, a moratorium on GM trials was announced in 2013 by the then Minister of MOEF & CC, (Ms) Jayanthi Natarajan. [from July 12, 2011 to December 21, 2013], though then Prime Minister Manmohan Singh had supported GM crops.

One main reason for the slowing in the progress, for the induction of GM technology in Indian commercial agriculture after 2006 seems to the author to be a landmark decision of the Government in 2010, by the Minister of MOEF & CC, Jairam Ramesh [May 28, 2009, to July 12, 2011]. The Genetic Engineering Approval Committee of the MOEF & CC (formerly MOE & F) was renamed as the Genetic Engineering Appraisal Committee (Geacindia 2010). This action appears to have resulted in the stripping off its wider powers to take independent decision on matters relating to the use of GM technologies in commercial agriculture based on sound scientific principles and scientifically generated experimental data. The powers for approval shifted from science-based decision making on a precautionary principal process to the decisions of the ruling political system. This situation is thought to be the beginning of the slowing down on the use of GM crops in Indian agriculture. After the change in the name, it implied that after scientific appraisal of proposals, the GEAC would have to obtain clearance from the ruling political system, and only after the consent and approval of the latter, the GEAC would be able to issue authorization for further actions on GM projects.

This is not the case when GMO approval is authorized in countries like USA. Regulation of GM crops in the United States is divided among three regulatory agencies (Biosecurity 2022) namely the Environmental Protection Agency (EPA), the Food and Drug Administration (FDA), and the U.S. Department of Agriculture (USDA), based on pre-existing laws, rules and procedures. The EPA ensures by scientific studies that the GMO does not do harm to the environment; if the plant produces any toxin [such as the Bt-toxins], it should neither be toxic nor be allergenic to animals and human, determined on sound scientific evaluation. The U.S. Food and Drug Administration (FDA) has been the primary regulatory agency (FAO 2022), which is responsible for ensuring the safety of GMO and non-GMO food for animals and human subjects. Based on current scientific knowledge, such foods are evaluated for safety. GM foods include plant-based as well as animal -based GMOs. US FDA examines GM foods to ensure that such foods are safe to animals, and those consumed by human do not contain any harmful substances. The Animal and Plant Health Inspection Service (APHIS) division of USDA sets regulations on the use of

GMOs in USA and ensures that such plants are not harmful (FDA 2022) to other plants, as also to non-target life forms. The three agencies work using pre-existing laws, rules and procedures. There is no political interference while applying the laws and rules; the authorization for rejection or for use of GMOs [and products thereof] is based on sound scientific information and documentation. Political interference may sprout only when the existing laws are found inadequate to address the public concerns of safety or other societal concerns.

The Indian regulatory system was almost brought to a deadlock after 2010 after India placed a moratorium on GM eggplant, although field trials of other GM crops were not formally halted. One major development had taken place thereafter when experimental data including limited field trials were completed on Transgenic Mustard Hybrid DMH-11, developed by the Delhi University. Though the GEAC had considered and cleared the case for environmental release of the product in its 133rd meeting held on May 11, 2017, the recommendations were placed before another Committee meeting held in May 2018 of the Union Ministry of Environment and Forest, Chaired by the Minister, and the Committee did not approve the commercial release of GM Mustard (Newindianexpress n.d.). GM Mustard was finally approved by the government in October 2022, as already stated (Reuters n.d.).

There had been considerable initial indications many years after March 2002, and especially after 2014 that the use of GM crops in Indian agriculture would soon be accelerated (CNBC 2015; Timesofindia n.d.). However, no new GM technology-based seeds had yet been approved in Indian commercial agriculture except those few based on transgenic Bt- cotton. Indian R&D consequently has been severely jeopardized in skill development in this critical area. In such situations, the young scientists did not get encouraged for long, to explore their career in research on GM technologies.

As regards GM drugs and pharmaceuticals, the products are non-living and therefore these do not attract environment protection laws. The GMOs used for their development or manufacture however require compliance with the Act and the current Rules-1989. More than thirty rDNA based products (Ghosh 2017) are presently being used in India which have all been approved for use under the Drugs Act. Several of these are produced in India from the recombinant DNA based microorganisms and cell lines. The recombinant microbes and cell lines have undergone evaluation of environmental safety assessment through RCGM and have obtained clearance from the GEAC. The first rDNA based drugs approved for use in India include recombinant HBV vaccine in 1997, followed by interferon alpha 2b, both introduced by Shantha Biotechniques, using recombinant *P. pastoris*; and thereafter several others were, introduced (Ghosh 2017), using many other recombinant microorganisms and cell lines.

Annexure-VIII

(Labeling of food products in India and the consequential issues)

1. Para 4 of Clause 2 of the Indian Rules-1989 read as under:
   "(4) These rules shall be applicable in the following specific cases;
   (a) sale, offers for sale, storage for the purpose of sale, offers and any kind of handling over with or without a consideration.
   (b) exportation and importation of genetically engineered cells or organisms.
   (c) production, manufacturing, processing, storage, import, drawing off, packaging and repacking of the Genetically Engineered Products.
   (d) Production, manufacture, etc. of drugs and pharmaceuticals and food stuffs distilleries and tanneries, etc. which make use of microorganisms genetically engineered microorganisms one way or the other."
2. The issue of labeling of food products is a controversial burning global issue. One section of the opinion makers is determined to ensure the labeling of such products. Consequently, there are presently 64 countries including India, which have developed rules and procedures for the packaged foods containing genetically modified, requiring labeling (Justlabelit 2022).
3. USA passed a law in 2016 to establish a national benchmark for labeling of GM foods. The law was effective from 01.01.2022. This law, called as the US National Bioengineered Food Disclosure Standard (AMS n.d.), requires mandatory disclosure on foods containing GMOs. GMO-derived foods are called as "bioengineered" foods in the USA. With the imposition of labeling of GM substance-derived foods, USA became the 65th nation in the world to adopt GM substance labeling country. Such an imposition is a costly affair. The compliance in the USA is estimated to cost (Allianceforscience 2022) the food industry close to US $569 million to $3.6 billion during the first year, and about US$ 68 million to $391 million annually, which eventually may be recovered from the consumers.
4. It is mentioned in this context that manufacture of packaged foods produced from genetically modified ingredients would require handling of such substances in bulk. Handling includes procurement, storage, processing, use as well as the disposal of wastes. If the finished packaged foods are to be labeled because these would contain GMOs or ingredients produced from GMOs, then precise segregation procedures are to be adopted, so that the use or consumption of the GMOs or products manufactured there from is quantifiable at each stage of use/consumption. Since in India, the labeling laws for the packaged foods containing GMOs are not implemented, such processes of handling are not yet in place. Implementation of such processes would increase costs of both bulk as well as consumer-friendly packaged foods considerably. The US estimates depicted above are indicators.
5. Besides packaged foods produced from genetically modified ingredients/organisms, approved agricultural GMOs grown in commercial agriculture in India and used or transported across the country or exported outside India raise multiple questions of handling at each stage, if such products are to be labeled. LMOs not approved in India shall not be permitted to enter the country in accordance with the existing environmental laws and regulations. However, when such LMOs shall be approved for cultivation in commercial agriculture in India, these would be required to be labeled after being packed and made ready for sale, if laws requiring labeling of such products, are in place in the country. Since there are

strong indications that packaged foods produced from genetically modified ingredients/organisms are to be labeled, the agricultural GMOs produced in bulk would also have to be labeled. Intent for labeling implies that the identity of such substances would have to be preserved. Preservation of identity in agricultural crops require implementation of a system of management and trade practices that allows identification of the source and the nature of materials at each stage as the material moves through a series of supply chain, starting from purchase of certified genetically modified seeds through cultivation, harvesting, transfer, processing and packaging for consumption. India being a vast country with cultivators of different economic and social status, it is too ambitious to think that a precise management system can be put in place to implement segregation and labeling of bulk GMOs / LMOs at all levels of cultivation, harvesting and sale.

6. In the meantime, the world over where a strong group of opinion makers exist, they are becoming stronger day by day, who exert pressure on the governments to implement labeling requirements for all kinds of GMOs used as food, feed and fodder. The minimum they want is that the consumers should have an option to choose, and if GMOs used as food are labeled the consumers can exercise their choice. This situation has become unilateral in favor of labeling GMOs used as food as multiple animal feeding studies, published from different parts of the world have shown adverse impacts on animals fed with GMO foods, although many other reports have found such foods as safe. In a critical review of the multiple feeding studies conducted by various groups all over the world on animals and human subjects, using genetically modified foods, published between the periods from 01.01.1983 to 11.06.2020, investigators from China had concluded that a good number of such studies had registered adverse events although such foods had already been authorized for use as safe by the regulatory authorities. Because of such results, the authors suggested that GM foods should be labeled (Shen et al. 2022) so that the consumers can make a choice.
7. Genetically Modified crops (GM crops) are developed for increasing the productivity of crops in agriculture and to reduce the costs of such crops, when compared with the corresponding non-GM crops. For some GM crops, the properties of their metabolic products, in terms of unique quality traits components, are improved. Worldwide, presently 26 numbers of GM crops incorporating various traits have been approved for commercial cultivation. In India, only one GM crop namely Bt-cotton was approved in 2002, and after over 20 years another one namely GM Mustard was approved in October 2022. GM Mustard is not yet in the market. It is anticipated that the GM labeling issue may wake up when GM mustard and its products are in the market.
8. Research and developmental work are being carried out on 21 number of GM-food crops in India (Anon 2015) as of the 2015 information. Over a period, many of such projects will mature and would be required to be examined for approval by the regulatory authorities. Commercial cultivation and sale of such GM plants in saleable packs or in bulk would have to be done after these are being labeled, if the labeling laws are in place, which is a stupendous task, and difficult to implement. The increased costs for doing so may make the finished GM food crops more expensive than the non-GM counterparts.

9. Agriculture is practiced in India by marginal, small, medium and large farmers. Farming by the marginal, small, medium farmers are driven more from the subsistence point of view than from industrial-temper point of view. It is recognized that implementation of labeling of GM foods must precede through the implementation of identity preservation procedures, and implementation of such procedures shall require developing standard operating practices (SOPs) and incurring of additional costs, besides education and training of the farmers. Also, people would have to be deployed for ensuring implementation, which also will incur additional costs. In one study conducted by the author, it was concluded that the labeling policy and the identity preservation procedures, if implemented would need to be framed by creating a rational balance (FEASIBILITY STUDY—GEAC n.d.) between and among the interests of the small and medium farmers, and the vocal public seeking compulsory labeling of LMOs, which is indeed a near-impossible task in Indian conditions. A way forward is to proceed with such safe products where there is no need to institute labeling of GMOs, and where safety is established on sound scientific basis.

References

Agritech (n.d.) GM Crops:: Bt Cotton—TNAU Agritech Portal:: Bio Technology. https://agritech.tnau.ac.in › bio-tech › biotech_btcotton_env

Allianceforscience (2022) Study suggests new US GMO labeling law likely to have little. https://allianceforscience.cornell.edu › Posts -https://allianceforscience.cornell.edu/blog/2022/01/study-suggests-new-us-gmo-labeling-law-likely-to-have-little-effect-on-consumers/

AMS (n.d.) BE Disclosure|Agricultural Marketing Service—USDA. https://www.ams.usda.gov › rules-regulations

Anon (2015) Document on frequently asked questions about genetically engineered (GE) plants—phase II capacity building projects on biosafety published by ministry of environment, forest and climate change in association with biotech consortium India Limited. p 15. Accessed 29 Jun 2022

BCH (n.d.) The cartagena protocol on biosafety. https://bch.cbd.int › protocol

Biosafety (2022) Rules-1989.pdf—ICAR biosafety portal. https://biosafety.icar.gov.in › uploads › 2015/11PDF. https://biosafety.icar.gov.in/wp-content/uploads/2015/11/Rules-1989.pdf. Accessed 25 May 2022

Biosecurity (2022) U.S. regulation of genetically modified crops. https://biosecurity.fas.org › 2.-agricultural-biotechnology, https://biosecurity.fas.org/education/dualuse-agriculture/2.-agricultural-biotechnology/us-regulation-of-genetically-engineered-crops.html. Accessed 24 Jun 2022

Britannica (2018) Britannica, The Editors of Encyclopaedia. "When was recombinant DNA technology invented?". Encyclopedia Britannica, 1 Nov. 2018. https://www.britannica.com/question/When-was-recombinant-DNA-technology-invented. Accessed 12 Jun 2022

Business (2022) Scientists oppose TECs recommendations on biotech. https://www.business-standard.com › ... › News. Accessed 24 Jun 2022

CBD (2001) INTERGOVERNMENTAL COMMITTEE FOR THE CARTAGENA PROTOCOL ON BIOSAFETY- iccp-02-15-en.pdf—Convention on Biological Diversity [October 05, 2001]. https://www.cbd.int/doc/meetings/bs/iccp-02/official/iccp-02-15-en.pdf

CBD (2002) INTERGOVERNMENTAL COMMITTEE FOR THE CARTAGENA PROTOCOL ON BIOSAFETY-REPORT -[The Hague, 22–26 April 2002]. https://www.cbd.int/doc/meetings/bs/iccp-03/official/iccp-03-10-en.pdf

CBD (2022) Nairobi Final Act of the Conference for the Adoption of the https://www.cbd.int › handbook › cbd-hb-09-enPDF. Accessed 27 May 2022

Chand S, Patidar OP, Meena VK, Shiv A (2018) Barnase -barstar system: an incredible technique to produce hybrid seeds in self-pollinated crops. Int J Crop Sci 8(2):109–113; barnase and barstar system. http://mmcmodinagar.ac.in › econtent › biotech PDF

Chimata MK, Bharti G (2019) Regulation of genome edited technologies in India. Transgenic Res 28:175–181. https://doi.org/10.1007/s11248-019-00148-z

CNBC (2015) Modi bets on GM crops for India's second green revolution. https://www.cnbc.com › 2015/02/21 › modi-bets-on-ge. https://www.cnbc.com/2015/02/21/modi-bets-on-genetically-modified-crops-for-indias-second-green-revolution.html

Committee (2022). http://164.100.47.5/committee_web/ReportFile/19/103/309_2018_6_17.pdf. Accessed 28 May 2022

Convention on Biological Diversity (1992) Download here. https://ibkp.dbtindia.gov.in/DBT_Content_Test/CMS/Guidelines/20181115122523619_cbd-en.pdf. Accesed 27 May 2022

Dbtindia (2022) Guidelines for safety assessment of genome edited plants. https://dbtindia.gov.in › latest-announcement › guidelin. https://dbtindia.gov.in/latest-announcement/guidelines-safety-assessment-genome-edited-plants2022

Dbtindia (n.d.) Guidelines for the safety assessment of foods derived from https://dbtindia.gov.in › default › files › upload files PDF—https://dbtindia.gov.in/sites/default/files/uploadfiles/Guidelines_for_the_Safety_Assessment_of_Foods_Derived_from_Genetically_Engineered_Plants%20%E2%80%93%202008.pdf

Downtoearth (2022) Wrong to say Bt cotton successful in India: environmental. https://www.downtoearth.org.in › news › agriculture. Accessed 24 Jun 2022

Environmental Movement (2022). Encyclopedia.com-https://www.encyclopedia.com › environmental-studies. Accessed 12 Jun 2022

Ereky K (1919) Biotechnologie der Fleisch-, Fett-, und Milcherzeugung im landwirtschaftlichen Grossbetriebe: für naturwissenschaftlich gebildete Landwirte verfasst. P. Parey—via Hathi Trust. https://en.wikipedia.org/wiki/K%C3%A1roly_Ereky. Wikipedia contributors. (2021). Károly Ereky. In: Wikipedia, the free encyclopedia. https://en.wikipedia.org/w/index.php?title=K%C3%A1roly_Ereky&oldid=1044445528. Accessed 28 May 2022

FAO (2022) How GMO crops impact our World|FDA. https://www.fda.gov. food agricultural-biotechnology. Accessed 24 Jun 2022

FAO (n.d.) Codex alimentarius commission—FAO. https://www.fao.org › gmfp › resources › al03_34ePDF. https://www.fao.org/fileadmin/user_upload/gmfp/resources/al03_34e.pdf

FDA (2022) How GMOs are regulated—US Food and Drug Administration. https://www.fda.gov. › food › agricultural-biotechnology. Accessed 24 Jun 2022

FEASIBILITY STUDY—GEAC (n.d.). https://geacindia.gov.in › resource-documents › 1...PDF

FOOD SAFETY AND STANDARDS ACT (2006). https://fssai.gov.in/cms/food-safety-and-standards-act2006.php#:~:text=It%20is%20an%20Act%20to,and%20wholesome%20food%20for%20humn

FSSAI (2020) Dated, the 21 August, 2020—FSSAI- https://fssai.gov.in › upload › advisories › 2020/08 PDF

FSSAI (2022) FSSAI-2022 Notification:Deaft Regulations for...-Food Safety Works, https//fssai.gov.in/uploadfiles/files/Draft_Notification_GMIs the_Food_21_11_2022.pdf

FSSAI (n.d.-a) Food Safety and Standards Authority of India's (FSSAI). http://www.fssai.gov.in/

FSSAI (n.d.-b) The Prevention of Food Adulteration Act & Rules (as on 1.10. https://fssai.gov.in › upload › files › pfa-acts-and-rules PDF). https://fssai.gov.in/upload/uploadfiles/files/pfa-acts-and-rules.pdf

Geacindia (1993) MOE&F SO 677(E) dated 13th September 1993. https://geacindia.gov.in/resource-documents/biosafety-regulations/acts-andrules/MoEF_gazette_notification_SO_677E_Rules_1989.pdf

Geacindia (2005) MOE&F GSR 493(E) dated 14.07.2005. https://geacindia.gov.in/resource-documents/biosafety-regulations/acts-and-rules/584-589E.pdf

Geacindia (2010) MOE&F GSR 613(E) dated 16.07.2010. http://geacindia.gov.in/resource-documents/biosafety-regulations/acts-and-rules/MoEF_gazette_notification_GSR_613E_Change_of_GEAC-Approval_to_Appraisal.pdf

Geacindia (2022) Relevant Acts and Rules—GEAC. https://geacindia.gov.in › acts-and-rules. Accessed 6 Jun 2022

Ghosh PK (1992) Biotechnology industries in India: investment opportunities. In: Malik V, Sand Sridhar V (eds) Book industrial biotechnology. Pub: Oxford & IBH Pub. Co. Pvt. Ltd, New Delhi; Research Papers—Prasanta K Ghosh. http://www.gandipsbio.com › articles, and http://www.gandipsbio.com/article/1992_Industrial_Biotechnology.pdf

Ghosh PK (1993) Biotechnology in India. Australas Biotechnol 3(4):214–218

Ghosh PK (2017) Similar biologics: global opportunities and issues. J Pharm Pharm Sci 19(4):552–596. https://doi.org/10.18433/J34K6B

Ghosh S, Ghosh PK (2007) SAFETY EVALUATION OF GENETICALLY MODIFIED FOOD: INDIAN REGULATORY OVERVIEW. VAK 2:8–22; GM Food Safety VAK-2007—Academia.edu—https://www.academia.edu ›

Government of India (Allocation of Business) Rules (1961). https://cabsec.gov.in › english › 1_Upload_1187PDF. Accessed 28 May 2022

History of the Convention (2022). https://www.cbd.int › history. Accessed 27 May 2022

IATP (2022a) India takes baby steps to GM crops, Greens fume|IATP. https://www.iatp.org › news › india-takes-baby-steps-to. Accessed 2 Jun 2022

IATP (2022b) What is wrong with Bt cotton. https://www.iatp.org › sites › default › files › What_Is_. Accessed 24 Jun 2022

IBKP (2007) MOE& F SO 1519€ dated 23.08.2007. https://ibkp.dbtindia.gov.in/DBT_Content_Test/CMS/Guidelines/20181115133508271_Authorization%20ICAR.pdf

Indiaenvironmentportal (n.d.) Operationalizing the regulation of genetically modified foods. http://www.indiaenvironmentportal.org.in › files › fs...PDF. http://www.indiaenvironmentportal.org.in/files/fssa_interim%20regulation%20on%20Operatonalising%20GM%20Food%20regulation%20in%20India.pdf

Indiagminfo (2012) June 5th 2012 notification of Dept. of Consumer Affairs, Legal Metrology (Packaged Commodities) Amendment Rules, 2012. http://indiagminfo.org/wp-content/uploads/2012/06/1st-amendment-of-2012-in-PCR.pdf

INVESTINDIA (2024) Invest India. https://www.investindia.gov.in/sector/biotechnology. Accessed 25 Jun 2024

Jonas K, Matin Q (2012) Economic impacts and impact dynamics of Bt (Bacillus thuringiensis) cotton in India. Proc Natl Acad Sci U S A 109(29):11652–11656. https://doi.org/10.1073/pnas.1203647109

Justlabelit (2022) Labeling around the world. https://www.justlabelit.org/right-to-know-center/labeling-around-the-world/. Accessed 28 June 2022

Legislative (2022a) THE CONSTITUTION OF INDIA—legislative department. https://legislative.gov.in › sites › default › files › COIPDF. Accessed 28 May 2022

Legislative (2022b) The constitution of India—legislative department. https://legislative.gov.in › default › files › COI...pdf. Accessed 29 May 2022

Lifegate (n.d.) 20 years of corruption and fraud is enough, says Vandana Shiva. https://www.lifegate.com › vandana-shiva-gm-mustard-bt-

Naik K (n.d.) Clinical trials in India: history, current regulations, and future considerations. https://www.emich.edu/chhs/health-sciences/programs/clinical-research-administration/documents/research/clinical-trials-in-india.pdf

NCBI (2022) Asilomar and recombinant DNA: the end of the beginning. https://www.ncbi.nlm.nih.gov › books › NBK234217. Accessed 25 May 2022

Newindianexpress (n.d.) No nod to commercial release of GM Mustard. https://www.newindianexpress.com › nation › may › no-n. https://www.newindianexpress.com/nation/2018/may/14/no-nod-to-commercial-release-of-gm-mustard-1814335.html

PARLIAMENT OF INDIA RAJYA SABHA (2022). http://164.100.47.5 › reports › English Committees PDF. Accessed 24 Jun 2022

PARLIAMENT OF INDIA RAJYA SABHA (n.d.). http://164.100.47.5 › committee_web › ReportFilePDF. http://164.100.47.5/committee_web/ReportFile/14/100/110_2018_8_16.pdf

Pharmaceutical Policy (2002). https://pharmaceuticals.gov.in › policy › pharmaceutical. Accessed 29 May 2022

PIB (2022) With collective efforts by entire cotton value chain, India will. https://pib.gov.in › PressReleasePage, dated 07 OCT 2021. Accessed 24 Jun 2022

Pingoud A, Wilson GG, Wende W (2014) Type II restriction endonucleases—a historical perspective and more. Nucleic Acids Res 42(12):7489–7527. https://doi.org/10.1093/nar/gku447; Accessed 12 Jun 2022

Plant Quarantine Order (2003) IBKP-Department of Biotechnology. https://ibkp.dbtindia.gov.in › CMS › Guidelines PDF and https://ibkp.dbtindia.gov.in/DBT_Content_Test/CMS/Guidelines/20181115121925276_PlantQuarantine_order_2003.pdf

Reuters (n.d.) India gives environmental approval for gene- modified mustard. https://www.reuters.com › business › environment › India

Reyes TF, Chen Y, Fraser RZ, Chan T, Li X (2021) Assessment of the potential allergenicity and toxicity of Pichia proteins in a novel leghemoglobin preparation. Regul Toxicol Pharmacol 119:104817. https://doi.org/10.1016/j.yrtph.2020.104817; Epub 2020 Nov 7. PMID: 33171209

Seed Control Order (Amendment) (2006). https://agricoop.nic.in › sites › default › files › Se

Seeds (Control) Amendment Order (2019). https://www.indianemployees.com › acts-rules › details-https://www.indianemployees.com/acts-rules/details/seeds-control-amendment-order-2019 and https://drive.google.com/file/d/1X0nbXQwMyxl1AbILPwBrNdPD3CtQ9YSc/view

Shen C, Yin X-C, Jiao B-Y, Li J, Jia P, Zhang X-W et al (2022) Evaluation of adverse effects/events of genetically modified food consumption: a systematic review of animal and human studies. Environ Sci Eur 34:8. https://doi.org/10.1186/s12302-021-00578-9

THE ENVIRONMENT (PROTECTION) ACT (1986) India Code https://www.indiacode.nic.in › ep_act_1986PDF. Accessed 25 May 2022

THE SEEDS ACT (1966) SeedNet India Portal. https://seednet.gov.in › Seed_Act_1966PDF-https://seednet.gov.in/PDFFILES/Seed_Act_1966.pd

Timesofindia (n.d.) Modi says genetically modified crops will benefit Indian. https://timesofindia.indiatimes.com › blogs › toi-editorials. https://timesofindia.indiatimes.com/blogs/toi-editorials/modi-says-genetically-modified-crops-will-benefit-indian-agriculture/

Treaties (2022) Convention on Biological Diversity—UNTC. https://treaties.un.org › pages › ViewDetails › src=TREATY. Accessed 27 May 2022

Wikipedia Contributors (2022) United Nations conference on the human environment. In Wikipedia, the free encyclopedia. https://en.wikipedia.org/w/index.php?title=United_Nations_Conference_on_the_Human_Environment&oldid=1088569029. Accessed 3 Jun 2022

Part VI

Human Resource Development

Biotechnology in India: Four Decades of Human Resource Development: Challenges and the Road Ahead

14

Krishnamoorthy Kannan

Abstract

Around the year 1982, the Government of India decided to focus on the newly emerging area of modern biology and biotechnology. This vision resulted in the creation of the National Biotechnology Board which later on became the Department of Biotechnology. At that point, the country had very few trained modern biotechnologists and they were scattered around some Indian Universities and National laboratories. It was decided to establish Biotechnology centres that were similar in concept to the IITs but differing in approach by identifying five already well established universities that had an ecosystem suitable for the growth of modern biotechnology. The National Biotechnology Strategies I, II, and III aimed to harness the potential of biotechnology as a premier precision tool for national development and well-being of society. The approach was to building and strengthening of the nationwide education, research, and translation ecosystem that would lead to India becoming Globally competitive. The New Education Policy (NEP 2020) has evolved to provide multiple entries and multiple exits in higher studies as the future jobs will require skilled workforce in a particular area of specialization rather than just a degree. The proposal of an 8 to 10-week internship as part of higher education allows the student to experience challenges in the career path. The World Economic Forum Report 2023 has suggested that the biotech sector should focus on big data analytics, climate change and environmental management technologies, and encryption and cybersecurity to not only protect innovation but also to incorporate these areas in biotechnol-

K. Kannan (✉)
KK Biotech, Gurgaon, India

Biotechnology, GGSIPU, New Delhi, India

Nagaland Central University, Lumami, Nagaland, India
e-mail: kannan@kkbiotech.net

G. John, A. K. Panda (eds.), *Biotechnology in India - Reworking A Strategy*, https://doi.org/10.1007/978-981-97-0896-3_14

ogy education. Empowering future biotechnologists in the areas of agriculture technologies, digital platforms and apps, e-commerce, digital trade, and AI will help them move forward in their careers. This article traces the evolution of the ecosystem to promote the workforce in biotechnology in India since 1982 and the role of the Department of Biotechnology (DBT) in providing the policy support and required approvals to recognize biotechnology as an industry.

14.1 Introduction

The dawn of modern biotechnology around 1970 ushered in an era when cells made products through processes unknown before. Making cells do things better at what they were good at by natural evolution was what classical biotechnology dealt with. On the other hand, modern biotechnology led to the coinage of two new techniques in the laboratories. One technique involved the tinkering of specific genes in a living cell called genetic engineering and the other technique involved the fusion of two cells, one a cell producing antibodies with a cancerous cell and making the immortal hybrid cell secrete antibodies (hybridoma technology). The technique of genetic engineering raised a storm in the public. Can man play God? This led to resolving the safety of the technique and a ban was in place to perform the technique. Through the 1975 Asilomar conference on recombinant DNA (Berg et al. 1975), regulations were put in place for one to practice genetic engineering in a laboratory. Hybridoma technology involved mammalian cells, and the technique was not patented. The technique of genetic engineering was patented. This led to the free use of Hybridoma Technology in 1976, leading to a sudden spurt in the production of a new type of antibodies called Monoclonal Antibodies (Köhler and Milstein 1975; Wikipedia 2023). The first FDA-approved therapeutic monoclonal antibody was in 1986 for in vivo use. Due to patent restrictions in practicing Genetic Engineering, the first few products appeared in the US market through two nascent biotechnology companies, Genentech and Amgen after due clinical trials in 1976 and 1980, respectively. Genentech was established in 1976, by the late venture capitalist Robert A. Swanson and biochemist Dr. Herbert W. Boyer. AMGen (Applied Molecular Genetics Inc.) was established in the year 1980, by George Rathmann a biological chemist, and venture capitalists William K. (Bill) Bowes and associates.

The products Erythropoietin (Amgen) and human Insulin (Genentech) made these companies achieve a billion-dollar earning in less than a decade for any pharma product till then. Both the nascent R&D companies took the support of multinational companies to manufacture and market their products.

Looking at the promise, potential, and reality, the Government of India decided to go ahead with promoting modern Biotechnology. Thus, began science-based technology as a tool to help the growth of the agriculture and healthcare sectors. To enable the practice of modern Biotechnology, new field, policies were needed to be put in place to approve Biotechnology as an Industry.

Figure 14.1 summarizes various interventions made to promote the growth of biotechnology over the last 4 decades. The policies were modified in 1986, 2002,

2007, 2014, 2015, 2016, and 2021. The National Knowledge Commission (2006) was an Indian think-tank that dwelled on considering policies that might sharpen India's comparative advantage in the knowledge-intensive sectors including Biotechnology. It advised the Prime Minister's Office on reforms in building excellence in the educational system to meet the knowledge challenges of the twenty-first century. This Commission was dissolved in 2014.

The policies of the government are reflected in the direction that Biotechnology took during the last four decades. The visionary leaders under the then Prime Minister created the National Biotechnology Board (NBTB) in 1982 with a Member Secretary under the Department of Science & Technology (DST) (Joshi et al. 2013). Creating a new workforce needed a paradigm shift in education to promote Biotechnology. NBTB in collaboration with the Ministry of Human Resource Development (MHRD) identified one department in each of the five universities, with the ecosystem of CSIR, ICAR, and ICMR laboratories. In the identified university department, a separate MSc Biotechnology training programme along with the existing MSc programme brought about a paradigm shift in education in that university (Sitaramam 1991, 1995, 1998; Modi 1991) integrating the Botany, Zoology, Microbiology, and Biochemistry departments. Furthermore, the NBTB identified and created a National Facility for Animal Tissue and Cell Culture, and facilities for Plant Tissue Culture, and Microbial Culture. The existing National Institute of Immunology (NII) was incorporated as a DBT autonomous institution. The MSc Biotechnology programme being an interdisciplinary subject also provided specific skill training for those interested in using plant, animal, or microbial cells. Looking at the potential, NBTB was made an independent Department of Biotechnology (DBT) in 1986 under the Ministry of Science & Technology. The DBT identified experts including Prof Ananda M Chakraborty, who held the first life patent in the world. Experts actively participated in promoting modern Biotechnology including

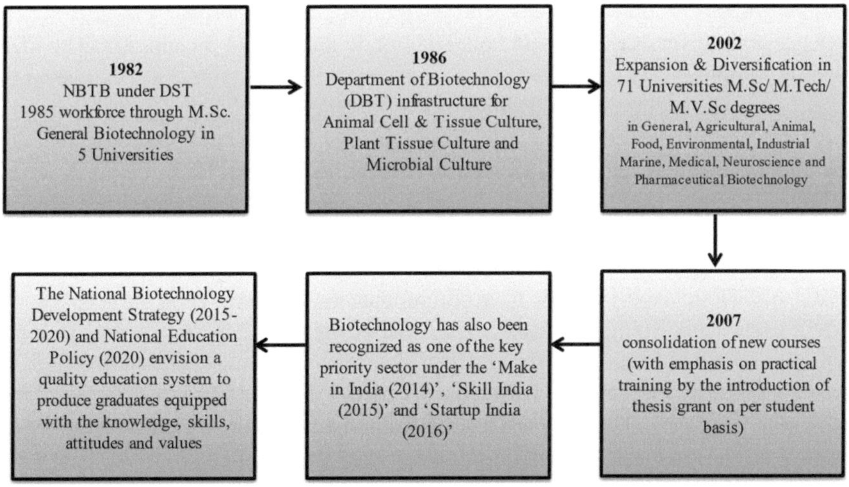

Fig. 14.1 Government strategy for biotechnology in India

Human Resource Development. The DBT also created overseas Biotechnology training fellowships for researchers and faculty. Students who completed the General Biotechnology programme with skills in Animal cell and Tissue culture, Plant tissue Culture, and Microbial culture went on to pursue higher studies. The Indian diagnostic industry also kept pace with Hybridoma Technology, and products were available in the 1980s.

Manufacturing was controlled through a government licensing policy popularly known as 'License Raj'. All this changed in the 1990s with the liberalization of the economy. The two sectors Information Technology (IT) and Biotechnology (BT) had a sudden spurt in growth for jobs and careers for students. IT grew independently till the government created the Ministry of IT in 1999 to bring a policy to enable the growth. This led to different growth curves for IT and BT. Interestingly bioinformatics has been included in DBT training programmes since 1985.

The University Grants Commission (UGC) and the All-India Council for Science & Technology (AICTE) were two accrediting bodies established in 1956 and 1987 by Acts of Parliament. The UGC was for the coordination, determination, and maintenance of standards of teaching, examination, and research in university education while AICTE was responsible for proper planning and coordinated development of the technical education and management education system in India. Both UGC and AICTE recognized biotechnology which led to BSc/MSc/MVSc courses in universities. BTech/MTech/MBA programmes were under the AICTE. The confusion is understandable as the word bio means life sciences which is a 'qualitative' subject with no mathematics and Technology being a 'quantitative' subject involving mathematics and statistics. This created a competition for jobs and careers.

As the general enrolment ratio (GER) in the universities was around 6% in 2000, the government pushed for increasing the GER in universities as only 2.5% enrolled in Post Graduate education, while doctoral programme enrolment was only 0.028%. Even with this small percentage, students interested in science were only around 20.45% while 45% opted for Arts. The government allowed private players to open colleges and universities in a self-financing mode throughout the country. The idea was to promote education in every state and minimize the migration of students to other states. Biotechnology Education also benefited and many private and state universities offered general and specialized areas of Biotechnology at postgraduate and undergraduate levels. The availability of qualified PG and Ph.D. degree holders in Biotechnology lagged behind leading to the demand for increased Faculty in AICTE-recognized programmes. This led to the recruitment through the National Entrance Test (NET) of persons qualified with a PG degree in life science areas in UGC. AICTE relaxed its rules to accommodate B. Tech-M. Tech Biotechnology qualified students to join as faculty at AICTE-recognized educational institutions. Probably this was necessary for technology area subjects where qualified PhDs were limited and the demand for IT professionals increased logarithmically. This relaxation resulted in lowering the teaching standards and increased the number of unemployable graduates in the country. With the less qualified students occupying faculty positions in the colleges, many UGC/CSIR NET-qualified students could not find placements in colleges as teaching faculty.

The Central Board of Secondary Education (CBSE) in the 2000s made efforts to introduce several electives along with existing choices to help school children make informed decisions for their future jobs/careers in higher education. The parents of these children also took a proactive interest. Due to the availability of the Internet, children were aware of emerging opportunity areas. Biotechnology was introduced in select schools that were willing to create a separate laboratory by investing at least rupees 5 lakhs in equipment and reagents. In the year 2002, 20 schools all over India introduced Biotechnology as an elective. Interestingly many students chose six electives as against the norm of five only. Many aspirant medical students chose a combination of biology and biotechnology as the Medical Council of India (MCI) did not recognize biotechnology, to appear for the medical entrance examination. The introduction of biotechnology as an elective introduced a threat perception that biology would lose its popularity and biology teachers felt they might become redundant. The positive aspect was that many biology teachers were ready to learn biotechnology during CBSE workshops conducted throughout the country at regular intervals including attending practical sessions. The students of the medical stream thus got another professional subject biotechnology as a career opportunity. BTech in biotechnology became a popular stream to pursue. Besides MSc Biotechnology, BTech/MTech Biotechnology grew in popularity.

The Kishore Vaigyanik Protsahan Yojana (KVPY) scholarship scheme, also known as the Young Scientist incentive plan, was initiated by DST in the year 1999 and aimed at encouraging students of class XII to take up research careers in the areas of basic science. This programme was administered by the prestigious Indian Institute of Sciences (IISc) Bangalore. Many students who pursued various branches of life sciences made a career in biotechnology.

Biotechnology became a business reality in the 1990s, and DBT in 2002 expanded and diversified human resource development programmes to 71 universities. Besides general biotechnology, MTech/M V Sc/M.Pharm in diverse subjects were introduced (Table 14.1). What started as creating a skilled workforce that could later introduce modern biotechnological approaches as an elective in universities was lost by this diversification. This led to competition with the pre-existing programmes in the diversified fields resulting in decreased job and career opportunities in India. Salaries were not attractive in the 2000s as the IT sector offered better salaries, leading to the perception buzzword that India was not yet ready for business in biotechnology. To remove this perception and provide career and job opportunities, to prevent brain drain, and to provide workforce in various emerging specialized areas of application in biotechnology, DBT created autonomous institutions and developed strategies.

The First National Biotechnology Development Strategy 2007 (Prsindia 2023) provided insight into the enormous opportunities in the country. Boundaries between various academic disciplines converged leading to newer opportunities and challenges. The decline in employable students in Biotechnology led DBT to relook at the course curricula. They began with a consolidation drive by introducing emphasis on practical training followed by the submission of a thesis as partial fulfilment

Table 14.1 Multipronged approaches of DBT-HRD

- MSc biotechnology started since 1985. Also MVSc MTech (medical, agricultural, marine, veterinary, industrial biotechnology, bioprocess technology, computational biology, food science, bioinformatics, and nutrition biology) started since 2002
  - Skill Vigyan State Science and Technology Partnership for Students, Technician, Faculty, and Entrepreneurship Training
  - Programme in Life Science and Biotechnology (Skill Development Programme)
  - Biotech Industrial Training Programme (BITP)
  - Foldscope Programme
- Scholarships and awards
  - Junior Research Fellowship (DBT-JRF)
  - Research Associateship Programme (DBT-RA)
  - Programmes for North eastern States (NER)
  - Har Gobind Khorana-Innovative Young Biotechnologist Fellowship (IYBF)
  - M K Bhan-Young Researchers Fellowship Program
  - Biotechnology Career Advancement and Re-orientation Programme (BioCARe)
  - National Biotechnology Innovation Award
  - Research Resource Service Facility and Platform (RRSFP)
  - Award for Institution: DBT-Boost to University Interdisciplinary Life Science Departments for Education and Research Program
  - Award for Institution: DBT-Scientific Infrastructure Access for Harnessing Academia University Research Joint Collaborations
  - DBT Outreach Programme
- Building a critical mass of science leaders
  - Ramalingaswami Re-entry Fellowship
  - Distinguished Biotechnology Research Professorship
  - Janaki Ammal National Women Bioscientist Fellowship (young and senior category)
  - S. Ramachandran-National Bioscience Fellowship for Career Development
  - Tata Innovation Fellowship
- International partnership programme
  - Khorana Program for Scholars
  - PhD Fellowship Programme under Indo-Australia Collaboration (Indian Institute of Technology, Bombay-Monash University)
  - DBT-TWAS Programme: 'The World Academy of Sciences' for the advancement of science in developing countries or more commonly known as 'TWAS'
  - DBT-Wellcome Trust Partnership/India Alliance (IA) Biomedical Research Career Programme (BRCP):
  - Partnership Programmes with EMBO, HFSPO

https://dbtindia.gov.in Annual Report 2022–23

for UG and PG degrees. To strengthen biology and biotechnology research, CSIR, ICMR, and DAE institutes joined hands.

Despite the multipronged approach (Table 14.1), there was a steady decline in the enrolment of youth in science not only in India but also globally. The challenges in basic science education were deliberated by a group of experts at a meeting organized in 2009 in Paris, by UNESCO to discuss education policies in science and mathematics. They concluded that science and mathematics education plays a very important role in developing critical thinking and creative thinking skills, thus enabling learners to understand and participate in public policy discussions, and encouraging behavioural changes that can initiate them to achieve the Millennium Development Goals.

The Star College Scheme was initiated by DBT in 2008 to support colleges and universities offering undergraduate education to improve science teaching across the country as suggested by UNESCO. The scheme provided the much-needed stimulus to encourage students and teachers to get inspired and pursue unsolved socio-economic problems of society and the country.

The DST started the INSPIRE MANAK (Million Minds Augmenting National Aspirations and Knowledge) under its flagship programme Innovation in Science Pursuit for Inspired Research (INSPIRE) in 2008 to create awareness among school students on the benefit of science in solving local problems through competitions. The objective was to strengthen the science and technology system and expand the R&D base of the country. They also provided INSPIRE fellowships to pursue higher studies up to Ph.D.

The creation of AcCSIR, an Academy in CSIR, parallel to a university, is contributing to skill education and technology development in the area of Biotechnology. AcCSIR was established by an Act of Parliament in 2010. The CSIR mandate in the biology and biotechnology education sector contributed to developing commercially viable products/technologies, carrying out research in genomics and molecular medicine, r-DNA technology, protein engineering, drugs/nutraceutical development, bioinformatics and systems biology, and immunodiagnostics to name a few. Created as a Hub and Spoke model, the Hub is AcCSIR (HQ) and spokes are located in 38 CSIR institutes/units and 18 non-CSIR institutes (Associate academic campuses) spread all over the country. Out of 38, only 19 are focused on a biotechnology training programme leading to a PhD.

The Department of Atomic Energy (DAE) has also focused on biotechnology training and employment in its institutions, TIFR, NCBS, BARC, etc. BARC has a division focused on Nuclear Agriculture and Biotechnology. Here, there is a broad-based R&D programme for various aspects of agriculture (barc.gov.in). TIFR, a public deemed university started the first molecular biology research group way back in 1960, which became the Department of Biological Sciences (DBS). The DBS established a new unit, the National Centre for Biological Sciences in 1970 at Bangalore. DBS has 14 labs covering various aspects of modern biology including biotechnology.

The aim of National Biotechnology Strategy II (PIB 2023) was to establish India as a world-class biomanufacturing hub, backed with significant investments, for the creation of new biotech products, creating a strong infrastructure for R&D and commercialization, and empowering human resources scientifically and technologically. DBT launched four major missions in education, healthcare, food and nutrition, and clean energy. This aspect of technology development and translation network was facilitated across India with global partnership, including five new clusters, 40 biotech incubators, 150 Technology Transfer Offices (TTOs), and 20 biotech connect centres. This ensured strategic and focused investments in building human capital by setting up the Life Sciences and Biotechnology Education Council. The DBT Regional Centre for Biotechnology (RCB) established with regional and global partnerships collaborated with the programmes of UNESCO as a category II centre for promoting Biotechnology education through an integrated MSc-PhD programme for advanced research and skill training. With focus on providing world-class education training and conducting innovative research at the interface of multiple disciplines, high-quality human resources in disciplinary and interdisciplinary areas of biotechnology needed in a globally competitive research milieu were created. In 2016, RCB was recognized as an institution of National Importance by the Parliament of India.

The multipronged approach (Table 14.1) of DBT in capacity building, competence building, and skill development in broad areas of biotechnology began in 1985. DBT's National Biotechnology Development Strategy III (NBDS) places schemes and programmes that are aligned with the efforts of the Government of India's flagship initiatives such as Skill India, Start-up India, and Digital India, among others.

In 2023, 675 candidates were admitted to the DBT-PG programme in Biotechnology in 63 programmes running in 56 participating host universities/institutes. This resulted in 1400 DBT-supported Postgraduate students from 2022 to 2023. AICTE recognized 305 engineering colleges which included 77% private, 22% public, and 1% public–private PG programmes in the country that were not supported under the umbrella of DBT.

The Star College Programme supports science departments in more than 300 colleges all over the country (Fig. 14.2). The scheme has ignited young minds (faculty and students) to engage in networking, exposure visits to research institutes and industries, and applying for research grants in order to prepare them for future challenges after the successful completion of their undergraduate courses.

Under the Science Setu Programme, DBT Autonomous Institutes have been linked with Star Colleges to expose undergraduate students and College Faculty to contemporary research. Virtual lab visits, webinars, lectures, and discussions were arranged on career opportunities for participating UG and PG students. The programme is organized under the aegis of the Atmanirbhar Bharat, India@75 campaign. The 'Star Status' colleges adopt a higher role of mentoring the colleges that have currently initiated their journey under the 'Strengthening Component'. As a result of these mentorship programmes, the number of applications for the Star College scheme has increased from the aspirational districts. In 2023, 23 new colleges received support under the Strengthening Component and 2 Colleges were supported under Star Status. The categorization of the scheme into urban and rural categories during 2018–2019 has ensured a level playing field for applicants hailing from rural and urban areas of the country. So far, 85 colleges from rural areas and 8 colleges from the aspirational districts are supported. To enhance the outreach of the Star College scheme in the Northeast Region (NER), a special call was issued only for NER Colleges, resulting in 35 proposal submissions.

Students are selected under DBT Junior Research Fellowship Programme for two categories through a national computer-based Biotechnology Eligibility Test (BET). Category I has 500 fellowships and Category II is for fellowship from extramural R&D projects. In 2023, 998 ongoing students were supported and 209 new fellows joined. A total of 142 research articles and reviews were published by DBT-JRF fellows, and 29 fellows were awarded doctoral degrees.

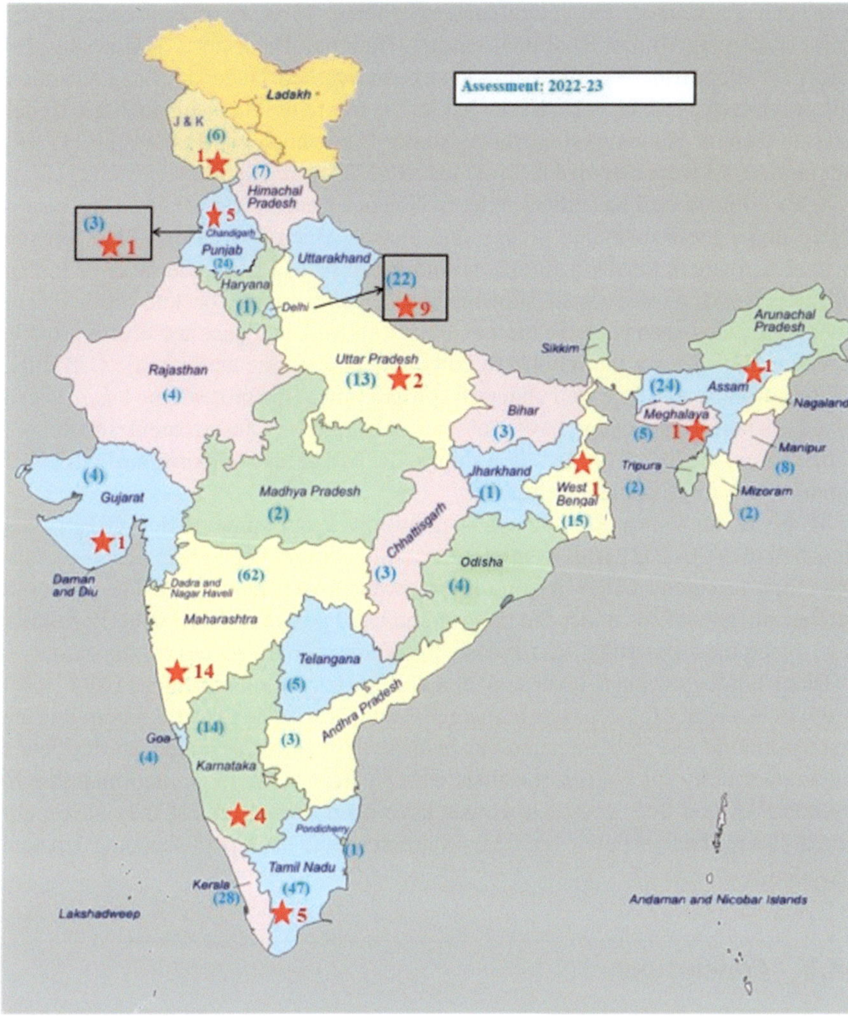

Fig. 14.2 Location of star colleges in India. (Source: https://dbtindia.gov.in)

DBT Research Associateship (DBT-RA) Programme objective of training post-doctoral students in frontier areas of research in Life Sciences and Biotechnology at premier institutions in the country and enhancing the post-doctoral culture in the country has been largely met. There is a provision for 100 fellowships per year. During 2022–2023, 107 ongoing fellows were supported under the programme; 87 research articles published, and 8 fellows employed/placed in universities/research institutions. The Distinguished Biotechnology Research Professorship utilizes the expertise of superannuated distinguished scientists in biological sciences, biotechnology, and related fields. The fellowships are limited to 5 at any point in time. Under the Janaki Ammal National Women Bio scientist Fellowship, 72 fellowships have been given under the young category (below 45 years of age) and 1 in the senior category. Under S. Ramachandran-National Bioscience Fellowship for Career Development, 182 fellowships have been awarded. Under the Tata Innovation Fellowship, ten candidates were selected for the award of fellowship in 2023. Under the Har Gobind Khorana—Innovative Young Biotechnologist Fellowship (IYBF) programme, 15 were awarded fellowships in 2023.

In an effort to build a critical mass of Science Leaders, 553 fellows were supported under the Ramanujan programme, out of which 339 fellows have already secured permanent faculty positions within India. Now a greater number of Indian scientists working overseas are showing an interest in coming back to India. Indian laboratories have also gained capacities, quantitatively and qualitatively, to provide excellent facilities for the fellows to pursue their scientific endeavours. Currently, 299 fellowships are supported under this programme. This programme has received mixed responses as many have not been absorbed in their parent institution. Furthermore, at the end of the fellowship, some had to take up positions with lower remuneration in private universities.

M K Bhan-Young Researchers Fellowship Programme (MKBYRFP) was launched in 2020–2021 with the aim to encourage bright young researchers to continue their research in DBT-Autonomous institutes after their Ph.D. The proposed number of fellowships under the programme is 50 per year. As of now, 32 fellows have joined the fellowship. It is too early to comment on the success of the initiative.

Biotechnology Career Advancement and Re-orientation Programme (BioCARe) provides research project grants to unemployed women scientists for whom it is the first extramural research grant and this helps in their career development. It offers a good opportunity for women scientists with a career break to get into mainstream research. Around 365 women scientists have been supported and they have published 450 research articles. In 2023, 15 articles have been published and 1 patent was filed.

14.2 Conclusion

The field of biotechnology education has shown tremendous growth in the last 4 decades both in terms of generating skilled manpower and updating curricula across the country. All these have been possible due to the tremendous efforts by government institutions like DBT, CSIR, ICMR, UGC, and AICTE. The biggest challenge is that even with such skilled manpower, the Indian industry is not yet ready to invest in R&D and shift to biotechnology as it involves time, funding, and creating a new product with a good shelf-life is challenging. It has to be followed by scaling up, demonstration, and dissemination. Hence, industry demand and supply are at variance. To bridge this gap, DBT has brought several policy changes to enable the growth of the biotechnology industry and enable startups to innovate by providing funds through BIRAC.

The first few batches saw an exodus of biotechnology-trained manpower to USA as biotechnology industry-based research were few in India. Currently, any pharma-based biotechnology company in the USA/Europe has the high probability of finding Indian national biotechnology-trained manpower at senior/middle management levels. Interestingly, a few of our M.Sc. biotechnology-trained manpower who joined the vaccine industry in India were lured by companies abroad to lead their vaccine initiatives. This reflects highly on the quality of curriculum and laboratory training skills. The training included imparting critical thinking, and analytical approach in solving problems. Many of them have contributed to various areas of emerging and cutting-edge research globally in leading multinational companies, be it R&D, Quality assurance, Manufacturing, or as patent lawyers. The year 2023 saw a few of them figuring in the list of Fellows of the Royal Society (FRS) London. In the first 20 years of the starting of the programme, there was a brain drain as students saw more opportunities abroad due to a global shortage of skilled manpower in the 1980s and 1990s. Since 2000 a few of our students started their own companies in India, the USA, France, and Switzerland. The exodus has certainly given visibility to our trained biotechnology students employed globally. The Government of India, through DBT, has framed attractive policies to reverse the brain drain, and some trained manpower are joining the country's reputed academic, research institutes and industries. The post-COVID-19 scenario has impacted the opportunities abroad and forced many to return to India.

NBDS III of DBT seems to be going along similar lines as the pharmaceutical sector in the 1990s. Consequent to economic liberalization in the 1990s, and with many drugs going off-patent the Indian pharma companies saw a huge opportunity in bulk drugs and formulations with a good export market globally. Many got their facilities upgraded to meet international manufacturing standards. They were accredited by the USA and Europe, the biggest generic medicine markets. Similarly, it is expected that India will become a manufacturing hub for generic products in the biotech sector.

With the growing population, food security is a major concern. The yields of major crops in India are lower than the global average. To increase yield, one needs an effective blending of conventional breeding with new plant breeding technologies, assisted breeding, genetic engineering, and gene and genome editing. The appropriate policy framework is critical in this regard.

The Biotechnology Consortium India Limited (BCIL) has played a big role in the growth of the biotechnology industry since 1990s by providing a platform for a trained biotechnology workforce in the public and private sectors. The 6 month industry internship has enabled many students to get jobs and some to become entrepreneurs. Association of Biotechnology Led Enterprises (ABLE) a not-for-profit organization has enabled trained biotech manpower to find suitable jobs and entrepreneurship in various sectors of biotechnology. BCIL and a few private players like FABA Academy and Biocon have created finishing schools for aspiring biotechnologists. They groom students for a better career in science, whether in academia or industry. They facilitate skill and professional development by identifying gaps in the existing university education system.

The biotechnology-trained workforce has started contributing to the Bioeconomy in four areas (Figs. 14.3 and 14.4) to the tune of over 80 billion $. The National

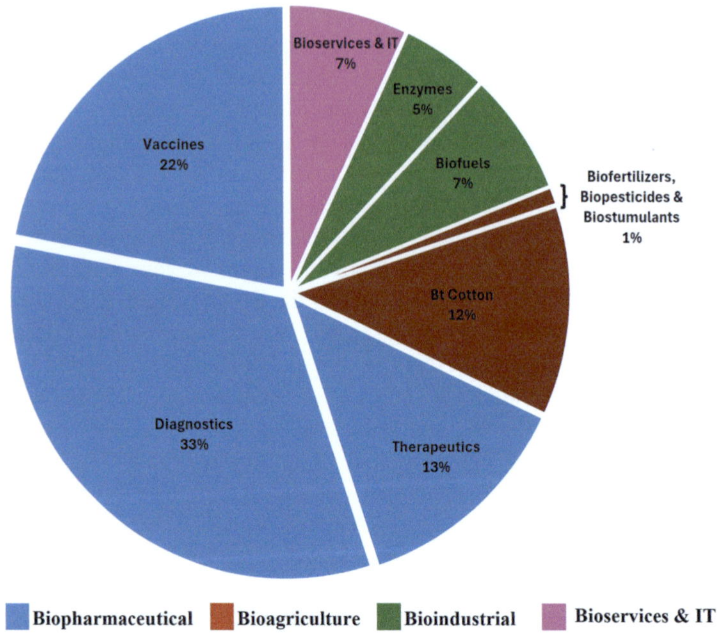

Fig. 14.3 Biotechnology share in bioeconomy (BIRAC 2022 Report) (Birac 2023)

Biotechnology Development Strategy III (Dbtindia 2023a, b) focuses on 'Knowledge and Innovation-driven Bioeconomy'. India is expected to make the top five list globally and be recognized as a Global Manufacturing Hub by 2025. The biotechnology sector is expected to contribute $ 150 billion to the anticipated $ 5 trillion Indian Economy by 2025.

14.3 The Road Ahead

The New Education Policy 2020 (NEP 2020) is a paradigm shift for teachers and students to look at the future and make informed decisions. It is a 4-year programme in Physical Sciences/Chemical Sciences/Biological Sciences/Engineering and Technology. The NEP 2020 provides flexibility for students as they can drop out after the first year and get a certificate after completing 8–10 weeks of internship. If they drop out after the second year, they get a diploma and a general bachelor's degree after the third year. These students can opt for a job. Students who wish to pursue the subject as a career can continue into the fourth year and be awarded a bachelor's degree in research. Universities may consider allowing some students to become entrepreneurs and allow flexibility to choose relevant courses and not be confined to a structured timetable, and also avoid courses that are not in line with their aspirations of growth in a desired field. Multiple entry, multiple exit policy implementation comes with many logistical issues.

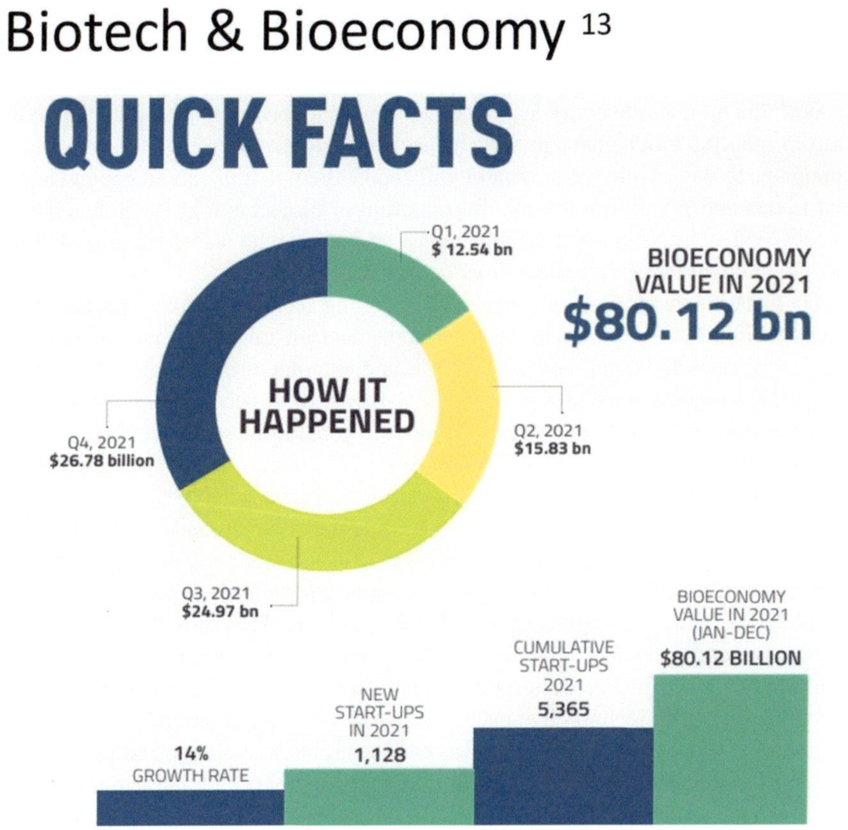

Fig. 14.4 India's bioeconomy has grown eight times in the last 8 years from $10 bn in 2014 to over $80 bn in 2022. Bioeconomy report 2022 by BIRAC (Birac 2023)

Currently, many Contract Research Organizations support preclinical trials without using animals. However, they work for drug testing for Caucasian populations. Unfortunately, we still use Caucasian values as standards to treat the Indian population. Because of our diet and food habits, we need to have our standards of normal values of biochemistry parameters, and little effort has been made in this direction. Probably, that may be the reason for reporting of a large number of diabetic cases and resulting in overmedication. The rampant use of antibiotics has resulted in multiple drug resistance in parts of our population. Furthermore, a large population of Indians are vegetarian and rice/wheat is their staple diet. There is a need to develop standards that are relevant to Indian conditions.

The current trend of workforce development has side lined or in some cases suppressed R&D work, critical to innovation and sustainability, in spite of the multi-pronged efforts by DBT and others, to meet the requirements aligned with the needs of the industry. The needs of the industry are more focused on talent that supports manufacturing over creative and innovative R&D. Focusing only on making India a manufacturing hub at the cost of R&D will lead to the import of know-how and perhaps outsourcing of technology and raw material. Contract manufacturing in the pharma sector has already taught us many lessons as R&D was neglected.

With the huge biodiversity available, we should involve school and college children to map the biodiversity in their respective regions. Orienting children from a young age to a scientific temperament will enable them to think about context based real issues and possible solutions. Introduction of biotechnology as an elective in select CBSE schools in 2002 created a new generation who wanted to contribute to the talented workforce to build a bioeconomy for the future.

There is no institution focusing on developing techniques and equipment for R&D/manufacturing. India has become dependent on importing various types of medical devices and equipment for analysis and manufacturing. A relevant example is that DNA sequencing equipment and the software tools necessary to operate them are imported. So, we have become a hub for generating sequence data through cheap qualified biotechies. In the early days of IT, the commonly used term was cyber support staff. To elaborate further, services were available for very nominal charges whereas the person delegating or leading the contract made many times that amount. It is time we stop looking at the West for a benchmark and create benchmark for our growth. Through R&D and a continuous pipeline of products, some of them may become big earners or blockbuster products. This should be the mission of biotechnology leaders. This is possible only through creative and analytical thinking in the workforce design rather than working like robots. Having been an early bird in the biotechnology space as a nation, we should strive hard by regular upgradation of skills that can be converted into usable knowledge and products by training the trained biotechnology workforce, say every 6 months, like continuing medical education for clinicians.

Looking at the future trend of jobs it seems that cheap manpower may no longer be an advantage as one needs specialized skills rather than degrees for employability. In the new era of big data, huge database repositories, and artificial intelligence, the biotechnology sector must adapt adequately and rapidly to these new areas (Tables 14.2 and 14.3). The future applications of developments like big data analytics, AI, IoT devices, etc. will greatly impact several areas of global and economic importance like climate change, agriculture, digital trade, etc. Given these developments, the curricula must also be revised along those lines to generate manpower ready to take advantage of these advancements. NEP2020 is such a step towards this direction and is a major paradigm shift not only for students but also for educators across the fields. NEP 2020 which has been implemented in most of the universities starting August 2023 provides an opportunity to strategize an academic programme with a focus on employability as well as entrepreneurship.

The balance between basic and translational research needs to be maintained to ensure that we have a robust pipeline of new knowledge, which helps us to take the translational work forward towards technology development. The strength of research and translation is further enhanced through strategic partnerships. In addition to innovation in process, technology, and product, there is a need for innovation in models of governance and partnerships. The vibrance of the ecosystem for the

Table 14.2 Future of Jobs: analysis[a]

Surveyed organizations predict 26 million fewer jobs by 2027 in record-keeping and administrative roles, including cashiers and ticket clerks; data entry, accounting, book keeping, and payroll clerks; and administrative and executive secretaries, driven mainly by digitalization and automation
Critical areas:
• Analytical thinking is considered a core skill
• Creative thinking, another cognitive skill, ranks second
• Self-efficacy skills—resilience, flexibility, and agility; motivation and self-awareness; and curiosity and lifelong learning—in recognition of the importance of workers' ability to adapt to disrupted workplaces
• Dependability and attention to detail, rank sixth, behind technological literacy
• Two attitudes relating to working with others—empathy and active listening and leadership and social influence—as well as quality control

[a] Future of Jobs, World Economic Report 2023

Table 14.3 Expected impact of technology adoption on jobs, 2023–2027. Share of organizations surveyed that expect each technology to create or displace jobs, ordered by the job creation net effect. (WEP 2023)

• Big-data analytics	58.0%
• Climate-change mitigation technology	49.5%
• Environmental management technologies	45.8%
• Encryption and cybersecurity	43.3%
• Biotechnology	42.9%
• Agriculture technologies	41.3%
• Digital platforms and apps	41.0%
• Health and care technologies	40.3%
• Education and workforce development technologies	39.8%
• Augmented and virtual reality	39.6%
• Power storage and generation	37.6%
• E-commerce and digital trade	36.6%

delivery of products needs strong regulatory support with key policy drivers and enablers. The vision of DBT is attaining new heights in biotechnology research, shaping biotechnology into a premier tool for the future, for the creation of wealth and ensuring social justice and welfare of the poor.

References

Berg P, Baltimore D, Brenner S, Roblin RO III, Singer MF (1975) Summary statement of the asilomar conference on recombinant DNA molecules. Proc Natl Acad Sci 72(6):1981–1984

Birac (2023). https://birac.nic.in/webcontent/1658318307_India_Bioeconomy_Report_2022.pdf. Accessed 3 Oct 2023

Dbtindia (2023a). https://dbtindia.gov.in/sites/default/files/Guidelines%20SAHAJ.pdf. Accessed 29 Sept 2023

Dbtindia (2023b). https://dbtindia.gov.in/sites/default/files/uploadfiles/Draft%20National%20Biotechnology%20Development%20Strategy%202020-25.pdf. Accessed 29 Sept 2023
Joshi K, Mehra K, Govil S, Singh N (2013) Biotechnology education in India: an overview. Policy Futures Educ 11(1):19–36
Köhler G, Milstein C (1975) Continuous cultures of fused cells secreting antibody of predefined specificity. Nature 256:495–497. https://doi.org/10.1038/256495a0
Modi VV (1991) Biotechnology training–innovative teaching and liaison with industry, MSU Baroda. Curr Sci 60(9/10):540–542
PIB (2023). https://pib.gov.in/newsite/printrelease.aspx?relid=134035#:~:text=The%20National%20Biotechnology%20Development%20Strategy,%2Dclass%20bio%2Dmanufacturing%20hub. Accessed 29 Sept 2023
Prsindia (2023). https://prsindia.org/files/bills_acts/bills_parliament/2004/bill26_2007010326_National_Biotechnology_Development_Strategy.pdf. Accessed 29 Sept 2023
Sitaramam V (1991) Biotechnology programme at Poona U- the beginning or the end? Curr Sci 60(9/10):537–540
Sitaramam V (1995) The Indian academy of sciences document on science education in universities: a criticism. Curr Sci 69(2):89–94
Sitaramam V (1998) On recommending students: the curious case of all students being in the top ten percent. Curr Sci 75(7):647–652
Wikipedia (2023). https://en.wikipedia.org/wiki/Monoclonal_antibody. Accessed 29 Sept 2023

MIX
Papier aus verantwortungsvollen Quellen
Paper from responsible sources
FSC® C105338

If you have any concerns about our products,
you can contact us on
ProductSafety@springernature.com

In case Publisher is established outside the EU,
the EU authorized representative is:
Springer Nature Customer Service Center GmbH
Europaplatz 3, 69115 Heidelberg, Germany

Printed by Libri Plureos GmbH
in Hamburg, Germany